UNDERWATER ACOUSTICS

A Linear Systems Theory Approach

Underwater Acoustics
A Linear Systems Theory Approach

LAWRENCE J. ZIOMEK

Department of Electrical and Computer Engineering
Naval Postgraduate School
Monterey, California

1985

ACADEMIC PRESS, INC.

(Harcourt Brace Jovanovich, Publishers)

Orlando San Diego New York London
Toronto Montreal Sydney Tokyo

ACADEMIC PRESS, INC.
Orlando, Florida 32887

United Kingdom Edition published by
ACADEMIC PRESS INC. (LONDON) LTD.
24–28 Oval Road, London NW1 7DX

Library of Congress Cataloging in Publication Data

Ziomek, Lawrence J.
Underwater acoustics.

Includes index.
1. Underwater acoustics. 2. Signal processing.
3. System analysis. I. Title.
QC242.2.Z56 1985 534'.24 84-18606
ISBN 0-12-781720-4 (alk. paper)

PRINTED IN THE UNITED STATES OF AMERICA

85 86 87 88 9 8 7 6 5 4 3 2 1

To my parents
CASIMIR and LOTTIE,
my wife
VIRGINIA,
and my daughters
NICOLE and KIRSTEN

Contents

Preface

This book is based, in part, upon lecture notes that I wrote for courses in Sonar Systems Engineering (in the Department of Electrical and Computer Engineering) and Radiation and Scattering of Waves in Fluids (in the Department of Physics) at the Naval Postgraduate School. I took an interdisciplinary approach in writing this book, involving the fields of electrical engineering (linear systems theory, statistical communication theory, and digital signal processing) and physics (underwater acoustics). However, the main approach was that of an electrical engineer with acoustic theory provided as needed.

The philosophy that I followed was to treat the ocean medium as a linear, random filter. This idea is by no means new; it has been part of the research literature since the mid-1960s. However, this book represents an attempt to write in the style of a textbook to make the material more easily approachable.

This book is rigorous primarily in the sense that its emphasis is on novel, general derivations of results from first principles whenever possible and practical, using a consistent and mainly standard notation from the first chapter to the last. However, it is not devoted to theory for theory's sake, since by the use of simplifying assumptions and examples, numerous practical results are obtained. It is hoped that by providing novel derivations whenever possible, a fresh point of view is expressed and duplication avoided.

The field of underwater acoustics is so broad and interdisciplinary that it is difficult to write a single book that treats in sufficient depth and clarity all the various topics, methods, and theories considered as part of the field. The danger is that one produces either an encyclopedic handbook, which treats many topics with elementary discussions and developments, or a research monograph, which reads like a several-hundred page journal article. While each has its merits and fills a definite need, it is my opinion that neither is totally suitable for use as a classroom textbook. It was therefore my philosophy to concentrate on developing in sufficient theoretical depth and clarity basic, fundamental results that are useful in a variety of underwater acoustic applications so that the overall level of the book lies between the two extremes of a handbook and a research monograph. The topics covered in this book indicate that it is just as important for the signal processor to understand the fundamentals of wave propagation as it is for the underwater acoustician to understand the fundamentals of array theory and signal processing. The vehicle I chose to bring together the areas of wave propagation and array theory and signal processing is *linear systems theory*. This book does not discuss the sonar equations or waveguides, not because I felt these topics were unimportant, but because there are already several books available that contain good discussions of these subjects.

This book could well be used by first-year graduate students and advanced college seniors in the fields of electrical engineering, ocean engineering, acoustics, and oceanography for a treatment of sonar systems engineering. It is quite suitable for self-study.

Chapter 1 provides a general overview of the book and background discussion concerning the treatment of the ocean medium as an underwater acoustic communication channel. Chapters 3 and 4 cover the fundamentals of complex apertures and arrays, respectively, while Chapter 5 covers basic topics in sonar signal processing. Complex aperture theory is discussed before array theory, analogous to the general practice of discussing the theory of continuous-time signals before discussing discrete-time signals. Once the general principles of aperture theory are developed, the array theory results can be derived quickly and easily. Chapters 3–5 contain many examples and problems with answers provided at the end of the book. Several of the problems at the end of Chapter 4 are well suited for computer projects, and I have used them for such purposes. Recommended prerequisites for this course include introductory-level courses in communication theory, probability, and random processes. An introductory-level course in acoustics would be desirable but is not absolutely necessary.

Another type of course that could be taught from this book is an advanced special topics course on mathematical models of the ocean medium. Chapters 1, 2 (Fundamentals of Linear, Time-Variant, Space-Variant Filters), 6 (Wave Propagation in Inhomogeneous Media), and 7 (Random Ocean Medium Transfer Functions) could form the foundation for such a course and could be supplemented by additional material of the instructor's choice. Chapters 2, 6, and 7 do contain examples and are written in a tutorial fashion suitable for classroom use. Chapter 7 is the essence of the book for it is here that most of the ideas contained in Chapters 1 through 6 are brought together.

I would like to thank my colleagues Profs. Donald E. Kirk and Robert D. Strum of the Department of Electrical and Computer Engineering at the Naval Postgraduate School for their early encouragement and support. My students also deserve special credit for their many thought-provoking questions and comments, which caused me to make many revisions for the better. I would also like to acknowledge the Naval Postgraduate School Foundation Research Program and the Defense Advanced Research Projects Agency (DARPA). Their support of my research efforts is responsible for the material appearing in Chapters 2 and 7 of this book. And finally, I would like to acknowledge Lily T. Nimri and Elaine R. Christian for their typing assistance and Alvin W. Lau who drew all of the artwork.

Lawrence J. Ziomek
Carmel, California

1

Introduction and Background Discussion

1.1 THE PROPAGATION OF ACOUSTIC SIGNALS IN THE OCEAN

The propagation of *small-amplitude* acoustic signals in the ocean can be described by the *linear*, inhomogeneous, scalar wave equation

$$\nabla^2\varphi(t,\mathbf{r}) - \frac{1}{c^2(\mathbf{r})}\frac{\partial^2}{\partial t^2}\varphi(t,\mathbf{r}) = x_M(t,\mathbf{r}), \tag{1.1-1}$$

where $\varphi(t,\mathbf{r})$ is the *velocity potential* at time t and position $\mathbf{r}$, $x_M(t,\mathbf{r})$ is the input acoustic signal to the medium or the *source distribution*, which represents the rate at which fluid volume is added (volume flow rate) at time t and position $\mathbf{r}$ per unit volume of fluid, and $c(\mathbf{r})$ is the speed of sound in the ocean in meters per second. If we assume that $c(\mathbf{r})$ is equal to a *constant* c, then the solution of Eq. (1.1-1) is given by

$$\varphi(t,\mathbf{r}) = -\frac{1}{4\pi}\int_{V_0}\frac{x_M(t-(|\mathbf{r}-\mathbf{r}_0|/c),\mathbf{r}_0)}{|\mathbf{r}-\mathbf{r}_0|}\,dV_0, \tag{1.1-2}$$

where the quantity

$$t-(|\mathbf{r}-\mathbf{r}_0|/c) \tag{1.1-3}$$

is known as the *retarded time* and the integration is performed (in general) over the volume occupied by the source (see Fig. 1.1-1). The *acoustic pressure* $p(t,\mathbf{r})$ in pascals (newtons per square meter) and the *acoustic fluid* (*particle*) *velocity*

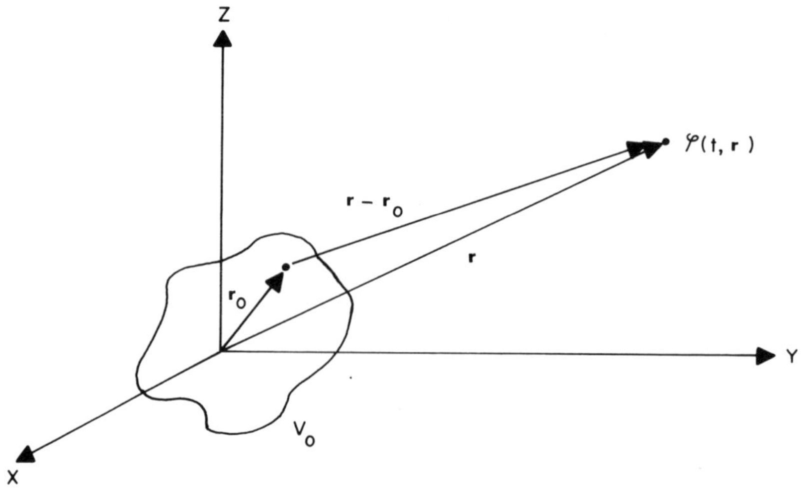

FIG. 1.1-1 Source distribution shown occupying a volume V_0.

vector $\mathbf{u}(t,\mathbf{r})$ in meters per second can be obtained from $\varphi(t,\mathbf{r})$ as follows:

$$p(t,\mathbf{r}) = -\rho_0 \frac{\partial}{\partial t}\varphi(t,\mathbf{r}), \tag{1.1-4}$$

where ρ_0 is the *equilibrium density* (assumed to be constant) of the ocean in kilograms per cubic meter and

$$\mathbf{u}(t,\mathbf{r}) = \nabla\varphi(t,\mathbf{r}). \tag{1.1-5}$$

Note that if we take the curl of both sides of Eq. (1.1-5), then

$$\nabla \times \mathbf{u}(t,\mathbf{r}) = \nabla \times \nabla\varphi(t,\mathbf{r}) = 0, \tag{1.1-6}$$

since the curl of the gradient of a scalar function [in this case, $\varphi(t,\mathbf{r})$] is equal to zero. Equation (1.1-6) indicates that the *vorticity* of the fluid, which is defined as $\nabla \times \mathbf{u}(t,\mathbf{r})$ and is proportional to the local angular velocity of the fluid, is equal to zero, and thus $\mathbf{u}(t,\mathbf{r})$ is said to be *irrotational.*

In addition to acoustic (sound) waves, there is water motion resulting from *internal waves.* Thus, in general, we must deal with the superposition of two wave motions: (1) acoustic (sound) waves, whose motions are *irrotational* and *compressional*; and (2) internal waves, whose motions are *rotational* and *incompressible.* The total fluid-velocity vector is therefore given by

$$\nabla\varphi(t,\mathbf{r}) + \nabla \times \mathbf{\Phi}(t,\mathbf{r}), \tag{1.1-7}$$

where $\nabla\varphi(t, \mathbf{r})$ is the fluid-velocity vector due to sound waves [see Eq. (1.1-5)] and $\nabla \times \mathbf{\Phi}(t, \mathbf{r})$ is the fluid-velocity vector due to internal waves. The expression $\mathbf{\Phi}(t, \mathbf{r})$ is called the *vector velocity potential*. Note that $\nabla \cdot \nabla \times \mathbf{\Phi}(t, \mathbf{r}) = 0$ (vector identity), and thus the wave motion $\nabla \times \mathbf{\Phi}(t, \mathbf{r})$ due to internal waves is said to be incompressible, as mentioned earlier. We will not be considering internal wave motion in this book, so Eq. (1.1-5) will be our defining relationship for the fluid-velocity vector.

Equation (1.1-2) is a representation of the acoustic field at distances close to the source, where the speed of sound can be considered constant. The analysis and results presented in Chapters 3 and 4 on the directivity functions of complex apertures and arrays, respectively, are based on Eq. (1.1-2). In addition, the various signal-processing topics considered in Chapter 5 also assume that c is constant. Of course, as is well known, the speed of sound in the ocean is not constant in general but is instead a function of temperature, depth, and salinity, that is, it is a function of position in the medium. Thus, different approximate solutions of Eq. (1.1-1) for variable speed-of-sound profiles $c(\mathbf{r})$ are discussed in Chapter 6. These solutions are then incorporated into the derivation of transfer functions of the random ocean medium, which is the subject of Chapter 7. Once a transfer function has been derived, it can then be *coupled* to the far-field directivity functions of the transmit and receive apertures (arrays) and to the frequency spectrum of the transmitted electrical signal. This systems approach is introduced next in more detail.

1.2 THE OCEAN MEDIUM AS AN UNDERWATER ACOUSTIC COMMUNICATION CHANNEL

Since the wave equation for small-amplitude acoustic signals is linear, we can represent the ocean medium (in general) as a *linear, time-variant, space-variant, random filter* (*system* or *communication channel*). With this interpretation in mind, refer to Fig. 1.2-1, which illustrates the geometry of a basic bistatic communication channel, and Fig. 1.2-2, which is a mathematical block-diagram representation of Fig. 1.2-1. With respect to Fig. 1.2-1, both the transmit and receive apertures (arrays) are, in general, volume apertures (arrays) and are in motion.

Figure 1.2-2 serves as a good introduction to the following concepts and associated notation:

(1) complex transmit and receive aperture functions, $A_T(f, \mathbf{r})$ and $A_R(\eta, \mathbf{r})$;

(2) transmit and receive far-field directivity functions, $D_T(f, \boldsymbol{\alpha})$ and $D_R(\eta, \boldsymbol{\beta})$;

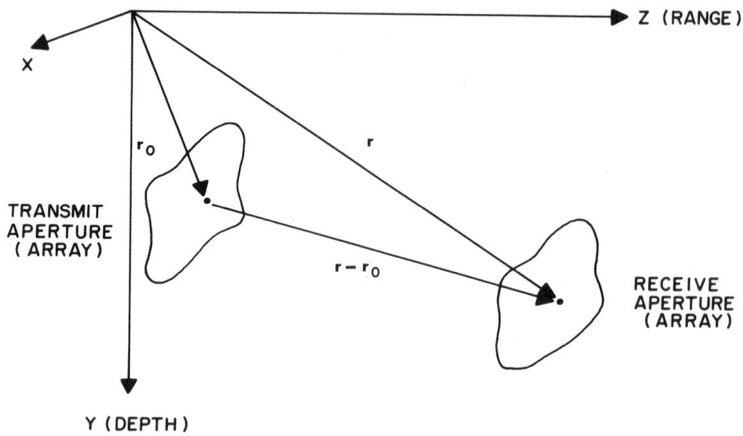

FIG. 1.2-1 Basic geometry of a bistatic communication channel.

(3) vectors whose components are spatial frequencies, $\boldsymbol{\alpha}$, $\mathbf{v}$, $\boldsymbol{\beta}$, and γ;
(4) frequency and angular spectra, $X(f, \boldsymbol{\alpha})$, $X_M(f, \mathbf{v})$, $Y_M(\eta, \boldsymbol{\beta})$, and $Y(\eta, \gamma)$;
(5) linear, time-variant, space-variant, random ocean medium impulse-response and transfer functions, $h_M(\tau, \mathbf{r}_0; t, \mathbf{r})$ and $H_M(f, \mathbf{v}; t, \mathbf{r})$.

Before proceeding further, a word of caution concerning Fig. 1.2-2 is in order. Note, for example, that $X_M \neq XD_T$, $Y_M \neq X_M H_M$, and $Y \neq Y_M D_R$, in general. The equations required for describing the linear, time-variant, space-variant, random filter's input–output relationships and for coupling the transmitted and received electrical signals to the medium via the transmit and receive apertures are developed in Chapters 2 and 3, respectively.

Let us now describe in more detail the notation used in Fig. 1.2-2. The position vectors $\mathbf{r}_0$ and $\mathbf{r}$ refer to the spatial coordinates (x_0, y_0, z_0) and (x, y, z), respectively, in meters, and t refers to time in seconds. The parameters f and η are frequencies in hertz where f represents input or transmitted frequencies and η represents output or received frequencies. Note that if $\eta \neq f$, then Doppler spread is implied.

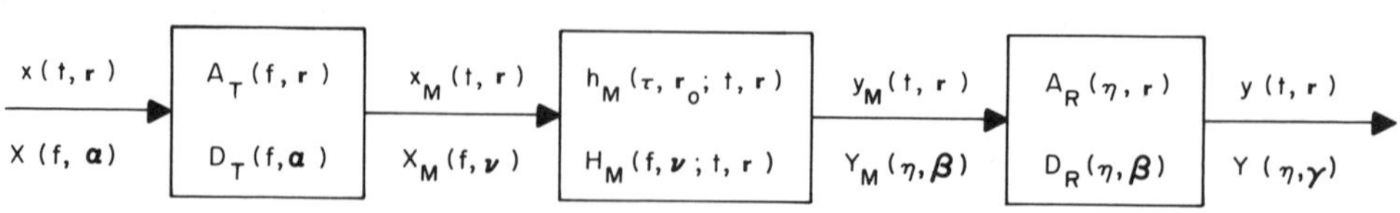

FIG. 1.2-2 Mathematical block-diagram representation of the bistatic communication channel depicted in Fig. 1.2-1.

The quantities $\boldsymbol{\alpha}$, $\mathbf{v}$, $\boldsymbol{\beta}$, and γ are vectors whose components are *spatial frequencies* in cycles per meter. Since spatial frequencies are related to both direction cosines and wavelength, and hence to wave-number components, they represent directions of wave propagation. The vector $\mathbf{v}$ represents input or transmitted spatial frequencies into the medium, as in $X_M(f, \mathbf{v})$, while $\boldsymbol{\beta}$ represents output or received spatial frequencies from the medium, as in $Y_M(\eta, \boldsymbol{\beta})$. Note that if $\boldsymbol{\beta} \neq \mathbf{v}$, then angular spread (scatter) is implied.

The remaining expressions found in Fig. 1.2-2 are further described in the following list:

$x(t, \mathbf{r})$	Input electrical signal to transmit electroacoustic transducer applied at time t and spatial location $\mathbf{r}$ of transducer
$X(f, \boldsymbol{\alpha})$	Frequency f and angular spectrum $\boldsymbol{\alpha}$ of input electrical signal
$A_T(f, \mathbf{r})$	Complex frequency response at spatial location $\mathbf{r}$ of transmit transducer. Also referred to as the complex transmit aperture function
$D_T(f, \boldsymbol{\alpha})$	Transmit far-field directivity function or beam pattern
$x_M(t, \mathbf{r})$	Input acoustic signal to the medium applied at time t and spatial location $\mathbf{r}$. Also, output acoustic signal from transmit electroacoustic transducer. Recall that $x_M(t, \mathbf{r})$ appeared as the source distribution in Eq. (1.1-1)
$X_M(f, \mathbf{v})$	Frequency f and angular spectrum $\mathbf{v}$ of input acoustic signal
$h_M(\tau, \mathbf{r}_0; t, \mathbf{r})$	Time-variant, space-variant, random impulse response or *Green's function* of the ocean medium. It describes the response of the medium at time t and spatial location $\mathbf{r}$ due to the application of a unit impulse at time $(t - \tau)$, or τ seconds ago, at a distance $\|\mathbf{r} - \mathbf{r}_0\|$ meters away (see Fig. 1.2-1)
$H_M(f, \mathbf{v}; t, \mathbf{r})$	Time-variant, space-variant, random transfer function of the ocean medium
$y_M(t, \mathbf{r})$	Output acoustic signal from the medium at time t and spatial location $\mathbf{r}$. Also, input acoustic signal to receive electroacoustic transducer
$Y_M(\eta, \boldsymbol{\beta})$	Frequency η and angular spectrum $\boldsymbol{\beta}$ of output acoustic signal
$A_R(\eta, \mathbf{r})$	Complex frequency response at spatial location $\mathbf{r}$ of receive transducer. Also referred to as the complex receive aperture function
$D_R(\eta, \boldsymbol{\beta})$	Receive far-field directivity function or beam pattern

$y(t, \mathbf{r})$	Output electrical signal from receive electroacoustic transducer at time t and spatial location $\mathbf{r}$ of transducer
$Y(\eta, \gamma)$	Frequency η and angular spectrum γ of output electrical signal

As mentioned previously, we can represent the ocean medium as a linear, time-variant, space-variant, random filter. The term "time-variant" implies motion among targets, the ocean surface, discrete point scatterers, and the transmit and receive apertures (arrays). Discrete point scatterers in the ocean may include, for example, gas bubbles, fish, and particulate matter. The time-variant property results in both Doppler spread and spread in time-delay values. If the filter is time-invariant, then no motion is implied. As a result, there will be no Doppler spread and no spread in time delay.

The term "space-variant" implies that the sound-speed profile (index of refraction) of the ocean is a function of position. This space-variant property results in scatter or angular spread due to refraction. If the filter is space-invariant, then an isospeed medium is implied. As a result, there will be no refraction, and hence no scatter or angular spread, since the sound rays will be traveling in straight lines.

In addition, since any motion and/or the index of refraction can be decomposed into a sum of deterministic (average) and random (fluctuating) components, these random components can be accounted for via a random filter representation as opposed to a deterministic filter representation. For example, by using a systems theory approach, surface, volume, and/or bottom reverberation returns can be modeled as the outputs from linear filters. In addition, target returns can also be modeled as filter outputs. Furthermore, different transmit signals and transmit and receive far-field directivity functions can easily be coupled to various models (i.e., transfer functions) of the random, inhomogeneous ocean medium in a straightforward and logical fashion in order to study problems in pulse propagation in random media, underwater acoustic communication, target detection, and parameter estimation using various space–time signal-processing algorithms.

BIBLIOGRAPHY

S. M. Flatté (ed.), *Sound Transmission Through a Fluctuating Ocean.* Cambridge Univ. Press, London and New York, 1979.

D. Middleton, A statistical theory of reverberation and similar first-order scattered fields. Part I: waveforms and the general process. *IEEE Trans. Inf. Theory* **IT-13**, 372–392 (1967).

S. Temkin, *Elements of Acoustics.* Wiley, New York, 1981.

L. J. Ziomek, *Linear Time-Variant Space-Variant Filters and the WKB Approximation.* Naval Postgraduate School Technical Report 62-83-058, October, 1983, 93 pages. Also available from DTIC, AD Number A 134886.

2

Fundamentals of Linear, Time-Variant, Space-Variant Filters

2.1 DETERMINISTIC FILTERS

2.1.1 Impulse Response and Transfer Functions

A linear, time-variant, space-variant filter, as depicted in Fig. 2.1-1, is characterized by its corresponding time-varying, space-varying impulse response $h(\tau, \mathbf{r}_0; t, \mathbf{r})$. The function $h(\tau, \mathbf{r}_0; t, \mathbf{r})$ describes the response of the filter at time t and spatial location $\mathbf{r} = (x, y, z)$ due to the application of a unit impulse at time $(t - \tau)$, or τ seconds ago, at a distance $|\mathbf{r} - \mathbf{r}_0|$ meters away, where $\mathbf{r}_0 = (x_0, y_0, z_0)$. Note that

$$h(\tau, \mathbf{r}_0; t, \mathbf{r}) \equiv h(t, \mathbf{r}; t - \tau, \mathbf{r} - \mathbf{r}_0). \tag{2.1-1}$$

The impulse response function is also called the *Green's function.*

The relationship between the input signal $x(t, \mathbf{r})$ and the output signal $y(t, \mathbf{r})$ is given by

$$y(t, \mathbf{r}) = \int_{-\infty}^{\infty} \int_{-\infty}^{\infty} x(t - \tau, \mathbf{r} - \mathbf{r}_0) h(\tau, \mathbf{r}_0; t, \mathbf{r}) \, d\tau \, d\mathbf{r}_0, \tag{2.1-2}$$

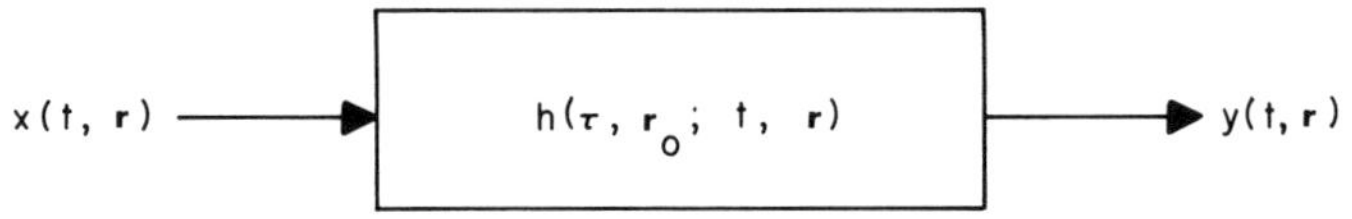

FIG. 2.1-1 Representation of a linear, time-variant, space-variant filter.

where it should be noted that both the input and output signals are functions of time and space.

EXAMPLE 2.1-1

Different forms of Eq. (2.1-2) can be obtained by making the following simplifying assumptions:

(1) If the linear filter h is time-invariant and space-invariant, then

$$h(\tau, \mathbf{r}_0; t, \mathbf{r}) = h(t - (t - \tau), \mathbf{r} - (\mathbf{r} - \mathbf{r}_0)) = h(\tau, \mathbf{r}_0), \tag{2.1-3}$$

and, as a result, Eq. (2.1-2) reduces to

$$y(t, \mathbf{r}) = \int_{-\infty}^{\infty} \int_{-\infty}^{\infty} x(t - \tau, \mathbf{r} - \mathbf{r}_0) h(\tau, \mathbf{r}_0) \, d\tau \, d\mathbf{r}_0, \tag{2.1-4}$$

which is a *multidimensional convolution integral*, as would be expected.

(2) If h is time-invariant but space-variant, then

$$\begin{aligned} h(\tau, \mathbf{r}_0; t, \mathbf{r}) &= h(t - (t - \tau), \mathbf{r}; \mathbf{r} - \mathbf{r}_0) \\ &= h(\tau, \mathbf{r}; \mathbf{r} - \mathbf{r}_0) = h(\tau, \mathbf{r}_0; \mathbf{r}), \end{aligned} \tag{2.1-5}$$

and, as a result, Eq. (2.1-2) reduces to

$$y(t, \mathbf{r}) = \int_{-\infty}^{\infty} \int_{-\infty}^{\infty} x(t - \tau, \mathbf{r} - \mathbf{r}_0) h(\tau, \mathbf{r}_0; \mathbf{r}) \, d\tau \, d\mathbf{r}_0. \tag{2.1-6}$$

(3) If h is time-variant but space-invariant, then

$$\begin{aligned} h(\tau, \mathbf{r}_0; t, \mathbf{r}) &= h(t, \mathbf{r} - (\mathbf{r} - \mathbf{r}_0); t - \tau) \\ &= h(t, \mathbf{r}_0; t - \tau) = h(\tau, \mathbf{r}_0; t), \end{aligned} \tag{2.1-7}$$

and, as a result, Eq. (2.1-2) reduces to

$$y(t, \mathbf{r}) = \int_{-\infty}^{\infty} \int_{-\infty}^{\infty} x(t - \tau, \mathbf{r} - \mathbf{r}_0) h(\tau, \mathbf{r}_0; t) \, d\tau \, d\mathbf{r}_0. \tag{2.1-8}$$

□

Note that if Eq. (2.1-2) is rewritten as

$$y(t, \mathbf{r}) = \int_{-\infty}^{\infty} \int_{-\infty}^{\infty} x(t - \alpha, \mathbf{r} - \zeta) h(\alpha, \zeta; t, \mathbf{r}) \, d\alpha \, d\zeta \tag{2.1-9}$$

and if the input is a unit impulse applied at time $\alpha = (t - \tau)$ seconds at a distance $|\zeta| = |\mathbf{r} - \mathbf{r}_0|$ meters away, that is, if

$$x(\alpha, \zeta) = \delta(\alpha - (t - \tau), \zeta - (\mathbf{r} - \mathbf{r}_0)), \tag{2.1-10}$$

then substituting Eq. (2.1-10) into Eq. (2.1-9) yields

$$y(t, \mathbf{r}) = \int_{-\infty}^{\infty} \int_{-\infty}^{\infty} \delta(\tau - \alpha, \mathbf{r}_0 - \zeta) h(\alpha, \zeta; t, \mathbf{r}) \, d\alpha \, d\zeta = h(\tau, \mathbf{r}_0; t, \mathbf{r}). \quad (2.1\text{-}11)$$

Analogous to the frequency response or transfer function $H(f)$ of linear, time-invariant systems is the time-varying, space-varying, frequency response or transfer function $H(f, \mathbf{v}; t, \mathbf{r})$ of linear, time-variant, space-variant systems. It is defined as follows:

$$H(f, \mathbf{v}; t, \mathbf{r}) \triangleq F_\tau F_{\mathbf{r}_0}\{h(\tau, \mathbf{r}_0; t, \mathbf{r})\} \quad (2.1\text{-}12)$$

or

$$H(f, \mathbf{v}; t, \mathbf{r}) \triangleq \int_{-\infty}^{\infty} \int_{-\infty}^{\infty} h(\tau, \mathbf{r}_0; t, \mathbf{r}) \exp(-j2\pi f \tau) \exp(+j2\pi \mathbf{v} \cdot \mathbf{r}_0) \, d\tau \, d\mathbf{r}_0, \quad (2.1\text{-}13)$$

where f corresponds to input frequencies in hertz and $\mathbf{v}$ is a vector whose components are input *spatial frequencies* in cycles per meter. As was previously mentioned in Chapter 1, spatial frequencies are related to wave-number components and, hence, they represent directions of wave propagation. Similarly,

$$h(\tau, \mathbf{r}_0; t, \mathbf{r}) = F_f^{-1} F_{\mathbf{v}}^{-1}\{H(f, \mathbf{v}; t, \mathbf{r})\} \quad (2.1\text{-}14)$$

or

$$h(\tau, \mathbf{r}_0; t, \mathbf{r}) = \int_{-\infty}^{\infty} \int_{-\infty}^{\infty} H(f, \mathbf{v}; t, \mathbf{r}) \exp(+j2\pi f \tau) \exp(-j2\pi \mathbf{v} \cdot \mathbf{r}_0) \, df \, d\mathbf{v}. \quad (2.1\text{-}15)$$

The choice of a plus sign in $\exp(+j2\pi \mathbf{v} \cdot \mathbf{r}_0)$ appearing in Eq. (2.1-13), which corresponds to the forward spatial Fourier transform with respect to $\mathbf{r}_0$, was *not* arbitrary. This choice of sign convention is meant to be consistent with that of the spatial Fourier transform relationship between a complex aperture function and its far-field directivity function (beam pattern) as developed later in Chapter 3. Besides, the integrand term $\exp(-j2\pi f \tau) \exp(+j2\pi \mathbf{v} \cdot \mathbf{r}_0)$ appearing in Eq. (2.1-13) has the nice physical interpretation of being a time-harmonic plane wave traveling in the direction of increasing $r_0 = |\mathbf{r}_0|$ with the sign convention as given.

EXAMPLE 2.1-2

If the linear filter h is time-invariant and space-invariant, then its corresponding transfer function can be obtained by substituting Eq. (2.1-3) into Eq. (2.1-13). Doing so yields the following result:

$$H(f,\mathbf{v}) = \int_{-\infty}^{\infty}\int_{-\infty}^{\infty} h(\tau,\mathbf{r}_0)\exp(-j2\pi f\tau)\exp(+j2\pi\mathbf{v}\cdot\mathbf{r}_0)\,d\tau\,d\mathbf{r}_0. \quad (2.1\text{-}16)$$

Similarly, from Eq. (2.1-15),

$$h(\tau,\mathbf{r}_0) = \int_{-\infty}^{\infty}\int_{-\infty}^{\infty} H(f,\mathbf{v})\exp(+j2\pi f\tau)\exp(-j2\pi\mathbf{v}\cdot\mathbf{r}_0)\,df\,d\mathbf{v}. \quad (2.1\text{-}17)$$

□

Calculating the transfer function according to Eq. (2.1-13) assumes that one knows the impulse response function. However, even if the impulse response is *not* known, the transfer function can still be obtained by using a time-harmonic plane wave as an input signal. This can be shown by representing a linear, time-varying, space-varying filter by the *linear operator* $L\{\cdot\}$, which operates on input signals that are functions of both time and space. The output of the filter $y(t,\mathbf{r})$ can then be expressed as

$$y(t,\mathbf{r}) = L\{x(t,\mathbf{r})\}. \quad (2.1\text{-}18)$$

If we let

$$x(t,\mathbf{r}) = \exp(+j2\pi ft)\exp(-j2\pi\mathbf{v}\cdot\mathbf{r}), \quad (2.1\text{-}19)$$

which is in the form of a time-harmonic plane wave traveling in the direction of increasing $r = |\mathbf{r}|$, then from Eqs. (2.1-2) and (2.1-18) we can write

$$L\{\exp(+j2\pi ft)\exp(-j2\pi\mathbf{v}\cdot\mathbf{r})\}$$
$$= \int_{-\infty}^{\infty}\int_{-\infty}^{\infty} \exp[+j2\pi f(t-\tau)]\exp[-j2\pi\mathbf{v}\cdot(\mathbf{r}-\mathbf{r}_0)]h(\tau,\mathbf{r}_0;t,\mathbf{r})\,d\tau\,d\mathbf{r}_0, \quad (2.1\text{-}20)$$

and, by using the definition of the transfer function given by Eq. (2.1-13), Eq. (2.1-20) can be expressed as

$$L\{\exp(+j2\pi ft)\exp(-j2\pi\mathbf{v}\cdot\mathbf{r})\} = H(f,\mathbf{v};t,\mathbf{r})\exp(+j2\pi ft)\exp(-j2\pi\mathbf{v}\cdot\mathbf{r}), \quad (2.1\text{-}21)$$

where $L\{\exp(+j2\pi ft)\exp(-j2\pi\mathbf{v}\cdot\mathbf{r})\}$ is the response of the filter to $\exp(+j2\pi ft)\exp(-j2\pi\mathbf{v}\cdot\mathbf{r})$ and $H(f,\mathbf{v};t,\mathbf{r})$ is the time-varying, space-varying transfer function of the filter evaluated at f and $\mathbf{v}$.

Equation (2.1-21) is a fundamental result that will be used in the derivation of ocean medium transfer functions in Chapter 7. In addition, Eq. (2.1-21) can be used to represent the output $y(t,\mathbf{r})$ in terms of the transfer function. For example, if we express the input $x(t,\mathbf{r})$ in terms of the following multidimensional inverse Fourier transform,

$$x(t, \mathbf{r}) = F_f^{-1} F_{\mathbf{v}}^{-1}\{X(f, \mathbf{v})\} \tag{2.1-22}$$

or

$$x(t, \mathbf{r}) = \int_{-\infty}^{\infty} \int_{-\infty}^{\infty} X(f, \mathbf{v}) \exp(+j2\pi f t) \exp(-j2\pi \mathbf{v} \cdot \mathbf{r})\, df\, d\mathbf{v}, \tag{2.1-23}$$

then, from Eqs. (2.1-18) and (2.1-21), we can write the output $y(t, \mathbf{r})$ as

$$y(t, \mathbf{r}) = \int_{-\infty}^{\infty} \int_{-\infty}^{\infty} X(f, \mathbf{v}) L\{\exp(+j2\pi f t) \exp(-j2\pi \mathbf{v} \cdot \mathbf{r})\}\, df\, d\mathbf{v} \tag{2.1-24}$$

or

$$y(t, \mathbf{r}) = \int_{-\infty}^{\infty} \int_{-\infty}^{\infty} X(f, \mathbf{v}) H(f, \mathbf{v}; t, \mathbf{r}) \exp(+j2\pi f t) \exp(-j2\pi \mathbf{v} \cdot \mathbf{r})\, df\, d\mathbf{v}. \tag{2.1-25}$$

Compare Eq. (2.1-25) with Eq. (2.1-2).

2.1.2 Spreading and Bifrequency Functions

Two additional filter functions will now be introduced, namely, the *spreading function* and the *bifrequency function*. The spreading function will be discussed first.

The spreading function $S(\tau, \mathbf{r}_0; \phi, \boldsymbol{\kappa})$ is defined as

$$S(\tau, \mathbf{r}_0; \phi, \boldsymbol{\kappa}) \triangleq F_t F_{\mathbf{r}}\{h(\tau, \mathbf{r}_0; t, \mathbf{r})\} \tag{2.1-26}$$

or

$$S(\tau, \mathbf{r}_0; \phi, \boldsymbol{\kappa}) \triangleq \int_{-\infty}^{\infty} \int_{-\infty}^{\infty} h(\tau, \mathbf{r}_0; t, \mathbf{r}) \exp(-j2\pi \phi t) \exp(+j2\pi \boldsymbol{\kappa} \cdot \mathbf{r})\, dt\, d\mathbf{r}, \tag{2.1-27}$$

where ϕ corresponds to the rate of change of the filter's impulse response in hertz and $\boldsymbol{\kappa}$ is a vector whose components are spatial frequencies that correspond to the rate of change of the impulse response in cycles per meter. Similarly,

$$h(\tau, \mathbf{r}_0; t, \mathbf{r}) = F_\phi^{-1} F_{\boldsymbol{\kappa}}^{-1}\{S(\tau, \mathbf{r}_0; \phi, \boldsymbol{\kappa})\} \tag{2.1-28}$$

or

$$h(\tau, \mathbf{r}_0; t, \mathbf{r}) = \int_{-\infty}^{\infty} \int_{-\infty}^{\infty} S(\tau, \mathbf{r}_0; \phi, \boldsymbol{\kappa}) \exp(+j2\pi \phi t) \exp(-j2\pi \boldsymbol{\kappa} \cdot \mathbf{r})\, d\phi\, d\boldsymbol{\kappa}. \tag{2.1-29}$$

An alternate representation of the output $y(t,\mathbf{r})$ can be obtained by substituting Eq. (2.1-29) into Eq. (2.1-2). Doing so yields

$$y(t,\mathbf{r}) = \int_{-\infty}^{\infty}\int_{-\infty}^{\infty}\int_{-\infty}^{\infty}\int_{-\infty}^{\infty} x(t-\tau,\mathbf{r}-\mathbf{r}_0)\exp(+j2\pi\phi t)\exp(-j2\pi\boldsymbol{\kappa}\cdot\mathbf{r}) \times S(\tau,\mathbf{r}_0;\phi,\boldsymbol{\kappa})\,d\phi\,d\boldsymbol{\kappa}\,d\tau\,d\mathbf{r}_0, \tag{2.1-30}$$

where the integrand is a time-, space-, and "frequency"- shifted (frequencies in hertz and spatial frequencies in cycles per meter) replica of the input signal $x(t,\mathbf{r})$ weighted by the spreading function $S(\tau,\mathbf{r}_0;\phi,\boldsymbol{\kappa})$. Therefore, for a given input (source) location as specified by $\mathbf{r}_0$, the spreading function determines the amount of spread in time delay τ (seconds), frequency ϕ(hertz), and spatial frequencies $\boldsymbol{\kappa}$ (cycles per meter) that an input signal will undergo as it passes through a linear, time-varying, space-varying channel. Equations (2.1-30) and (2.1-2) both indicate that the output $y(t,\mathbf{r})$ at some receiver location $\mathbf{r}$ is dependent on the location $\mathbf{r}_0$ of the input (source) via the spreading function $S(\tau,\mathbf{r}_0;\ \phi,\boldsymbol{\kappa})$ with respect to Eq. (2.1-30) and via the impulse response $h(\tau,\mathbf{r}_0;t,\mathbf{r})$ with respect to Eq. (2.1-2). For example, the existence and extent of SOFAR transmission depends on the depth of both the source and receiver.

The last filter function to be discussed is the bifrequency function. The bifrequency function $B(f,\mathbf{v};\phi,\boldsymbol{\kappa})$ is defined as

$$B(f,\mathbf{v};\phi,\boldsymbol{\kappa}) \triangleq F_t F_{\mathbf{r}}\{H(f,\mathbf{v};t,\mathbf{r})\} \tag{2.1-31}$$

or

$$B(f,\mathbf{v};\phi,\boldsymbol{\kappa}) \triangleq \int_{-\infty}^{\infty}\int_{-\infty}^{\infty} H(f,\mathbf{v};t,\mathbf{r})\exp(-j2\pi\phi t)\exp(+j2\pi\boldsymbol{\kappa}\cdot\mathbf{r})\,dt\,d\mathbf{r}. \tag{2.1-32}$$

Similarly,

$$H(f,\mathbf{v};t,\mathbf{r}) = F_{\phi}^{-1}F_{\boldsymbol{\kappa}}^{-1}\{B(f,\mathbf{v};\phi,\boldsymbol{\kappa})\} \tag{2.1-33}$$

or

$$H(f,\mathbf{v};t,\mathbf{r}) = \int_{-\infty}^{\infty}\int_{-\infty}^{\infty} B(f,\mathbf{v};\phi,\boldsymbol{\kappa})\exp(+j2\pi\phi t)\exp(-j2\pi\boldsymbol{\kappa}\cdot\mathbf{r})\,d\phi\,d\boldsymbol{\kappa}. \tag{2.1-34}$$

The term "bifrequency function" was originally used to denote the appearance of the two frequency variables f and ϕ in linear, time-varying systems theory. We can generalize this notion by using the bifrequency function to denote the appearance of the input "frequency" pair f and $\mathbf{v}$ and the filter variation "frequency" pair ϕ and $\boldsymbol{\kappa}$ in linear, time-varying, space-varying systems theory.

Just as the spreading function gives an indication of how rapidly the impulse response changes with time and space, the bifrequency function gives an indication of how rapidly the transfer function changes with time and space. The bifrequency function can also be obtained by taking the Fourier transform of the spreading function with respect to τ and $\mathbf{r}_0$, that is,

$$B(f, \mathbf{v}; \phi, \boldsymbol{\kappa}) = F_\tau F_{\mathbf{r}_0}\{S(\tau, \mathbf{r}_0; \phi, \boldsymbol{\kappa})\} \tag{2.1-35}$$

or

$$B(f, \mathbf{v}; \phi, \boldsymbol{\kappa}) = \int_{-\infty}^{\infty}\int_{-\infty}^{\infty} S(\tau, \mathbf{r}_0; \phi, \boldsymbol{\kappa}) \exp(-j2\pi f\tau) \exp(+j2\pi \mathbf{v} \cdot \mathbf{r}_0)\, d\tau\, d\mathbf{r}_0, \tag{2.1-36}$$

and, similarly,

$$S(\tau, \mathbf{r}_0; \phi, \boldsymbol{\kappa}) = F_f^{-1} F_\mathbf{v}^{-1}\{B(f, \mathbf{v}; \phi, \boldsymbol{\kappa})\} \tag{2.1-37}$$

or

$$S(\tau, \mathbf{r}_0; \phi, \boldsymbol{\kappa}) = \int_{-\infty}^{\infty}\int_{-\infty}^{\infty} B(f, \mathbf{v}; \phi, \boldsymbol{\kappa}) \exp(+j2\pi f\tau) \exp(-j2\pi \mathbf{v} \cdot \mathbf{r}_0)\, df\, d\mathbf{v}. \tag{2.1-38}$$

In addition, the impulse response and bifrequency functions form a Fourier transform pair, that is,

$$B(f, \mathbf{v}; \phi, \boldsymbol{\kappa}) = F_\tau F_{\mathbf{r}_0} F_t F_\mathbf{r}\{h(\tau, \mathbf{r}_0; t, \mathbf{r})\} \tag{2.1-39}$$

or

$$B(f, \mathbf{v}; \phi, \boldsymbol{\kappa}) = \int_{-\infty}^{\infty}\int_{-\infty}^{\infty}\int_{-\infty}^{\infty}\int_{-\infty}^{\infty} h(\tau, \mathbf{r}_0; t, \mathbf{r}) \exp(-j2\pi f\tau) \exp(+j2\pi \mathbf{v} \cdot \mathbf{r}_0) \times \exp(-j2\pi\phi t) \exp(+j2\pi \boldsymbol{\kappa} \cdot \mathbf{r})\, d\tau\, d\mathbf{r}_0\, dt\, d\mathbf{r}, \tag{2.1-40}$$

and, similarly,

$$h(\tau, \mathbf{r}_0; t, \mathbf{r}) = F_f^{-1} F_\mathbf{v}^{-1} F_\phi^{-1} F_\boldsymbol{\kappa}^{-1}\{B(f, \mathbf{v}; \phi, \boldsymbol{\kappa})\} \tag{2.1-41}$$

or

$$h(\tau, \mathbf{r}_0; t, \mathbf{r}) = \int_{-\infty}^{\infty}\int_{-\infty}^{\infty}\int_{-\infty}^{\infty}\int_{-\infty}^{\infty} B(f, \mathbf{v}; \phi, \boldsymbol{\kappa}) \exp(+j2\pi f\tau) \exp(-j2\pi \mathbf{v} \cdot \mathbf{r}_0) \times \exp(+j2\pi\phi t) \exp(-j2\pi \boldsymbol{\kappa} \cdot \mathbf{r})\, df\, d\mathbf{v}\, d\phi\, d\boldsymbol{\kappa}. \tag{2.1-42}$$

The interdependence that exists among the four filter functions is illustrated in Fig. 2.1-2. Any one of these functions may be used to define completely a linear, time-varying, space-varying communication channel.

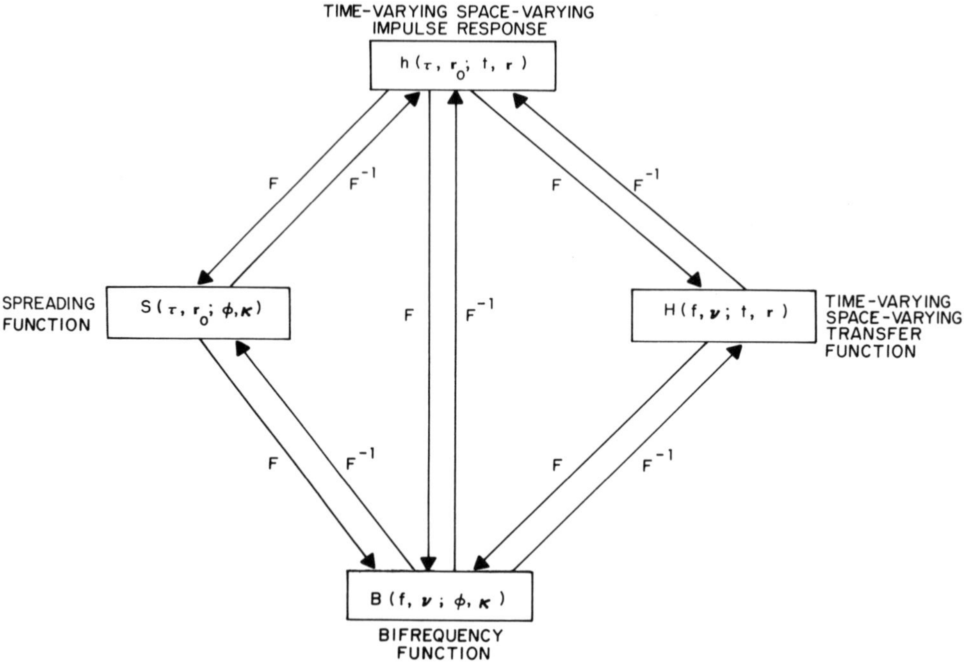

FIG. 2.1-2 Interdependence among the four filter functions that characterize linear, time-varying, space-varying communication channels.

2.1.3 Output Frequency and Angular Spectrum

The output frequency η and angular ($\boldsymbol{\beta}$) spectrum $Y(\eta, \boldsymbol{\beta})$ is defined as

$$Y(\eta, \boldsymbol{\beta}) \triangleq F_t F_{\mathbf{r}}\{y(t, \mathbf{r})\} \tag{2.1-43}$$

or

$$Y(\eta, \boldsymbol{\beta}) \triangleq \int_{-\infty}^{\infty} \int_{-\infty}^{\infty} y(t, \mathbf{r}) \exp(-j2\pi\eta t) \exp(+j2\pi\boldsymbol{\beta} \cdot \mathbf{r})\, dt\, d\mathbf{r}, \tag{2.1-44}$$

where η corresponds to output frequencies in hertz and $\boldsymbol{\beta}$ is a vector whose components are output spatial frequencies with units of cycles per meter. Substituting Eq. (2.1-25) into Eq. (2.1-44) yields

$$Y(\eta, \boldsymbol{\beta}) = \int_{-\infty}^{\infty} \int_{-\infty}^{\infty} X(f, \mathbf{v}) \int_{-\infty}^{\infty} \int_{-\infty}^{\infty} H(f, \mathbf{v}; t, \mathbf{r}) \exp[-j2\pi(\eta - f)t] \times \exp[+j2\pi(\boldsymbol{\beta} - \mathbf{v}) \cdot \mathbf{r}]\, dt\, d\mathbf{r}\, df\, d\mathbf{v}, \tag{2.1-45}$$

where, from Eq. (2.1-32), it can be seen that the inner multidimensional integral is equal to $B(f, \mathbf{v}; \eta - f, \boldsymbol{\beta} - \mathbf{v})$, so that

$$Y(\eta, \boldsymbol{\beta}) = \int_{-\infty}^{\infty} \int_{-\infty}^{\infty} X(f, \mathbf{v}) B(f, \mathbf{v}; \eta - f, \boldsymbol{\beta} - \mathbf{v})\, df\, d\mathbf{v}, \qquad (2.1\text{-}46)$$

which is in the form of a *multidimensional "frequency" domain convolution integral.* Note that $\phi \equiv \eta - f$ or $\eta \equiv f + \phi$ and that $\boldsymbol{\kappa} \equiv \boldsymbol{\beta} - \mathbf{v}$ or $\boldsymbol{\beta} \equiv \mathbf{v} + \boldsymbol{\kappa}$. Thus, the output frequencies η in hertz are equal to the sum of the input frequencies f and the variations in frequency ϕ due to the time-varying property of the filter. Similarly, the output spatial frequencies $\boldsymbol{\beta}$ (directions of wave propagation) in cycles per meter are equal to the sum of the input spatial frequencies $\mathbf{v}$ (directions of propagation of transmitted plane waves) and the variations in spatial frequency (directions of wave propagation) $\boldsymbol{\kappa}$ due to the space-varying property of the filter. Equation (2.1-46) demonstrates that a linear, time-variant, space-variant filter will spread the input frequency and angular spectrum $X(f, \mathbf{v})$ in both frequencies in hertz and spatial frequencies in cycles per meter.

EXAMPLE 2.1-3

If the linear filter h is time-invariant and space-invariant, then $H(f, \mathbf{v}; t, \mathbf{r}) = H(f, \mathbf{v})$ [see Eq. (2.1-16)] and, as a result, Eq. (2.1-32) reduces to

$$B(f, \mathbf{v}; \phi, \boldsymbol{\kappa}) = H(f, \mathbf{v}) \int_{-\infty}^{\infty} \exp(-j2\pi\phi t)\, dt \int_{-\infty}^{\infty} \exp(+j2\pi\boldsymbol{\kappa} \cdot \mathbf{r})\, d\mathbf{r}$$

or

$$B(f, \mathbf{v}; \phi, \boldsymbol{\kappa}) = H(f, \mathbf{v})\, \delta(\phi)\, \delta(\boldsymbol{\kappa}) \qquad (2.1\text{-}47)$$

since

$$F_t\{1\} = \delta(\phi) \qquad \text{and} \qquad F_{\mathbf{r}}\{1\} = \delta(\boldsymbol{\kappa}).$$

Substituting Eq. (2.1-47) into Eq. (2.1-46) yields

$$Y(\eta, \boldsymbol{\beta}) = \int_{-\infty}^{\infty} \int_{-\infty}^{\infty} X(f, \mathbf{v}) H(f, \mathbf{v})\, \delta(\eta - f)\, \delta(\boldsymbol{\beta} - \mathbf{v})\, df\, d\mathbf{v},$$

which simplifies to

$$Y(\eta, \boldsymbol{\beta}) = X(\eta, \boldsymbol{\beta}) H(\eta, \boldsymbol{\beta}),$$

and by replacing η with f and $\boldsymbol{\beta}$ with $\mathbf{v}$, we finally obtain

$$Y(f, \mathbf{v}) = X(f, \mathbf{v}) H(f, \mathbf{v}), \qquad (2.1\text{-}48)$$

which is the output frequency and angular spectrum from a linear, time-invariant, space-invariant filter. Note that the output frequency f and the output spatial-frequency vector $\mathbf{v}$ are identical with those of the input $X(f, \mathbf{v})$. Hence, as would be expected, there is *no* "frequency" spreading. This would correspond to the physical situation of transmitting a signal via a transmit aperture (array) to a receive aperture (array) when the platforms containing the apertures (arrays) are *not* in motion and the intervening ocean medium has a *constant* speed of sound (index of refraction) and no discrete point scatterers in motion. No motion implies no Doppler (frequency) spread, and a constant speed of sound implies that the sound rays travel in straight lines, that is, there is no angular spread (scatter) or change in the direction of propagation of the transmitted sound rays. □

2.2 RANDOM FILTERS

2.2.1 Filter Autocorrelation Functions

In Section 2.1 four system functions were introduced that are used to characterize linear, time-variant, space-variant filters. However, if the filter is random, then each of these system functions must be considered as random functions. As a result, we must work with the respective system autocorrelation functions, which are defined as follows:

$$R_h(\tau, \tau', \mathbf{r}_0, \mathbf{r}'_0; t, t', \mathbf{r}, \mathbf{r}') \triangleq \mathrm{E}\{h(\tau, \mathbf{r}_0; t, \mathbf{r})h^*(\tau', \mathbf{r}'_0; t', \mathbf{r}')\}, \tag{2.2-1}$$

$$R_H(f, f', \mathbf{v}, \mathbf{v}'; t, t', \mathbf{r}, \mathbf{r}') \triangleq \mathrm{E}\{H(f, \mathbf{v}; t, \mathbf{r})H^*(f', \mathbf{v}'; t', \mathbf{r}')\}, \tag{2.2-2}$$

$$R_S(\tau, \tau', \mathbf{r}_0, \mathbf{r}'_0; \phi, \phi', \boldsymbol{\kappa}, \boldsymbol{\kappa}') \triangleq \mathrm{E}\{S(\tau, \mathbf{r}_0; \phi, \boldsymbol{\kappa})S^*(\tau', \mathbf{r}'_0; \phi', \boldsymbol{\kappa}')\}, \tag{2.2-3}$$

and

$$R_B(f, f', \mathbf{v}, \mathbf{v}'; \phi, \phi', \boldsymbol{\kappa}, \boldsymbol{\kappa}') \triangleq \mathrm{E}\{B(f, \mathbf{v}; \phi, \boldsymbol{\kappa})B^*(f', \mathbf{v}'; \phi', \boldsymbol{\kappa}')\}, \tag{2.2-4}$$

where $\mathrm{E}\{\cdot\}$ is the expectation or ensemble average operator and the asterisk denotes complex conjugate.

If we use the sign convention that *forward* transforms with respect to τ, t, $\mathbf{r}'_0$, and $\mathbf{r}'$ are defined with a minus sign in the exponent of the complex exponential (*inverse* transforms with respect to f, ϕ, $\mathbf{v}'$, and $\boldsymbol{\kappa}'$ are defined with a plus sign) and *forward* transforms with respect to τ', t', $\mathbf{r}_0$, and $\mathbf{r}$ are defined with a plus sign (*inverse* transforms with respect to f', ϕ', $\mathbf{v}$, and $\boldsymbol{\kappa}$ are defined with a minus sign), then it can be shown that the four system autocorrelation functions are related by the following Fourier transform pairs:

$$R_H(f, f', \mathbf{v}, \mathbf{v}'; t, t', \mathbf{r}, \mathbf{r}') = F_\tau F_{\tau'} F_{\mathbf{r}_0} F_{\mathbf{r}'_0}\{R_h(\tau, \tau', \mathbf{r}_0, \mathbf{r}'_0; t, t', \mathbf{r}, \mathbf{r}')\} \tag{2.2-5}$$

and

$$R_h(\tau,\tau',\mathbf{r}_0,\mathbf{r}'_0;t,t',\mathbf{r},\mathbf{r}') = F_f^{-1}F_{f'}^{-1}F_{\mathbf{v}}^{-1}F_{\mathbf{v}'}^{-1}\{R_H(f,f',\mathbf{v},\mathbf{v}';t,t',\mathbf{r},\mathbf{r}')\}; \tag{2.2-6}$$

$$R_B(f,f',\mathbf{v},\mathbf{v}';\phi,\phi',\boldsymbol{\kappa},\boldsymbol{\kappa}') = F_tF_{t'}F_{\mathbf{r}}F_{\mathbf{r}'}\{R_H(f,f',\mathbf{v},\mathbf{v}';t,t',\mathbf{r},\mathbf{r}')\} \tag{2.2-7}$$

and

$$R_H(f,f',\mathbf{v},\mathbf{v}';t,t',\mathbf{r},\mathbf{r}') = F_\phi^{-1}F_{\phi'}^{-1}F_{\boldsymbol{\kappa}}^{-1}F_{\boldsymbol{\kappa}'}^{-1}\{R_B(f,f',\mathbf{v},\mathbf{v}';\phi,\phi',\boldsymbol{\kappa},\boldsymbol{\kappa}')\}; \tag{2.2-8}$$

$$R_S(\tau,\tau',\mathbf{r}_0,\mathbf{r}'_0;\phi,\phi',\boldsymbol{\kappa},\boldsymbol{\kappa}') = F_tF_{t'}F_{\mathbf{r}}F_{\mathbf{r}'}\{R_h(\tau,\tau',\mathbf{r}_0,\mathbf{r}'_0;t,t',\mathbf{r},\mathbf{r}')\} \tag{2.2-9}$$

and

$$R_h(\tau,\tau',\mathbf{r}_0,\mathbf{r}'_0;t,t',\mathbf{r},\mathbf{r}') = F_\phi^{-1}F_{\phi'}^{-1}F_{\boldsymbol{\kappa}}^{-1}F_{\boldsymbol{\kappa}'}^{-1}\{R_S(\tau,\tau',\mathbf{r}_0,\mathbf{r}'_0;\phi,\phi',\boldsymbol{\kappa},\boldsymbol{\kappa}')\}; \tag{2.2-10}$$

$$R_B(f,f',\mathbf{v},\mathbf{v}';\phi,\phi',\boldsymbol{\kappa},\boldsymbol{\kappa}') = F_\tau F_{\tau'}F_{\mathbf{r}_0}F_{\mathbf{r}'_0}\{R_S(\tau,\tau',\mathbf{r}_0,\mathbf{r}'_0;\phi,\phi',\boldsymbol{\kappa},\boldsymbol{\kappa}')\} \tag{2.2-11}$$

and

$$R_S(\tau,\tau',\mathbf{r}_0,\mathbf{r}'_0;\phi,\phi',\boldsymbol{\kappa},\boldsymbol{\kappa}') = F_f^{-1}F_{f'}^{-1}F_{\mathbf{v}}^{-1}F_{\mathbf{v}'}^{-1}\{R_B(f,f',\mathbf{v},\mathbf{v}';\phi,\phi',\boldsymbol{\kappa},\boldsymbol{\kappa}')\}; \tag{2.2-12}$$

and finally,

$$R_B(f,f',\mathbf{v},\mathbf{v}';\phi,\phi',\boldsymbol{\kappa},\boldsymbol{\kappa}') = F_\tau F_{\tau'}F_{\mathbf{r}_0}F_{\mathbf{r}'_0}F_tF_{t'}F_{\mathbf{r}}F_{\mathbf{r}'}\{R_h(\tau,\tau',\mathbf{r}_0,\mathbf{r}'_0;t,t',\mathbf{r},\mathbf{r}')\} \tag{2.2-13}$$

and

$$R_h(\tau,\tau',\mathbf{r}_0,\mathbf{r}'_0;t,t',\mathbf{r},\mathbf{r}') = F_f^{-1}F_{f'}^{-1}F_{\mathbf{v}}^{-1}F_{\mathbf{v}'}^{-1}F_\phi^{-1}F_{\phi'}^{-1}F_{\boldsymbol{\kappa}}^{-1}F_{\boldsymbol{\kappa}'}^{-1} \times \{R_B(f,f',\mathbf{v},\mathbf{v}';\phi,\phi',\boldsymbol{\kappa},\boldsymbol{\kappa}')\}. \tag{2.2-14}$$

The interdependence that exists among the four filter autocorrelation functions is illustrated in Fig. 2.2-1.

2.2.2 Uncorrelated Spreading and the Scattering Function

In this section we shall examine the consequences of assuming that the spreading function $S(\tau,\mathbf{r}_0;\phi,\boldsymbol{\kappa})$ is uncorrelated with $S(\tau',\mathbf{r}'_0;\phi',\boldsymbol{\kappa}')$ for all values of $\tau' \neq \tau$, $\mathbf{r}'_0 \neq \mathbf{r}_0$, $\phi' \neq \phi$, and $\boldsymbol{\kappa}' \neq \boldsymbol{\kappa}$.

The assumption of *uncorrelated spreading* is mathematically equivalent to stating that the *autocovariance* of $S(\tau,\mathbf{r}_0;\phi,\boldsymbol{\kappa})$ and $S(\tau',\mathbf{r}'_0;\phi',\boldsymbol{\kappa}')$ is zero for all

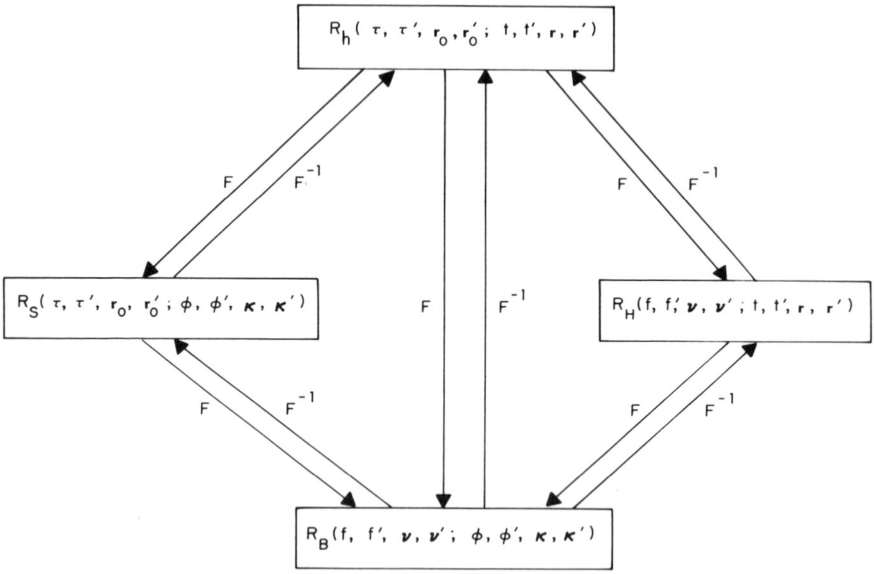

FIG. 2.2-1 Interdependence among the four filter autocorrelation functions that characterize linear, time-varying, space-varying, random communication channels.

values of $\tau' \neq \tau$, $\mathbf{r}'_0 \neq \mathbf{r}_0$, $\phi' \neq \phi$, and $\boldsymbol{\kappa}' \neq \boldsymbol{\kappa}$, that is,

$$C_S(\tau, \tau', \mathbf{r}_0, \mathbf{r}'_0; \phi, \phi', \boldsymbol{\kappa}, \boldsymbol{\kappa}') = R_S(\tau, \tau', \mathbf{r}_0, \mathbf{r}'_0; \phi, \phi', \boldsymbol{\kappa}, \boldsymbol{\kappa}') - \mu_S(\tau, \mathbf{r}_0; \phi, \boldsymbol{\kappa})\mu_S^*(\tau', \mathbf{r}'_0; \phi', \boldsymbol{\kappa}') = 0; \quad (2.2\text{-}15)$$

$\tau' \neq \tau, \mathbf{r}'_0 \neq \mathbf{r}_0, \phi' \neq \phi$, and $\boldsymbol{\kappa}' \neq \boldsymbol{\kappa}$, where C_S is the *autocovariance function* and $\mu_S(\tau, \mathbf{r}_0; \phi, \boldsymbol{\kappa}) = \mathrm{E}\{S(\tau, \mathbf{r}_0; \phi, \boldsymbol{\kappa})\}$. If it is assumed that $\mu_S(\tau, \mathbf{r}_0; \phi, \boldsymbol{\kappa}) = 0$, then Eq. (2.2-15) is equivalent to

$$R_S(\tau, \tau', \mathbf{r}_0, \mathbf{r}'_0; \phi, \phi', \boldsymbol{\kappa}, \boldsymbol{\kappa}') = R_S(\tau, \mathbf{r}_0; \phi, \boldsymbol{\kappa})\,\delta(\tau - \tau') \times \delta(\mathbf{r}_0 - \mathbf{r}'_0)\,\delta(\phi - \phi')\,\delta(\boldsymbol{\kappa} - \boldsymbol{\kappa}'), \quad (2.2\text{-}16)$$

where

$$R_S(\tau, \mathbf{r}_0; \phi, \boldsymbol{\kappa}) = \mathrm{E}\{|S(\tau, \mathbf{r}_0, \phi, \boldsymbol{\kappa})|^2\} \quad (2.2\text{-}17)$$

is called the *scattering function* and is equal to the mean-square value of the spreading function. The scattering function can be thought of as an average power density function that determines the average amount of spread that an input signal's power will undergo as a function of time delay τ, frequency ϕ, and spatial frequencies $\boldsymbol{\kappa}$ for a given input (source) location $\mathbf{r}_0$. Note that the scattering function is a real, positive-valued function.

Equation (2.2-16) is the result of the assumption that the spreading function is zero mean. However, if the spreading function is nonzero mean, it is convenient to do the analysis with the *centered process*:

$$S_C(\tau, \mathbf{r}_0; \phi, \boldsymbol{\kappa}) = S(\tau, \mathbf{r}_0; \phi, \boldsymbol{\kappa}) - \mu_S(\tau, \mathbf{r}_0; \phi, \boldsymbol{\kappa}).$$

The random process $S_C(\tau, \mathbf{r}_0; \phi, \boldsymbol{\kappa})$ has zero mean and, as a result, its autocovariance function is equal to

$$\begin{aligned} C_{S_C}(\tau, \tau', \mathbf{r}_0, \mathbf{r}_0'; \phi, \phi', \boldsymbol{\kappa}, \boldsymbol{\kappa}') &= \mathrm{E}\{S_C(\tau, \mathbf{r}_0; \phi, \boldsymbol{\kappa}) S_C^*(\tau', \mathbf{r}_0'; \phi', \boldsymbol{\kappa}')\} \\ &= R_{S_C}(\tau, \tau', \mathbf{r}_0, \mathbf{r}_0'; \phi, \phi', \boldsymbol{\kappa}, \boldsymbol{\kappa}') \\ &= C_S(\tau, \tau', \mathbf{r}_0, \mathbf{r}_0'; \phi, \phi', \boldsymbol{\kappa}, \boldsymbol{\kappa}'), \end{aligned}$$

that is, the autocorrelation function of the zero-mean centered process is equal to the autocovariance function of the original nonzero-mean spreading function.

Let us next examine the effect that the assumption of uncorrelated spreading has on the remaining three system autocorrelation functions. As Fig. 2.2-1 indicates, R_H can be obtained from R_S by performing the following transformations:

$$R_H(f, f', \mathbf{v}, \mathbf{v}'; t, t', \mathbf{r}, \mathbf{r}') = F_\phi^{-1} F_{\phi'}^{-1} F_{\boldsymbol{\kappa}}^{-1} F_{\boldsymbol{\kappa}'}^{-1} F_\tau F_{\tau'} F_{\mathbf{r}_0} F_{\mathbf{r}_0'} \{R_S(\tau, \tau', \mathbf{r}_0, \mathbf{r}_0'; \phi, \phi', \boldsymbol{\kappa}, \boldsymbol{\kappa}')\} \tag{2.2-18}$$

or

$$\begin{aligned} R_H(f, f', \mathbf{v}, \mathbf{v}'; t, t', \mathbf{r}, \mathbf{r}') = \int_{-\infty}^{\infty} \cdots \int_{-\infty}^{\infty} & R_S(\tau, \tau', \mathbf{r}_0, \mathbf{r}_0'; \phi, \phi', \boldsymbol{\kappa}, \boldsymbol{\kappa}') \\ & \times \exp[+j2\pi(\phi t - \phi' t')] \exp[-j2\pi(\boldsymbol{\kappa} \cdot \mathbf{r} - \boldsymbol{\kappa}' \cdot \mathbf{r}')] \\ & \times \exp[-j2\pi(f\tau - f'\tau')] \exp[+j2\pi(\mathbf{v} \cdot \mathbf{r}_0 - \mathbf{v}' \cdot \mathbf{r}_0')] \\ & \times d\phi\, d\phi'\, d\boldsymbol{\kappa}\, d\boldsymbol{\kappa}'\, d\tau\, d\tau'\, d\mathbf{r}_0\, d\mathbf{r}_0'. \end{aligned} \tag{2.2-19}$$

If Eq. (2.2-16) is substituted into Eq. (2.2-19), then

$$R_H(f, f', \mathbf{v}, \mathbf{v}'; t, t', \mathbf{r}, \mathbf{r}') = R_H(\Delta f, \Delta\mathbf{v}; \Delta t, \Delta\mathbf{r}), \tag{2.2-20}$$

where

$$\begin{aligned} R_H(\Delta f, \Delta\mathbf{v}; \Delta t, \Delta\mathbf{r}) = \int_{-\infty}^{\infty} \int_{-\infty}^{\infty} \int_{-\infty}^{\infty} \int_{-\infty}^{\infty} & R_S(\tau, \mathbf{r}_0; \phi, \boldsymbol{\kappa}) \exp(+j2\pi\phi\Delta t) \\ & \times \exp(-j2\pi\boldsymbol{\kappa} \cdot \Delta\mathbf{r}) \exp(-j2\pi\,\Delta f \tau) \\ & \times \exp(+j2\pi\,\Delta\mathbf{v} \cdot \mathbf{r}_0)\, d\phi\, d\boldsymbol{\kappa}\, d\tau\, d\mathbf{r}_0, \end{aligned} \tag{2.2-21}$$

$\Delta f = f - f', \Delta\mathbf{v} = \mathbf{v} - \mathbf{v}', \Delta t = t - t'$, and $\Delta\mathbf{r} = \mathbf{r} - \mathbf{r}'$. It can be seen from Eq. (2.2-20) that when uncorrelated spreading is assumed, the autocorrelation function R_H becomes a function of the differences Δf, $\Delta\mathbf{v}$, Δt, and $\Delta\mathbf{r}$ only. This implies that the random process $H(f, \mathbf{v}; t, \mathbf{r})$ is wide-sense stationary in frequency, spatial frequencies, time, and space. An additional requirement for $H(f, \mathbf{v}; t, \mathbf{r})$ to be wide-sense stationary is that the mean value

$$\mu_H(f, \mathbf{v}; t, \mathbf{r}) = \mathrm{E}\{H(f, \mathbf{v}; t, \mathbf{r})\}$$

is a constant. Since the four filter functions are related by linear transformations (see Fig. 2.1-2), and since it was assumed that $\mu_S(\tau, \mathbf{r}_0; \phi, \boldsymbol{\kappa}) = 0$, then $\mu_h(\tau, \mathbf{r}_0; t, \mathbf{r}) = 0$, $\mu_H(f, \mathbf{v}; t, \mathbf{r}) = 0$, and $\mu_B(f, \mathbf{v}; \phi, \boldsymbol{\kappa}) = 0$, which are constant, zero-mean values. Therefore, the condition of uncorrelated spreading in time delay τ, input (source) location $\mathbf{r}_0$, frequency ϕ, and spatial frequencies $\boldsymbol{\kappa}$ is equivalent to a condition of wide-sense stationarity in frequency Δf, spatial frequencies $\Delta\mathbf{v}$, time Δt, and output (receiver) location $\Delta\mathbf{r}$, respectively. If uncorrelated spreading in τ, $\mathbf{r}_0$, ϕ, and $\boldsymbol{\kappa}$ occur together, then we have a *wide-sense stationary uncorrelated spreading* (WSSUS) communication channel.

Consider the autocorrelation function R_h next. If Eq. (2.2-16) is substituted into Eq. (2.2-10) and the indicated transformations are performed, then

$$R_h(\tau, \tau', \mathbf{r}_0, \mathbf{r}_0'; t, t', \mathbf{r}, \mathbf{r}') = R_h(\tau, \mathbf{r}_0; \Delta t, \Delta\mathbf{r})\, \delta(\tau - \tau')\, \delta(\mathbf{r}_0 - \mathbf{r}_0'), \quad (2.2\text{-}22)$$

where

$$R_h(\tau, \mathbf{r}_0; \Delta t, \Delta\mathbf{r}) = \int_{-\infty}^{\infty}\int_{-\infty}^{\infty} R_S(\tau, \mathbf{r}_0; \phi, \boldsymbol{\kappa}) \exp(+j2\pi\phi\,\Delta t) \times \exp(-j2\pi\boldsymbol{\kappa}\cdot\Delta\mathbf{r})\, d\phi\, d\boldsymbol{\kappa}. \quad (2.2\text{-}23)$$

Equation (2.2-22) indicates that the random process $h(\tau, \mathbf{r}_0; t, \mathbf{r})$ is wide-sense stationary in time and space because of the Δt and $\Delta\mathbf{r}$ dependence and because $\mu_h(\tau, \mathbf{r}_0; t, \mathbf{r}) = 0$. Equation (2.2-22) also indicates that $h(\tau, \mathbf{r}_0; t, \mathbf{r})$ is uncorrelated for all values of $\tau' \neq \tau$ and $\mathbf{r}_0' \neq \mathbf{r}_0$.

Finally, if Eq. (2.2-16) is substituted into Eq. (2.2-11) and the indicated transformations are performed, then

$$R_B(f, f', \mathbf{v}, \mathbf{v}'; \phi, \phi', \boldsymbol{\kappa}, \boldsymbol{\kappa}') = R_B(\Delta f, \Delta\mathbf{v}; \phi, \boldsymbol{\kappa})\, \delta(\phi - \phi')\, \delta(\boldsymbol{\kappa} - \boldsymbol{\kappa}'), \quad (2.2\text{-}24)$$

where

$$R_B(\Delta f, \Delta\mathbf{v}; \phi, \boldsymbol{\kappa}) = \int_{-\infty}^{\infty}\int_{-\infty}^{\infty} R_S(\tau, \mathbf{r}_0; \phi, \boldsymbol{\kappa}) \exp(-j2\pi\,\Delta f\tau) \times \exp(+j2\pi\,\Delta\mathbf{v}\cdot\mathbf{r}_0)\, d\tau\, d\mathbf{r}_0. \quad (2.2\text{-}25)$$

Equation (2.2-24) indicates that the random process $B(f, \mathbf{v}; \phi, \boldsymbol{\kappa})$ is wide-sense stationary in frequency and spatial frequencies because of the Δf and $\Delta\mathbf{v}$ dependence and because $\mu_B(f, \mathbf{v}; \phi, \boldsymbol{\kappa}) = 0$. Equation (2.2-24) also indicates that $B(f, \mathbf{v}; \phi, \boldsymbol{\kappa})$ is uncorrelated for all values of $\phi' \neq \phi$ and $\boldsymbol{\kappa}' \neq \boldsymbol{\kappa}$.

Therefore, in summary, under the assumption of *uncorrelated spreading* the four filter autocorrelation functions reduce as follows:

$$R_S(\tau,\tau',\mathbf{r}_0,\mathbf{r}_0';\phi,\phi',\boldsymbol{\kappa},\boldsymbol{\kappa}') = R_S(\tau,\mathbf{r}_0;\phi,\boldsymbol{\kappa})\,\delta(\tau-\tau')\,\delta(\mathbf{r}_0-\mathbf{r}_0') \times \delta(\phi-\phi')\,\delta(\boldsymbol{\kappa}-\boldsymbol{\kappa}'), \tag{2.2-16}$$

$$R_H(f,f',\mathbf{v},\mathbf{v}';t,t',\mathbf{r},\mathbf{r}') = R_H(\Delta f,\Delta\mathbf{v};\Delta t,\Delta\mathbf{r}), \tag{2.2-20}$$

$$R_h(\tau,\tau',\mathbf{r}_0,\mathbf{r}_0';t,t',\mathbf{r},\mathbf{r}') = R_h(\tau,\mathbf{r}_0;\Delta t,\Delta\mathbf{r})\,\delta(\tau-\tau')\,\delta(\mathbf{r}_0-\mathbf{r}_0'), \tag{2.2-22}$$

and

$$R_B(f,f',\mathbf{v},\mathbf{v}';\phi,\phi',\boldsymbol{\kappa},\boldsymbol{\kappa}') = R_B(\Delta f,\Delta\mathbf{v};\phi,\boldsymbol{\kappa})\,\delta(\phi-\phi')\,\delta(\boldsymbol{\kappa}-\boldsymbol{\kappa}'), \tag{2.2-24}$$

where $R_S(\tau,\mathbf{r}_0;\phi,\boldsymbol{\kappa})$ is the scattering function.

2.2.3 The Scattering Function and Its Fourier Transforms

By inspecting Eq. (2.2-23), we can write that

$$R_h(\tau,\mathbf{r}_0;\Delta t,\Delta\mathbf{r}) = F_\phi^{-1}F_{\boldsymbol{\kappa}}^{-1}\{R_S(\tau,\mathbf{r}_0;\phi,\boldsymbol{\kappa})\} \tag{2.2-26}$$

and

$$R_S(\tau,\mathbf{r}_0;\phi,\boldsymbol{\kappa}) = F_{\Delta t}F_{\Delta\mathbf{r}}\{R_h(\tau,\mathbf{r}_0;\Delta t,\Delta\mathbf{r})\}, \tag{2.2-27}$$

or

$$R_S(\tau,\mathbf{r}_0;\phi,\boldsymbol{\kappa}) = \int_{-\infty}^{\infty}\int_{-\infty}^{\infty} R_h(\tau,\mathbf{r}_0;\Delta t,\Delta\mathbf{r})\exp(-j2\pi\phi\,\Delta t) \times \exp(+j2\pi\boldsymbol{\kappa}\cdot\Delta\mathbf{r})\,d\Delta t\,d\Delta\mathbf{r}. \tag{2.2-28}$$

Also, by inspecting Eq. (2.2-25), we can write that

$$R_B(\Delta f,\Delta\mathbf{v};\phi,\boldsymbol{\kappa}) = F_\tau F_{\mathbf{r}_0}\{R_S(\tau,\mathbf{r}_0;\phi,\boldsymbol{\kappa})\} \tag{2.2-29}$$

and

$$R_S(\tau,\mathbf{r}_0;\phi,\boldsymbol{\kappa}) = F_{\Delta f}^{-1}F_{\Delta\mathbf{v}}^{-1}\{R_B(\Delta f,\Delta\mathbf{v};\phi,\boldsymbol{\kappa})\}, \tag{2.2-30}$$

or

$$R_S(\tau,\mathbf{r}_0;\phi,\boldsymbol{\kappa}) = \int_{-\infty}^{\infty}\int_{-\infty}^{\infty} R_B(\Delta f,\Delta\mathbf{v};\phi,\boldsymbol{\kappa})\exp(+j2\pi\,\Delta f\tau) \times \exp(-j2\pi\,\Delta\mathbf{v}\cdot\mathbf{r}_0)\,d\Delta f\,d\Delta\mathbf{v}. \tag{2.2-31}$$

Additional transform pairs can be obtained as follows. With the use of Eq. (2.2-23), Eq. (2.2-21) can be expressed as

$$R_H(\Delta f, \Delta \mathbf{v}; \Delta t, \Delta \mathbf{r}) = F_\tau F_{\mathbf{r}_0}\{R_h(\tau, \mathbf{r}_0; \Delta t, \Delta \mathbf{r})\} \tag{2.2-32}$$

or

$$R_H(\Delta f, \Delta \mathbf{v}; \Delta t, \Delta \mathbf{r}) = \int_{-\infty}^{\infty}\int_{-\infty}^{\infty} R_h(\tau, \mathbf{r}_0; \Delta t, \Delta \mathbf{r}) \exp(-j2\pi\,\Delta f \tau) \times \exp(+j2\pi\,\Delta \mathbf{v}\cdot\mathbf{r}_0)\,d\tau\,d\mathbf{r}_0 \tag{2.2-33}$$

and

$$R_h(\tau, \mathbf{r}_0; \Delta t, \Delta \mathbf{r}) = F_{\Delta f}^{-1} F_{\Delta \mathbf{v}}^{-1}\{R_H(\Delta f, \Delta \mathbf{v}; \Delta t, \Delta \mathbf{r})\} \tag{2.2-34}$$

or

$$R_h(\tau, \mathbf{r}_0; \Delta t, \Delta \mathbf{r}) = \int_{-\infty}^{\infty}\int_{-\infty}^{\infty} R_H(\Delta f, \Delta \mathbf{v}; \Delta t, \Delta \mathbf{r}) \exp(+j2\pi\,\Delta f \tau) \times \exp(-j2\pi\,\Delta \mathbf{v}\cdot\mathbf{r}_0)\,d\Delta f\,d\Delta \mathbf{v}. \tag{2.2-35}$$

Next, if Eq. (2.2-28) is substituted into Eq. (2.2-25), then

$$R_B(\Delta f, \Delta \mathbf{v}; \phi, \boldsymbol{\kappa}) = F_\tau F_{\mathbf{r}_0} F_{\Delta t} F_{\Delta \mathbf{r}}\{R_h(\tau, \mathbf{r}_0; \Delta t, \Delta \mathbf{r})\} \tag{2.2-36}$$

or

$$R_B(\Delta f, \Delta \mathbf{v}; \phi, \boldsymbol{\kappa}) = \int_{-\infty}^{\infty}\int_{-\infty}^{\infty}\int_{-\infty}^{\infty}\int_{-\infty}^{\infty} R_h(\tau, \mathbf{r}_0; \Delta t, \Delta \mathbf{r}) \exp(-j2\pi\,\Delta f \tau) \times \exp(+j2\pi\,\Delta \mathbf{v}\cdot\mathbf{r}_0)\exp(-j2\pi\phi\,\Delta t) \times \exp(+j2\pi\boldsymbol{\kappa}\cdot\Delta\mathbf{r})\,d\tau\,d\mathbf{r}_0\,d\Delta t\,d\Delta\mathbf{r} \tag{2.2-37}$$

and

$$R_h(\tau, \mathbf{r}_0; \Delta t, \Delta \mathbf{r}) = F_{\Delta f}^{-1} F_{\Delta \mathbf{v}}^{-1} F_{\phi}^{-1} F_{\boldsymbol{\kappa}}^{-1}\{R_B(\Delta f, \Delta \mathbf{v}; \phi, \boldsymbol{\kappa})\} \tag{2.2-38}$$

or

$$R_h(\tau, \mathbf{r}_0; \Delta t, \Delta \mathbf{r}) = \int_{-\infty}^{\infty}\int_{-\infty}^{\infty}\int_{-\infty}^{\infty}\int_{-\infty}^{\infty} R_B(\Delta f, \Delta \mathbf{v}; \phi, \boldsymbol{\kappa}) \exp(+j2\pi\,\Delta f \tau) \times \exp(-j2\pi\,\Delta \mathbf{v}\cdot\mathbf{r}_0)\exp(+j2\pi\phi\,\Delta \mathrm{t}) \times \exp(-j2\pi\boldsymbol{\kappa}\cdot\Delta\mathbf{r})\,d\Delta f\,d\Delta\mathbf{v}\,d\phi\,d\boldsymbol{\kappa}. \tag{2.2-39}$$

Finally, with the use of Eq. (2.2-33), Eq. (2.2-37) can be expressed as

$$R_B(\Delta f, \Delta \mathbf{v}; \phi, \boldsymbol{\kappa}) = F_{\Delta t} F_{\Delta \mathbf{r}} \{ R_H(\Delta f, \Delta \mathbf{v}; \Delta t, \Delta \mathbf{r}) \} \tag{2.2-40}$$

or

$$R_B(\Delta f, \Delta \mathbf{v}; \phi, \boldsymbol{\kappa}) = \int_{-\infty}^{\infty} \int_{-\infty}^{\infty} R_H(\Delta f, \Delta \mathbf{v}; \Delta t, \Delta \mathbf{r}) \exp(-j2\pi\phi\, \Delta t) \times \exp(+j2\pi \boldsymbol{\kappa} \cdot \Delta \mathbf{r})\, d\Delta t\, d\Delta \mathbf{r} \tag{2.2-41}$$

and

$$R_H(\Delta f, \Delta \mathbf{v}; \Delta t, \Delta \mathbf{r}) = F_{\phi}^{-1} F_{\boldsymbol{\kappa}}^{-1} \{ R_B(\Delta f, \Delta \mathbf{v}; \phi, \boldsymbol{\kappa}) \} \tag{2.2-42}$$

or

$$R_H(\Delta f, \Delta \mathbf{v}; \Delta t, \Delta \mathbf{r}) = \int_{-\infty}^{\infty} \int_{-\infty}^{\infty} R_B(\Delta f, \Delta \mathbf{v}; \phi, \boldsymbol{\kappa}) \exp(+j2\pi\phi\, \Delta t) \times \exp(-j2\pi \boldsymbol{\kappa} \cdot \Delta \mathbf{r})\, d\phi\, d\boldsymbol{\kappa}. \tag{2.2-43}$$

The various Fourier transform pairs are summarized in Fig. 2.2-2. The scattering function, or any of its Fourier transforms, is a complete second-order statistical description of a WSSUS communication channel.

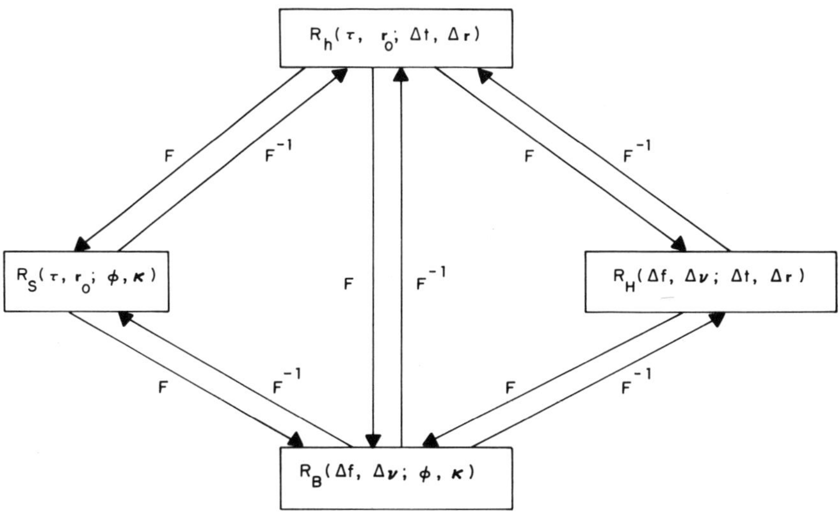

FIG. 2.2-2 The scattering function $R_S(\tau, \mathbf{r}_0; \phi, \boldsymbol{\kappa})$ and its Fourier transforms.

Another very important Fourier transform relationship can be obtained by substituting Eq. (2.2-35) into Eq. (2.2-28), which yields

$$R_S(\tau, \mathbf{r}_0; \phi, \boldsymbol{\kappa}) = F_{\Delta t} F_{\Delta \mathbf{r}} F_{\Delta f}^{-1} F_{\Delta \mathbf{v}}^{-1}\{R_H(\Delta f, \Delta \mathbf{v}; \Delta t, \Delta \mathbf{r})\} \tag{2.2-44}$$

or

$$R_S(\tau, \mathbf{r}_0; \phi, \boldsymbol{\kappa}) = \int_{-\infty}^{\infty}\int_{-\infty}^{\infty}\int_{-\infty}^{\infty}\int_{-\infty}^{\infty} R_H(\Delta f, \Delta \mathbf{v}; \Delta t, \Delta \mathbf{r}) \exp(-j2\pi\phi\, \Delta t) \times \exp(+j2\pi\boldsymbol{\kappa} \cdot \Delta \mathbf{r}) \exp(+j2\pi\, \Delta f \tau) \times \exp(-j2\pi\, \Delta \mathbf{v} \cdot \mathbf{r}_0)\, d\Delta t\, d\Delta \mathbf{r}\, d\Delta f\, d\Delta \mathbf{v} \tag{2.2-45}$$

and

$$R_H(\Delta f, \Delta \mathbf{v}; \Delta t, \Delta \mathbf{r}) = F_{\phi}^{-1} F_{\boldsymbol{\kappa}}^{-1} F_{\tau} F_{\mathbf{r}_0}\{R_S(\tau, \mathbf{r}_0; \phi, \boldsymbol{\kappa})\} \tag{2.2-46}$$

since

$$R_H(\Delta f, \Delta \mathbf{v}; \Delta t, \Delta \mathbf{r}) = \int_{-\infty}^{\infty}\int_{-\infty}^{\infty}\int_{-\infty}^{\infty}\int_{-\infty}^{\infty} R_S(\tau, \mathbf{r}_0; \phi, \boldsymbol{\kappa}) \exp(+j2\pi\phi\, \Delta t) \times \exp(-j2\pi\boldsymbol{\kappa} \cdot \Delta \mathbf{r}) \exp(-j2\pi\, \Delta f \tau) \times \exp(+j2\pi\, \Delta \mathbf{v} \cdot \mathbf{r}_0)\, d\phi\, d\boldsymbol{\kappa}\, d\tau\, d\mathbf{r}_0. \tag{2.2-21}$$

2.2.4 Input–Output Relations

We shall now proceed to derive an expression for the autocorrelation function of the output from a linear, time-varying, space-varying, random filter. The output autocorrelation function $R_y(t, t', \mathbf{r}, \mathbf{r}')$ is defined as follows:

$$R_y(t, t', \mathbf{r}, \mathbf{r}') \triangleq \mathrm{E}\{y(t, \mathbf{r}) y^*(t', \mathbf{r}')\}. \tag{2.2-47}$$

If Eq. (2.1-30) is substituted into Eq. (2.2-47), then the output autocorrelation function in time and space is given by

$$R_y(t, t', \mathbf{r}, \mathbf{r}') = \int_{-\infty}^{\infty} \cdots \int_{-\infty}^{\infty} x(t - \tau, \mathbf{r} - \mathbf{r}_0) x^*(t' - \tau', \mathbf{r}' - \mathbf{r}_0') \times \exp[+j2\pi(\phi t - \phi' t')] \exp[-j2\pi(\boldsymbol{\kappa} \cdot \mathbf{r} - \boldsymbol{\kappa}' \cdot \mathbf{r}')] \times R_s(\tau, \tau', \mathbf{r}_0, \mathbf{r}_0'; \phi, \phi', \boldsymbol{\kappa}, \boldsymbol{\kappa}')\, d\phi\, d\boldsymbol{\kappa}\, d\tau\, d\mathbf{r}_0\, d\phi'\, d\boldsymbol{\kappa}'\, d\tau'\, d\mathbf{r}_0'. \tag{2.2-48}$$

If the random filter is a WSSUS communication channel, then the autocorrelation function of the spreading function is given by Eq. (2.2-16). Substituting Eq. (2.2-16) into Eq. (2.2-48) yields

$$R_y(t,t',\mathbf{r},\mathbf{r}') = \int_{-\infty}^{\infty}\int_{-\infty}^{\infty}\int_{-\infty}^{\infty}\int_{-\infty}^{\infty} x(t-\tau,\mathbf{r}-\mathbf{r}_0)x^*(t'-\tau,\mathbf{r}'-\mathbf{r}_0) \times \exp(+j2\pi\phi\,\Delta t)\exp(-j2\pi\boldsymbol{\kappa}\cdot\Delta\mathbf{r}) \times R_s(\tau,\mathbf{r}_0;\phi,\boldsymbol{\kappa})\,d\phi\,d\boldsymbol{\kappa}\,d\tau\,d\mathbf{r}_0. \quad (2.2\text{-}49)$$

The mean-square value of $y(t,\mathbf{r})$ is given by

$$\mathrm{E}\{|y(t,\mathbf{r})|^2\} = R_y(t,t,\mathbf{r},\mathbf{r}),$$

where, from Eq. (2.2-48), we have in general that

$$R_y(t,t,\mathbf{r},\mathbf{r}) = \int_{-\infty}^{\infty}\cdots\int_{-\infty}^{\infty} x(t-\tau,\mathbf{r}-\mathbf{r}_0)x^*(t-\tau',\mathbf{r}-\mathbf{r}_0') \times \exp(+j2\pi\,\Delta\phi t)\exp(-j2\pi\,\Delta\boldsymbol{\kappa}\cdot\mathbf{r}) \times R_s(\tau,\tau',\mathbf{r}_0,\mathbf{r}_0';\phi,\phi',\boldsymbol{\kappa},\boldsymbol{\kappa}')\,d\phi\,d\boldsymbol{\kappa}\,d\tau\,d\mathbf{r}_0 \times d\phi'\,d\boldsymbol{\kappa}'\,d\tau'\,d\mathbf{r}_0', \quad (2.2\text{-}50)$$

where $\Delta\phi = \phi - \phi'$ and $\Delta\boldsymbol{\kappa} = \boldsymbol{\kappa} - \boldsymbol{\kappa}'$. In the case of a WSSUS communication channel, we have from Eq. (2.2-49) that

$$R_y(t,t,\mathbf{r},\mathbf{r}) = \int_{-\infty}^{\infty}\int_{-\infty}^{\infty}\int_{-\infty}^{\infty}\int_{-\infty}^{\infty} |x(t-\tau,\mathbf{r}-\mathbf{r}_0)|^2 R_S(\tau,\mathbf{r}_0;\phi,\boldsymbol{\kappa})\,d\phi\,d\boldsymbol{\kappa}\,d\tau\,d\mathbf{r}_0. \quad (2.2\text{-}51)$$

If we define the energy of the output signal E_y as

$$E_y \triangleq \int_{-\infty}^{\infty}\int_{-\infty}^{\infty} |y(t,\mathbf{r})|^2\,dt\,d\mathbf{r}, \quad (2.2\text{-}52)$$

then the average output energy $\bar{E}_y$ can be expressed as

$$\bar{E}_y \triangleq \mathrm{E}\{E_y\} = \int_{-\infty}^{\infty}\int_{-\infty}^{\infty} \mathrm{E}\{|y(t,\mathbf{r})|^2\}\,dt\,d\mathbf{r} \quad (2.2\text{-}53)$$

or

$$\bar{E}_y = \int_{-\infty}^{\infty}\int_{-\infty}^{\infty} R_y(t,t,\mathbf{r},\mathbf{r})\,dt\,d\mathbf{r}. \quad (2.2\text{-}54)$$

If Eq. (2.2-51) is substituted into Eq. (2.2-54), then

$$\bar{E}_y/E_x = \int_{-\infty}^{\infty}\int_{-\infty}^{\infty}\int_{-\infty}^{\infty}\int_{-\infty}^{\infty} R_S(\tau,\mathbf{r}_0;\phi,\boldsymbol{\kappa})\,d\phi\,d\boldsymbol{\kappa}\,d\tau\,d\mathbf{r}_0, \quad (2.2\text{-}55)$$

where

$$E_x = \int_{-\infty}^{\infty} \int_{-\infty}^{\infty} |x(t, \mathbf{r})|^2 \, dt \, d\mathbf{r} \tag{2.2-56}$$

is the energy of the input signal. Equation (2.2-55) indicates that the ratio of the output (received) average energy to the input (transmitted) energy for a WSSUS communication channel can be obtained by integrating the scattering function of the channel. Also, note that the average output energy is not a function of the input signal's shape.

Alternate expressions for the output autocorrelation function can be obtained from Eq. (2.1-25). If Eq. (2.1-25) is substituted into Eq. (2.2-47), then

$$\begin{aligned} R_y(t, t', \mathbf{r}, \mathbf{r}') &= \int_{-\infty}^{\infty} \int_{-\infty}^{\infty} \int_{-\infty}^{\infty} \int_{-\infty}^{\infty} X(f, \mathbf{v}) X^*(f', \mathbf{v}') R_H(f, f', \mathbf{v}, \mathbf{v}'; t, t', \mathbf{r}, \mathbf{r}') \\ &\quad \times \exp[+j2\pi(ft - f't')] \exp[-j2\pi(\mathbf{v} \cdot \mathbf{r} - \mathbf{v}' \cdot \mathbf{r}')] \, df \, d\mathbf{v} \, df' \, d\mathbf{v}'. \end{aligned} \tag{2.2-57}$$

If the random filter is a WSSUS communication channel, then the autocorrelation function of the transfer function is given by Eq. (2.2-20). Substituting Eq. (2.2-20) into Eq. (2.2-57) yields

$$\begin{aligned} R_y(t, t', \mathbf{r}, \mathbf{r}') &= \int_{-\infty}^{\infty} \int_{-\infty}^{\infty} \int_{-\infty}^{\infty} \int_{-\infty}^{\infty} X(f, \mathbf{v}) X^*(f', \mathbf{v}') R_H(\Delta f, \Delta \mathbf{v}; \Delta t, \Delta \mathbf{r}) \\ &\quad \times \exp[+j2\pi(ft - f't')] \exp[-j2\pi(\mathbf{v} \cdot \mathbf{r} - \mathbf{v}' \cdot \mathbf{r}')] \, df \, d\mathbf{v} \, df' \, d\mathbf{v} \end{aligned} \tag{2.2-58}$$

A relationship between the input and output power spectral densities will be obtained next.

Let us assume that the input signal $x(t, \mathbf{r})$ is a zero-mean, wide-sense stationary (in time and space), random process that is uncorrelated with the transfer function. Under these assumptions, the output autocorrelation function given by Eq. (2.2-57) becomes

$$\begin{aligned} R_y(t, t', \mathbf{r}, \mathbf{r}') &= \int_{-\infty}^{\infty} \int_{-\infty}^{\infty} \int_{-\infty}^{\infty} \int_{-\infty}^{\infty} \mathrm{E}\{X(f, \mathbf{v}) X^*(f', \mathbf{v}')\} R_H(f, f', \mathbf{v}, \mathbf{v}'; t, t', \mathbf{r}, \mathbf{r}') \\ &\quad \times \exp[+j2\pi(ft - f't')] \exp[-j2\pi(\mathbf{v} \cdot \mathbf{r} - \mathbf{v}' \cdot \mathbf{r}')] \, df \, d\mathbf{v} \, df' \, d\mathbf{v}'. \end{aligned} \tag{2.2-59}$$

Since $x(t, \mathbf{r})$ was assumed to be wide-sense stationary in both time and space, it can be shown that

$$E\{X(f, \mathbf{v})X^*(f', \mathbf{v}')\} = S_x(f, \mathbf{v})\,\delta(f - f')\,\delta(\mathbf{v} - \mathbf{v}'), \tag{2.2-60}$$

where

$$X(f, \mathbf{v}) = F_t F_{\mathbf{r}}\{x(t, \mathbf{r})\},$$
$$S_x(f, \mathbf{v}) \triangleq F_{\Delta t} F_{\Delta \mathbf{r}}\{R_x(\Delta t, \Delta \mathbf{r})\} \tag{2.2-61}$$

is the power spectral density of the input, and

$$R_x(\Delta t, \Delta \mathbf{r}) = E\{x(t, \mathbf{r})x^*(t', \mathbf{r}')\},$$

where $\Delta t = t - t'$ and $\Delta \mathbf{r} = \mathbf{r} - \mathbf{r}'$. If Eq. (2.2-60) is substituted into Eq. (2.2-59), then

$$R_y(t, t', \mathbf{r}, \mathbf{r}') = \int_{-\infty}^{\infty}\int_{-\infty}^{\infty} S_x(f, \mathbf{v}) R_H(f, f, \mathbf{v}, \mathbf{v}; t, t', \mathbf{r}, \mathbf{r}') \times \exp(+j2\pi f\,\Delta t)\exp(-j2\pi \mathbf{v}\cdot\Delta \mathbf{r})\,df\,d\mathbf{v}, \tag{2.2-62}$$

and if it is further assumed that the transfer function is wide-sense stationary in both time and space, then Eq. (2.2-62) reduces to

$$R_y(\Delta t, \Delta \mathbf{r}) = \int_{-\infty}^{\infty}\int_{-\infty}^{\infty} S_x(f, \mathbf{v}) R_H(f, f, \mathbf{v}, \mathbf{v}; \Delta t, \Delta \mathbf{r}) \times \exp(+j2\pi f\,\Delta t)\exp(-j2\pi \mathbf{v}\cdot\Delta \mathbf{r})\,df\,d\mathbf{v}, \tag{2.2-63}$$

which implies that the output $y(t, \mathbf{r})$ is also wide-sense stationary in both time and space. If we define the output power spectral density $S_y(\eta, \boldsymbol{\beta})$ as

$$S_y(\eta, \boldsymbol{\beta}) \triangleq F_{\Delta t} F_{\Delta \mathbf{r}}\{R_y(\Delta t, \Delta \mathbf{r})\}, \tag{2.2-64}$$

then substituting Eq. (2.2-63) into the right-hand side of Eq. (2.2-64) yields

$$S_y(\eta, \boldsymbol{\beta}) = \int_{-\infty}^{\infty}\int_{-\infty}^{\infty} S_x(f, \mathbf{v}) R_B(f, f, \mathbf{v}, \mathbf{v}; \eta - f, \boldsymbol{\beta} - \mathbf{v})\,df\,d\mathbf{v}, \tag{2.2-65}$$

where, from Eq. (2.2-41),

$$R_B(f, f, \mathbf{v}, \mathbf{v}; \eta - f, \boldsymbol{\beta} - \mathbf{v}) = \int_{-\infty}^{\infty}\int_{-\infty}^{\infty} R_H(f, f, \mathbf{v}, \mathbf{v}; \Delta t, \Delta \mathbf{r}) \times \exp[-j2\pi(\eta - f)\,\Delta t] \times \exp[+j2\pi(\boldsymbol{\beta} - \mathbf{v})\cdot\Delta \mathbf{r}]\,d\Delta t\,d\Delta \mathbf{r}. \tag{2.2-66}$$

Note that Eq. (2.2-65) is in the form of a *multidimensional convolution integral* analogous to Eq. (2.1-46) for the deterministic case. The convolution process accounts for the "frequency" spreading (frequencies in hertz and spatial frequencies in cycles per meter) of the input power spectral density.

EXAMPLE 2.2-1

Using Eqs. (2.2-65) and (2.2-66), let us compute the output power spectral density for the case when the transfer function is time-invariant, space-invariant, and *deterministic*.

First, note that

$$R_H(f, f, \mathbf{v}, \mathbf{v}; \Delta t, \Delta \mathbf{r}) = \mathrm{E}\{H(f, \mathbf{v}; t, \mathbf{r})H^*(f, \mathbf{v}; t', \mathbf{r}')\}.$$

If $H(f, \mathbf{v}; t, \mathbf{r})$ is both time-invariant and space-invariant, then [see Eq. (2.1-16)]

$$H(f, \mathbf{v}; t, \mathbf{r}) = H(f, \mathbf{v}),$$

and if the transfer function is also deterministic, then

$$R_H(f, f, \mathbf{v}, \mathbf{v}; \Delta t, \Delta \mathbf{r}) = \mathrm{E}\{|H(f, \mathbf{v})|^2\} = |H(f, \mathbf{v})|^2.$$

Therefore, Eq. (2.2-66) becomes

$$\begin{aligned} R_B(f, f, \mathbf{v}, \mathbf{v}; \eta - f, \boldsymbol{\beta} - \mathbf{v}) &= |H(f, \mathbf{v})|^2 \int_{-\infty}^{\infty} \exp[-j2\pi(\eta - f)\Delta t]\, d\Delta t \\ &\quad \times \int_{-\infty}^{\infty} \exp[+j2\pi(\boldsymbol{\beta} - \mathbf{v}) \cdot \Delta \mathbf{r}]\, d\Delta \mathbf{r} \\ &= |H(f, \mathbf{v})|^2\, \delta(\eta - f)\, \delta(\boldsymbol{\beta} - \mathbf{v}), \end{aligned}$$

and substituting this expression into Eq. (2.2-65) yields the desired result

$$S_y(\eta, \boldsymbol{\beta}) = |H(\eta, \boldsymbol{\beta})|^2 S_x(\eta, \boldsymbol{\beta}).$$

By replacing η with f and $\boldsymbol{\beta}$ with $\mathbf{v}$, we finally obtain [compare with Eq. (2.1-48)]

$$S_y(f, \mathbf{v}) = |H(f, \mathbf{v})|^2 S_x(f, \mathbf{v}). \quad \square$$

BIBLIOGRAPHY

L. J. Ziomek, *A Scattering Function Approach to Underwater Acoustic Detection and Signal Design*. Ph. D. Dissertation, The Pennsylvania State University, University Park, Pennsylvania, Chap. 2, pp. 10–49, (1981).

L. J. Ziomek, *Linear Time-Variant Space-Variant Filters and the WKB Approximation*. Naval Postgraduate School Technical Report 62-83-058, October 1983, 93 pp. Also available from DTIC, AD Number A 134886.

3

Complex Apertures

3.1 COUPLING THE TRANSMITTED AND RECEIVED SIGNALS TO THE MEDIUM

Let us consider the transmit electroacoustic transducer shown in Fig. 3.1-1. The transducer occupies a volume V in general. This physical situation is known as a *volume aperture*. Imagine applying an electrical input signal $x(t, \mathbf{r})$ to the transducer at time t and spatial location $\mathbf{r}$. If we treat the infinitesimal region dV of the transducer as a *linear filter* with *impulse response* $\alpha_T(t, \mathbf{r})$, then

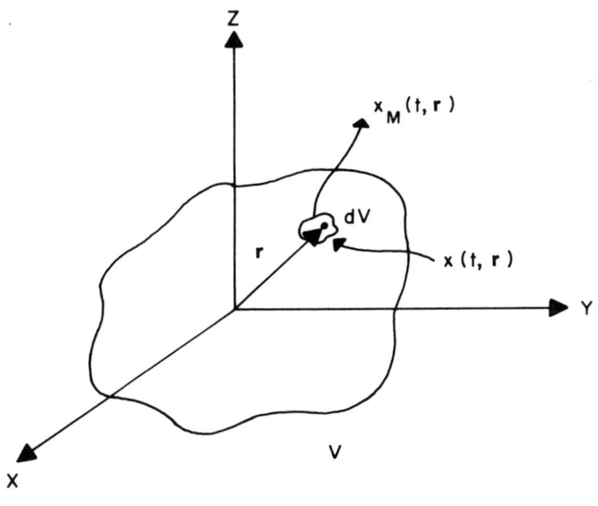

FIG. 3.1-1 Transmit electroacoustic transducer shown occupying a volume V.

the output acoustic signal $x_M(t, \mathbf{r})$ from region dV of the transducer at time t and spatial location $\mathbf{r}$ can be expressed as a time-domain convolution integral where $\mathbf{r}$ is a parameter, as follows:

$$x_M(t, \mathbf{r}) = \int_{-\infty}^{\infty} x(\tau, \mathbf{r}) \alpha_T(t - \tau, \mathbf{r}) \, d\tau \tag{3.1-1}$$

or

$$x_M(t, \mathbf{r}) = x(t, \mathbf{r}) * \alpha_T(t, \mathbf{r}), \tag{3.1-2}$$

where the asterisk $(*)$ denotes convolution with respect to t in this case. Taking the Fourier transform of both sides of Eq. (3.1-2) with respect to t yields

$$X_M(f, \mathbf{r}) = X(f, \mathbf{r}) A_T(f, \mathbf{r}), \tag{3.1-3}$$

where

$$X_M(f, \mathbf{r}) = F_t\{x_M(t, \mathbf{r})\} = \int_{-\infty}^{\infty} x_M(t, \mathbf{r}) \exp(-j2\pi f t) \, dt \tag{3.1-4}$$

is the frequency spectrum (in hertz) of the source distribution,

$$X(f, \mathbf{r}) = F_t\{x(t, \mathbf{r})\} = \int_{-\infty}^{\infty} x(t, \mathbf{r}) \exp(-j2\pi f t) \, dt \tag{3.1-5}$$

is the frequency spectrum of the input electrical signal, and

$$A_T(f, \mathbf{r}) = F_t\{\alpha_T(t, \mathbf{r})\} = \int_{-\infty}^{\infty} \alpha_T(t, \mathbf{r}) \exp(-j2\pi f t) \, dt \tag{3.1-6}$$

is the *complex frequency response* (transfer function) or *complex aperture function* at spatial location $\mathbf{r}$ of the transmit transducer. Since

$$x_M(t, \mathbf{r}) = F_f^{-1}\{X_M(f, \mathbf{r})\}, \tag{3.1-7}$$

substituting Eq. (3.1-3) into Eq. (3.1-7) yields

$$x_M(t, \mathbf{r}) = \int_{-\infty}^{\infty} X(f, \mathbf{r}) A_T(f, \mathbf{r}) \exp(+j2\pi f t) \, df. \tag{3.1-8}$$

Compared with Eq. (3.1-1), Eq. (3.1-8) is a more useful representation of the source distribution for our purposes.

A similar situation exists at the receive transducer, shown in Fig. 3.1-2. As before, the transducer occupies a volume V in general. By following the same reasoning used in the development of Eqs. (3.1-1) through (3.1-8), we can write

$$y(t, \mathbf{r}) = \int_{-\infty}^{\infty} y_M(\tau, \mathbf{r}) \alpha_R(t - \tau, \mathbf{r}) \, d\tau \tag{3.1-9}$$

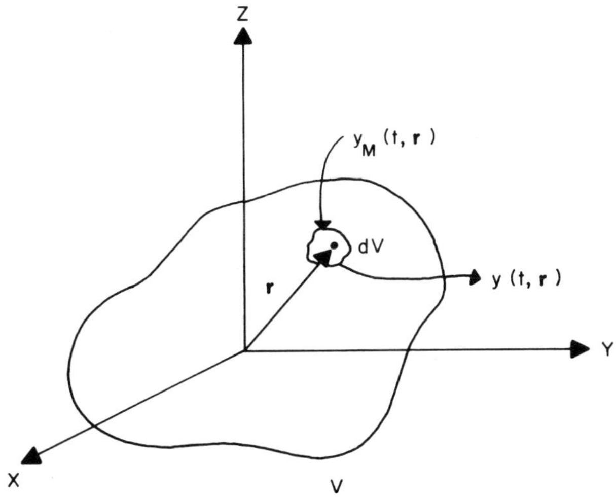

FIG. 3.1-2 Receive electroacoustic transducer shown occupying a volume V.

or

$$y(t,\mathbf{r}) = y_{\mathrm{M}}(t,\mathbf{r}) * \alpha_{\mathrm{R}}(t,\mathbf{r}), \tag{3.1-10}$$

where $y_{\mathrm{M}}(t,\mathbf{r})$ is the input acoustic signal to the transducer at time t and spatial location $\mathbf{r}$, $\alpha_{\mathrm{R}}(t,\mathbf{r})$ is the *impulse response* of the infinitesimal region dV of the transducer located at $\mathbf{r}$, and $y(t,\mathbf{r})$ is the output electrical signal from region dV of the transducer at time t and spatial location $\mathbf{r}$. Taking the Fourier transform of both sides of Eq. (3.1-10) with respect to t yields

$$Y(\eta,\mathbf{r}) = Y_{\mathrm{M}}(\eta,\mathbf{r})A_{\mathrm{R}}(\eta,\mathbf{r}), \tag{3.1-11}$$

and taking the inverse Fourier transform of both sides of Eq. (3.1-11) with respect to η yields

$$y(t,\mathbf{r}) = \int_{-\infty}^{\infty} Y_{\mathrm{M}}(\eta,\mathbf{r})A_{\mathrm{R}}(\eta,\mathbf{r})\exp(+j2\pi\eta t)\,d\eta, \tag{3.1-12}$$

where $A_{\mathrm{R}}(\eta,\mathbf{r})$ is the *complex frequency response* (transfer function) or *complex aperture function* at spatial location $\mathbf{r}$ of the receive transducer. Note that Eq. (3.1-12) is analogous to Eq. (3.1-8). Whereas Eq. (3.1-8) represents the coupling of the input electrical signal $x(t,\mathbf{r})$ to the medium, that is, the production of the input acoustic signal $x_{\mathrm{M}}(t,\mathbf{r})$ via the transmit aperture function $A_{\mathrm{T}}(f,\mathbf{r})$, Eq. (3.1-12) represents the coupling of the medium, that is, the output acoustic signal $y_M(t, \mathbf{r})$, to the output electrical signal $y(t, \mathbf{r})$ via the receive aperture function $A_R(\eta, \mathbf{r})$.

Recall that the expression for the velocity potential given by Eq. (1.1-2) is the solution of the linear, inhomogeneous wave equation given by Eq. (1.1-1) when $c(\mathbf{r}) = c$. If Eq. (3.1-8) is substituted into Eq. (1.1-2), then

$$\varphi(t, \mathbf{r}) = -\frac{1}{4\pi} \int_{-\infty}^{\infty} \int_{V_0} X(f, \mathbf{r}_0) A_T(f, \mathbf{r}_0) g(\mathbf{r} \,|\, \mathbf{r}_0) \, dV_0 \exp(+j2\pi ft) \, df, \quad (3.1\text{-}13)$$

where

$$g(\mathbf{r} \,|\, \mathbf{r}_0) = \frac{\exp(-jk|\mathbf{r} - \mathbf{r}_0|)}{|\mathbf{r} - \mathbf{r}_0|} \quad (3.1\text{-}14)$$

is the *free-space Green's function* and

$$k = 2\pi f/c = 2\pi/\lambda \quad \text{rad/m} \quad (3.1\text{-}15)$$

is the wave number (see Fig. 1.1-1).

3.2 DIRECTIVITY FUNCTIONS

3.2.1 Near-Field Directivity Functions

The solution of the wave equation as given by Eq. (3.1-13) for $c(\mathbf{r}) = c$ is an exact expression. However, by approximating the free-space Green's function $g(\mathbf{r}|\mathbf{r}_0)$ via *Fresnel* and *Fraunhofer expansions*, both the near-field and far-field directivity functions or beam patterns of the transmit aperture can be derived, respectively.

Let us begin the analysis by defining the following vectors: let

$$\mathbf{r}_0 = x_0\hat{x} + y_0\hat{y} + z_0\hat{z}, \quad (3.2\text{-}1)$$

where

$$r_0 = |\mathbf{r}_0| = (x_0^2 + y_0^2 + z_0^2)^{1/2}, \quad (3.2\text{-}2)$$

and

$$\mathbf{r} = x\hat{x} + y\hat{y} + z\hat{z} = r\hat{a}_r, \quad (3.2\text{-}3)$$

where

$$r = |\mathbf{r}| = (x^2 + y^2 + z^2)^{1/2}. \quad (3.2\text{-}4)$$

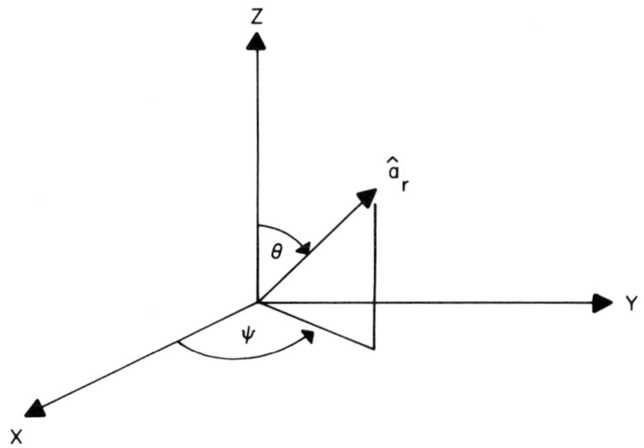

FIG. 3.2-1 The unit vector $\hat{a}_r$ and the spherical angles θ and ψ. Note that angle ψ is measured in a counterclockwise direction.

Let $\hat{a}_r$ be a unit vector given by

$$\hat{a}_r = u\hat{x} + v\hat{y} + w\hat{z}, \tag{3.2-5}$$

where

$$u = \sin\theta\cos\psi, \tag{3.2-6}$$

$$v = \sin\theta\sin\psi, \tag{3.2-7}$$

and

$$w = \cos\theta \tag{3.2-8}$$

are the *direction cosines* with respect to the positive X, Y, and Z axes, respectively (see Fig. 3.2-1). Note that

$$u^2 + v^2 + w^2 = 1. \tag{3.2-9}$$

Next, assume that the observation point $\mathbf{r}$ is in the *near-field*, that is,

$$r_0 < r < \pi R^2/\lambda, \tag{3.2-10}$$

where R represents the *maximum radial dimension* of the transmit transducer and λ is the wavelength (in meters) in the medium. The criterion given by Eq. (3.2-10) will be derived shortly, but first refer back to Eq. (3.1-14) for $g(\mathbf{r}|\mathbf{r}_0)$ and consider the range term $|\mathbf{r} - \mathbf{r}_0|$, which appears both as an amplitude and phase factor. Noting that

$$|\mathbf{r} - \mathbf{r}_0| = [(\mathbf{r} - \mathbf{r}_0)\cdot(\mathbf{r} - \mathbf{r}_0)]^{1/2}, \tag{3.2-11}$$

it can be shown that

$$|\mathbf{r} - \mathbf{r}_0| = r(1 + b)^{1/2}, \tag{3.2-12}$$

where

$$b = (r_0/r)^2 - 2(\hat{a}_r \cdot \mathbf{r}_0)/r. \tag{3.2-13}$$

We are now in a position to approximate $g(\mathbf{r}|\mathbf{r}_0)$.

First, consider the spherical spreading amplitude term $1/|\mathbf{r} - \mathbf{r}_0|$. Since $r > r_0$, we can assume that $b \approx 0$, so that

$$1/|\mathbf{r} - \mathbf{r}_0| \approx 1/r. \tag{3.2-14}$$

While this approximation is sufficient for the amplitude term, it is *not* satisfactory for the phase term,

$$\exp(-jk|\mathbf{r} - \mathbf{r}_0|), \tag{3.2-15}$$

since small changes in range $|\mathbf{r} - \mathbf{r}_0|$ can lead to large changes in its value.

EXAMPLE 3.2-1

Let

$$|\mathbf{r} - \mathbf{r}_0| = n\lambda/2, \qquad n = 0, 2, 4, \ldots,$$

where n is an even integer. Therefore,

$$k|\mathbf{r} - \mathbf{r}_0| = (2\pi/\lambda)(n\lambda/2) = n\pi, \qquad n = 0, 2, 4, \ldots,$$

so that

$$\exp(-jk|\mathbf{r} - \mathbf{r}_0|) = 1.$$

Now let

$$|\mathbf{r} - \mathbf{r}_0| = (n + 1)\lambda/2, \qquad n = 0, 2, 4, \ldots,$$

which represents a change in range of only one-half of a wavelength. As a result,

$$\exp(-jk|\mathbf{r} - \mathbf{r}_0|) = -1. \quad \square$$

Therefore, in order to approximate Eq. (3.2-15), we use a *binomial expansion* of the square root in Eq. (3.2-12), that is,

$$|\mathbf{r} - \mathbf{r}_0| \approx r[1 + (b/2) - (b^2/8) + \cdots], \qquad |b| < 1, \tag{3.2-16}$$

assuming that $|b| < 1$ with b given by Eq. (3.2-13). If the first three terms of the binomial expansion are used as shown in Eq. (3.2-16) and if terms involving r_0

raised to powers greater than two are neglected, then Eq. (3.2-16) becomes

$$|\mathbf{r} - \mathbf{r}_0| \approx r - \hat{a}_r \cdot \mathbf{r}_0 + (r_0^2 - [\hat{a}_r \cdot \mathbf{r}_0]^2)/(2r). \tag{3.2-17}$$

Thus, substituting Eqs. (3.2-14) and (3.2-17) into Eq. (3.1-14) yields

$$g(\mathbf{r}\,|\,\mathbf{r}_0) \approx [\exp(-jkr)/r]\exp(+jk\hat{a}_r \cdot \mathbf{r}_0)\exp\{-jk[r_0^2 - (\hat{a}_r \cdot \mathbf{r}_0)^2]/(2r)\}, \tag{3.2-18}$$

which is a near-field expansion of the free-space Green's function involving *all* terms up to second power in r_0. A *Fresnel approximation* or expansion of $g(\mathbf{r}\,|\,\mathbf{r}_0)$ can also be obtained from Eq. (3.2-16) by using only the first two terms instead of the first three. A Fresnel approximation is equivalent to setting the dot product term $(\hat{a}_r \cdot \mathbf{r}_0)^2$ equal to zero in Eqs. (3.2-17) and (3.2-18). Some authors claim that a Fresnel approximation is one that includes second-order terms, for example, r_0^2. This is *not* an accurate statement, since if one is interested in retaining all second-order terms involving r_0^2, for example, then $(\hat{a}_r \cdot \mathbf{r}_0)^2$ must be included. For the moment, we will work with Eq. (3.2-18) in its present form, so that later in the analysis we can demonstrate why a Fresnel approximation is commonly used.

First, let us examine the last complex exponential term in Eq. (3.2-18). Since $k = 2\pi/\lambda$, this term can be written as

$$\exp\{-j\pi[r_0^2 - (\hat{a}_r \cdot \mathbf{r}_0)^2]/(\lambda r)\}. \tag{3.2-19}$$

Equation (3.2-19) will be a significant phase term if

$$\pi|r_0^2 - (\hat{a}_r \cdot \mathbf{r}_0)^2|/(\lambda r) > 1$$

or

$$r < \pi|r_0^2 - (\hat{a}_r \cdot \mathbf{r}_0)^2|/\lambda \tag{3.2-20}$$

since, if r is greater than the right-hand side of Eq. (3.2-20), then Eq. (3.2-19) reduces to $\exp(j0) = 1$, a constant. In order to obtain a conservative criterion based on Eq. (3.2-20), we consider the worst case, which is when the numerator becomes the largest. This corresponds to setting $(\hat{a}_r \cdot \mathbf{r}_0)^2 = 0$ and the *radial distance* r_0 equal to its maximum value, say, R. Therefore, if

$$r < \pi R^2/\lambda,$$

then Eq. (3.2-19) is a significant phase term and we say that the observation point is in the near-field [compare with Eq. (3.2-10)].

Now that we have a near-field expansion of $g(\mathbf{r}\,|\,\mathbf{r}_0)$ as given by Eq. (3.2-18), we can proceed with the derivation of the near-field directivity function. Substituting Eq. (3.2-18) into the velocity potential expression of Eq. (3.1-13) yields

$$\varphi(t,\mathbf{r}) \approx -\frac{1}{4\pi r}\int_{-\infty}^{\infty}\int_{V_0} X_M(f,\mathbf{r}_0)\exp\left\{\frac{-jk[r_0^2 - (\hat{a}_r\cdot\mathbf{r}_0)^2]}{2r}\right\}$$
$$\times \exp(+jk\hat{a}_r\cdot\mathbf{r}_0)\,dV_0 \exp\{+j2\pi f[t-(r/c)]\}\,df, \tag{3.2-21}$$

where [see Eq. (3.1-3)]

$$X_M(f,\mathbf{r}_0) = X(f,\mathbf{r}_0)A_T(f,\mathbf{r}_0) \tag{3.2-22}$$

and $t-(r/c)$ is the retarded time. With the use of Eqs. (3.1-15), (3.2-1), and (3.2-5), we can write

$$\exp(+jk\hat{a}_r\cdot\mathbf{r}_0) = \exp[+j2\pi(f_X x_0 + f_Y y_0 + f_Z z_0)], \tag{3.2-23}$$

where

$$f_X = u/\lambda, \tag{3.2-24}$$

$$f_Y = v/\lambda, \tag{3.2-25}$$

and

$$f_Z = w/\lambda \tag{3.2-26}$$

are called *spatial frequencies* with units of cycles per meter and u, v, and w are the direction cosines given by Eqs. (3.2-6) through (3.2-8), respectively. Note that the components of the propagation vector $\mathbf{k} = k\hat{a}_r$ in the X, Y, and Z directions are given by $k_X = 2\pi f_X$, $k_Y = 2\pi f_Y$, and $k_Z = 2\pi f_Z$, respectively. If Eq. (3.2-23) is substituted into Eq. (3.2-21) and if we designate the volume integral by I, then

$$I = \int_{-\infty}^{\infty}\int_{-\infty}^{\infty}\int_{-\infty}^{\infty} X_M(f,\mathbf{r}_0)\exp\left\{\frac{-jk[r_0^2 - (\hat{a}_r\cdot\mathbf{r}_0)^2]}{2r}\right\}$$
$$\times \exp[+j2\pi(f_X x_0 + f_Y y_0 + f_Z z_0)]\,dx_0\,dy_0\,dz_0. \tag{3.2-27}$$

At this point in the analysis we have two options. We can stop with the form of Eq. (3.2-27) and define it as being the near-field directivity function of the source distribution $X_M(f,\mathbf{r}_0)$ or we can proceed further by employing the Fresnel approximation. We shall choose the latter course of action.

If we let $(\hat{a}_r\cdot\mathbf{r}_0)^2 = 0$ in Eq. (3.2-27), then

$$I = \int_{-\infty}^{\infty}\int_{-\infty}^{\infty}\int_{-\infty}^{\infty} X_M(f,\mathbf{r}_0)\exp[-jkr_0^2/(2r)]$$
$$\times \exp[+j2\pi(f_X x_0 + f_Y y_0 + f_Z z_0)]\,dx_0\,dy_0\,dz_0, \tag{3.2-28}$$

which can be interpreted as being a *three-dimensional spatial Fourier transform* of the product of two functions. The question now is "the product of which two functions" since $X_M(f, \mathbf{r}_0)$ itself represents a product [see Eq. (3.2-22)]. If Eq. (3.2-22) is substituted into Eq. (3.2-28) and the transmitted electrical signal is separated from the aperture function, then Eq. (3.2-28) can be written as

$$I = F_{\mathbf{r}_0}\left\{X(f, \mathbf{r}_0)\left[A_T(f, \mathbf{r}_0)\exp\left(\frac{-jkr_0^2}{2r}\right)\right]\right\}, \qquad (3.2\text{-}29)$$

where $F_{\mathbf{r}_0}$ represents a spatial Fourier transform with respect to $\mathbf{r}_0 = (x_0, y_0, z_0)$. Since a product in the spatial domain implies a convolution in the spatial frequency domain, Eq. (3.2-29) finally becomes

$$I = X(f, \boldsymbol{\alpha}) * D_T(f, r, \boldsymbol{\alpha}) \triangleq X_M(f, r, \mathbf{v}), \qquad (3.2\text{-}30)$$

where

$$X_M(f, r, \mathbf{v}) \triangleq \int_{-\infty}^{\infty} X(f, \boldsymbol{\alpha}) D_T(f, r, \mathbf{v} - \boldsymbol{\alpha})\, d\boldsymbol{\alpha} \qquad (3.2\text{-}31)$$

and

$$X(f, \boldsymbol{\alpha}) = F_{\mathbf{r}_0}\{X(f, \mathbf{r}_0)\}, \qquad (3.2\text{-}32)$$

or

$$X(f, \boldsymbol{\alpha}) = \int_{V_0} X(f, \mathbf{r}_0)\exp(+j2\pi\boldsymbol{\alpha}\cdot\mathbf{r}_0)\, dV_0 \qquad (3.2\text{-}33)$$

and

$$D_T(f, r, \boldsymbol{\alpha}) \triangleq F_{\mathbf{r}_0}\left\{A_T(f, \mathbf{r}_0)\exp\left(\frac{-jkr_0^2}{2r}\right)\right\}, \qquad (3.2\text{-}34)$$

or

$$D_T(f, r, \boldsymbol{\alpha}) \triangleq \int_{V_0} A_T(f, \mathbf{r}_0)\exp\left(\frac{-jkr_0^2}{2r}\right)\exp(+j2\pi\boldsymbol{\alpha}\cdot\mathbf{r}_0)\, dV_0, \qquad (3.2\text{-}35)$$

where

$$\boldsymbol{\alpha} = (f_X, f_Y, f_Z). \qquad (3.2\text{-}36)$$

Equation (3.2-35) is the *near-field directivity function* of the complex aperture function $A_T(f, \mathbf{r}_0)$. Recall that the aperture function is the frequency response of the transducer. Note that *the near-field directivity function is a function of the range r to the observation point.* The form of Eq. (3.2-35) is also known as a *Fresnel diffraction integral* in optics or as a *Fresnel transform.*

Finally, if Eq. (3.2-30) is substituted into Eq. (3.2-21) in place of the volume integral, then the expression for the velocity potential becomes

$$\varphi(t,\mathbf{r}) \equiv \varphi(t,r,\theta,\psi) \approx -\frac{1}{4\pi r}\int_{-\infty}^{\infty} X_{\mathrm{M}}(f,r,\mathbf{v})\exp\left\{+j2\pi f\left(t-\frac{r}{c}\right)\right\}df, \tag{3.2-37}$$

where $X_{\mathrm{M}}(f,r,\mathbf{v})$ is given by Eq. (3.2-31) and $\mathbf{v} = (f_X, f_Y, f_Z)$.

3.2.2 Far-Field Directivity Functions

An observation point $\mathbf{r}$ will be considered to be in the *far-field* if [see Eq. (3.2-10)]

$$r > \pi R^2/\lambda. \tag{3.2-38}$$

As a result, the *Fraunhofer approximation* or expansion of the free-space Green's function can be obtained from Eq. (3.2-18) by neglecting all those terms containing r_0^2. Therefore, the Fraunhofer or far-field approximation of $g(\mathbf{r}|\mathbf{r}_0)$ is given by

$$g(\mathbf{r}|\mathbf{r}_0) \approx [\exp(-jkr)/r]\exp(+jk\hat{a}_r \cdot \mathbf{r}_0) \tag{3.2-39}$$

or, substituting Eq. (3.2-23) into Eq. (3.2-39),

$$g(\mathbf{r}|\mathbf{r}_0) \approx [\exp(-jkr)/r]\exp(+j2\pi\boldsymbol{\alpha} \cdot \mathbf{r}_0). \tag{3.2-40}$$

Similarly, the far-field directivity function can be obtained from the near-field directivity function given by Eq. (3.2-35) by neglecting the term containing r_0^2. Therefore, the *far-field directivity function* or *beam pattern* of the complex aperture function $A_{\mathrm{T}}(f,\mathbf{r}_0)$ is

$$D_{\mathrm{T}}(f,\boldsymbol{\alpha}) \triangleq F_{\mathbf{r}_0}\{A_{\mathrm{T}}(f,\mathbf{r}_0)\} \tag{3.2-41}$$

or

$$D_{\mathrm{T}}(f,\boldsymbol{\alpha}) \triangleq \int_{V_0} A_{\mathrm{T}}(f,\mathbf{r}_0)\exp(+j2\pi\boldsymbol{\alpha} \cdot \mathbf{r}_0)\,dV_0. \tag{3.2-42}$$

Similarly,

$$A_{\mathrm{T}}(f,\mathbf{r}_0) = F_{\boldsymbol{\alpha}}^{-1}\{D_{\mathrm{T}}(f,\boldsymbol{\alpha})\} \tag{3.2-43}$$

or

$$A_{\mathrm{T}}(f,\mathbf{r}_0) = \int_{-\infty}^{\infty} D_{\mathrm{T}}(f,\boldsymbol{\alpha})\exp(-j2\pi\boldsymbol{\alpha} \cdot \mathbf{r}_0)\,d\boldsymbol{\alpha}. \tag{3.2-44}$$

Therefore, *the aperture function and the far-field beam pattern form a spatial*

Fourier transform pair. Note that the far-field beam pattern is *not* a function of the range r to the observation point, as the near-field directivity function is. The form of Eq. (3.2-42) is also known as a *Fraunhofer diffraction integral* in optics.

Recall that

$$X_M(f, \mathbf{r}_0) = X(f, \mathbf{r}_0) A_T(f, \mathbf{r}_0) \tag{3.2-22}$$

was derived *without* making any near-field versus far-field approximations. However, if the spatial Fourier transform is taken of both sides of Eq. (3.2-22), a far-field approximation is indirectly being made since

$$F_{\mathbf{r}_0}\{X_M(f, \mathbf{r}_0)\} = X(f, \boldsymbol{\alpha}) * D_T(f, \boldsymbol{\alpha}), \tag{3.2-45}$$

where $D_T(f, \boldsymbol{\alpha})$ is the far-field directivity function [see Eq. (3.2-41)]. If we let

$$X_M(f, \mathbf{v}) = F_{\mathbf{r}_0}\{X_M(f, \mathbf{r}_0)\}, \tag{3.2-46}$$

then

$$X_M(f, \mathbf{v}) = \int_{-\infty}^{\infty} X(f, \boldsymbol{\alpha}) D_T(f, \mathbf{v} - \boldsymbol{\alpha}) \, d\boldsymbol{\alpha}, \tag{3.2-47}$$

which is the far-field version of Eq. (3.2-31). Therefore, in the far-field the velocity potential can be written as [compare with Eq. (3.2-37)]

$$\varphi(t, \mathbf{r}) \equiv \varphi(t, r, \theta, \psi) \approx -\frac{1}{4\pi r} \int_{-\infty}^{\infty} X_M(f, \mathbf{v}) \exp\left\{+j2\pi f\left(t - \frac{r}{c}\right)\right\} df. \tag{3.2-48}$$

Similarly, since

$$Y(\eta, \mathbf{r}) = Y_M(\eta, \mathbf{r}) A_R(\eta, \mathbf{r}) \tag{3.1-11}$$

at the receive transducer,

$$F_{\mathbf{r}}\{Y(\eta, \mathbf{r})\} = Y_M(\eta, \boldsymbol{\beta}) * D_R(\eta, \boldsymbol{\beta}), \tag{3.2-49}$$

where $D_R(\eta, \boldsymbol{\beta})$ is the far-field directivity function of the complex receive aperture function $A_R(\eta, \mathbf{r})$. If we let

$$Y(\eta, \boldsymbol{\gamma}) = F_{\mathbf{r}}\{Y(\eta, \mathbf{r})\}, \tag{3.2-50}$$

then

$$Y(\eta, \boldsymbol{\gamma}) = \int_{-\infty}^{\infty} Y_M(\eta, \boldsymbol{\beta}) D_R(\eta, \boldsymbol{\gamma} - \boldsymbol{\beta}) \, d\boldsymbol{\beta}. \tag{3.2-51}$$

Equation (3.2-51) will be used in conjunction with our discussion of FFT beamforming in Chapter 5.

EXAMPLE 3.2-2

Assume that an identical input electrical signal is applied at all spatial locations $\mathbf{r}_0$ of the transmit transducer, that is, $X(f, \mathbf{r}_0)$ is *not* a function of $\mathbf{r}_0$, so that $X(f, \mathbf{r}_0) \equiv X(f)$. As a result, Eq. (3.2-33) becomes

$$X(f, \boldsymbol{\alpha}) = X(f) \int_{V_0} \exp(+j2\pi\boldsymbol{\alpha} \cdot \mathbf{r}_0)\, dV_0$$

or

$$X(f, \boldsymbol{\alpha}) = X(f)\, \delta(\boldsymbol{\alpha})$$

since $F_{\mathbf{r}_0}\{1\} = \delta(\boldsymbol{\alpha})$. Substituting this result into Eq. (3.2-31) yields

$$X_{\mathrm{M}}(f, r, \mathbf{v}) = X(f) D_{\mathrm{T}}(f, r, \mathbf{v}),$$

and upon substituting this relationship into Eq. (3.2-37), the near-field velocity-potential expression reduces to

$$\varphi(t, r, \theta, \psi) \approx -\frac{1}{4\pi r} \int_{-\infty}^{\infty} X(f) D_{\mathrm{T}}(f, r, \mathbf{v}) \exp\left\{+j2\pi f\left(t - \frac{r}{c}\right)\right\} df.$$

Similarly, with $X(f, \boldsymbol{\alpha}) = X(f)\, \delta(\boldsymbol{\alpha})$, Eq. (3.2-47) becomes

$$X_{\mathrm{M}}(f, \mathbf{v}) = X(f) D_{\mathrm{T}}(f, \mathbf{v}),$$

and the far-field velocity-potential expression given by Eq. (3.2-48) reduces to

$$\varphi(t, r, \theta, \psi) \approx -\frac{1}{4\pi r} \int_{-\infty}^{\infty} X(f) D_{\mathrm{T}}(f, \mathbf{v}) \exp\left\{+j2\pi f\left(t - \frac{r}{c}\right)\right\} df.$$

If it is further assumed that the input electrical signal is time-harmonic, that is, if

$$x(t, \mathbf{r}_0) \equiv x(t) = \exp(+j2\pi f_0 t),$$

then

$$X(f, \mathbf{r}_0) \equiv X(f) = \delta(f - f_0),$$

and, as a result,

$$\varphi(t, r, \theta, \psi) \approx -D_{\mathrm{T}}(f_0, r, \mathbf{v}_0) \frac{\exp\{+j2\pi f_0[t - (r/c)]\}}{4\pi r}$$

in the near-field and

$$\varphi(t, r, \theta, \psi) \approx -D_{\mathrm{T}}(f_0, \mathbf{v}_0) \frac{\exp\{+j2\pi f_0[t - (r/c)]\}}{4\pi r}$$

in the far-field, where $\mathbf{v}_0 = (f_{X_0}, f_{Y_0}, f_{Z_0})$, $f_{X_0} = u/\lambda_0$, $f_{Y_0} = v/\lambda_0$, and $f_{Z_0} = w/\lambda_0$. The wavelength $\lambda_0 = c/f_0$ is the wavelength in the medium and u, v, and w are the direction cosines given by Eqs. (3.2-6) through (3.2-8), respectively. □

Example 3.2-2 raises an interesting point. If we decide that any electronics used for purposes of amplitude shading and/or beam steering are to be considered as part of the transducer, then the assumption that an identical input electrical signal is applied at all spatial locations of the transmit transducer is true in most practical cases. Because of this assumption, we have seen that the convolution integral expression for $X_M(f, \mathbf{v})$ as given by Eq. (3.2-47), for example, reduced to a product. However, the situation is different at the receive transducer. Since the input acoustic field $y_M(t, \mathbf{r})$ incident upon the receive transducer is *not*, in general, identical at all spatial locations across the face of the transducer, the convolution integral given by Eq. (3.2-51), for example, must be used.

3.3 LINEAR APERTURES AND FAR-FIELD DIRECTIVITY FUNCTIONS

3.3.1 Amplitude Windows

The complex frequency response or complex aperture function of a transducer can be represented as

$$A(f, \mathbf{r}) = a(f, \mathbf{r}) \exp[+j\theta(f, \mathbf{r})], \tag{3.3-1}$$

where $a(f, \mathbf{r})$ is the magnitude of the frequency response and $\theta(f, \mathbf{r})$ is the phase at spatial location $\mathbf{r}$ of the transducer. Both $a(f, \mathbf{r})$ and $\theta(f, \mathbf{r})$ are real functions. The function $a(f, \mathbf{r})$ is also known as the *amplitude window*.

Consider the case of a thin, elongated transducer of length L meters lying along the X axis, as shown in Fig. 3.3-1. This physical situation is known as a *linear aperture*. Also shown in Fig. 3.3-1 is an observation point with spherical coordinates (ρ, θ, ψ). The observation point is assumed to be in the far-field region of the aperture, that is, at a range $\rho > \pi R^2/\lambda$ meters, where $R = L/2$ meters is the maximum radial extent of the transducer.

Recall that the far-field directivity function of a general, volume aperture is [see Eq. (3.2-42)]

$$D(f, \boldsymbol{\alpha}) = \int_V A(f, \mathbf{r}) \exp(+j2\pi\boldsymbol{\alpha} \cdot \mathbf{r})\, dV. \tag{3.3-2}$$

Since the linear aperture illustrated in Fig. 3.3-1 is shown lying along the

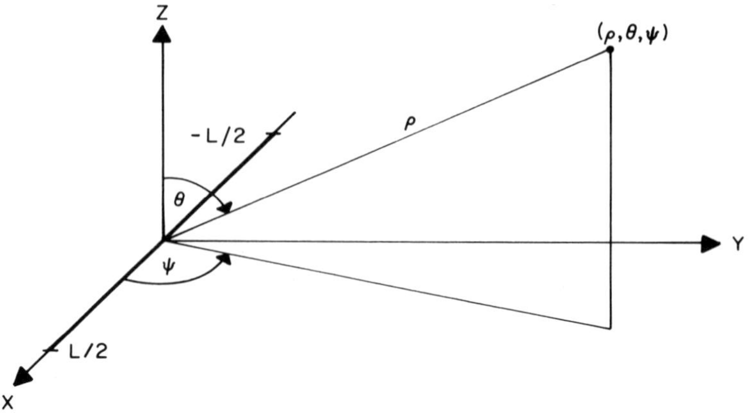

FIG. 3.3-1 Thin, elongated transducer of length L meters lying along the X axis. Also shown is a far-field observation point with spherical coordinates (ρ, θ, ψ).

X axis,

$$\mathbf{r} = (x, 0, 0), \tag{3.3-3}$$

and, as a result,

$$A(f, \mathbf{r}) = A(f, x). \tag{3.3-4}$$

Also, we can write that

$$dV \to dx \tag{3.3-5}$$

and [see Eq. (3.2-36)]

$$\boldsymbol{\alpha} \cdot \mathbf{r} = (f_X, f_Y, f_Z) \cdot (x, 0, 0) = f_X x. \tag{3.3-6}$$

Therefore, substituting Eqs. (3.3-4) through (3.3-6) into Eq. (3.3-2) yields

$$D(f, f_X) = \int_{-L/2}^{L/2} A(f, x) \exp(+j2\pi f_X x)\, dx, \tag{3.3-7}$$

which is the far-field directivity function of a linear aperture, where

$$f_X = u/\lambda = \sin\theta \cos\psi/\lambda. \tag{3.3-8}$$

Equation (3.3-7) can also be written as

$$D(f, f_X) = F_x\{A(f, x)\}. \tag{3.3-9}$$

Since the spatial frequency f_X can be expressed in terms of θ and ψ, Eq. (3.3-7) specifies the far-field beam pattern of a linear aperture as a function of the operating frequency f and the spherical angles θ and ψ, that is,

$$D(f, f_X) \equiv D(f, \theta, \psi). \tag{3.3-10}$$

With the use of Eq. (3.3-7), we shall compute closed form expressions for the far-field beam patterns of several hypothetical amplitude windows.

While the various amplitude windows to be discussed next are hypothetical in the sense of representing actual frequency responses of electroacoustic transducers, they are very common functions and will serve our purposes of demonstrating far-field beam-pattern calculations, beamwidth calculations, and the relationship between beamwidth and sidelobe levels of far-field beam patterns very well. Besides, sampled versions of these continuous amplitude windows are, in fact, used to amplitude-weight arrays of electroacoustic transducers, as will be discussed in Chapter 4. In addition, time-domain versions of these continuous amplitude windows can also be used to reduce the sidelobe levels of the auto-ambiguity functions of common transmitted electrical signals, as will be discussed in Chapter 5.

In the examples that follow it is assumed that the phase function $\theta(f, x)$ is equal to zero, that is, $A(f, x) = a(f, x)$. We shall examine the effects of a nonzero $\theta(f, x)$ later in this chapter.

EXAMPLE 3.3-1: THE RECTANGULAR WINDOW

The rectangular amplitude window is given by

$$a(f, x) = \text{rect}(L), \tag{3.3-11}$$

where (see Fig. 3.3-2)

$$\text{rect}(L) = 1, \qquad |x| \leq L/2. \tag{3.3-12}$$

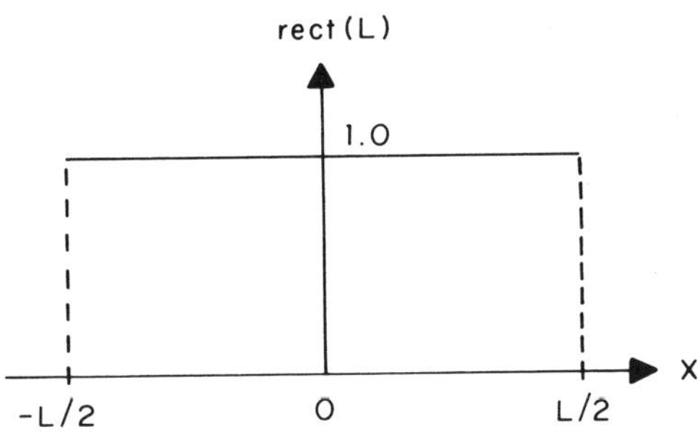

FIG. 3.3-2 Rectangular amplitude window.

Substituting Eq. (3.3-11) and Eq. (3.3-12) into Eq. (3.3-7) yields

$$D(f, f_X) = \int_{-L/2}^{L/2} \exp(+j2\pi f_X x)\,dx \qquad (3.3\text{-}13)$$

or

$$D(f, f_X) = \sin(\pi L f_X)/(\pi f_X), \qquad (3.3\text{-}14)$$

which is the *unnormalized* far-field directivity function of the rectangular amplitude window. The *normalized* directivity function, $D_N(f, f_X)$, is defined as follows:

$$D_N(f, f_X) \triangleq D(f, f_X)/D(f, 0), \qquad (3.3\text{-}15)$$

where

$$D(f, 0) = D(f, f_X)\Big|_{f_X=0} \qquad (3.3\text{-}16)$$

is the *normalization factor*. If we set $f_X = 0$ in Eq. (3.3-14), we obtain

$$D(f, 0) = L. \qquad (3.3\text{-}17)$$

Substituting Eqs. (3.3-14) and (3.3-17) into Eq. (3.3-15) yields

$$D_N(f, f_X) = \sin(\pi L f_X)/(\pi L f_X), \qquad (3.3\text{-}18)$$

which is the normalized far-field beam pattern of the rectangular window. Note that

$$D_N(f, 0) = 1. \qquad (3.3\text{-}19)$$

□

The normalization factor given by Eq. (3.3-17) was based on the knowledge that the unnormalized beam pattern of the rectangular window has its maximum value at the origin, that is, at $f_X = 0$. Thus, *the normalization factor is simply the maximum value of the unnormalized beam pattern*, whether this value occurs at $f_X = 0$ or not. Equation (3.3-18) can also be written as

$$D_N(f, u/\lambda) = \sin(\pi L u/\lambda)/(\pi L u/\lambda) \qquad (3.3\text{-}20)$$

since $f_X = u/\lambda$ [see Eq. (3.3-8)]. The magnitude of Eq. (3.3-20) is plotted as a function of the direction cosine u in Fig. 3.3-3. The plot is shown only for positive values of u since the magnitude of the beam pattern is symmetric. The level of the first sidelobe is approximately -13 dB.

From Fig. 3.3-3 it can be seen that the width of the main lobe of the far-field beam pattern is dependent upon the ratio λ/L. Thus, the beamwidth is directly proportional to the wavelength λ (inversely proportional to the operating frequency f) and inversely proportional to the length L of the transducer. Beamwidth will be discussed in detail in Sections 3.3.2 and 3.3.3.

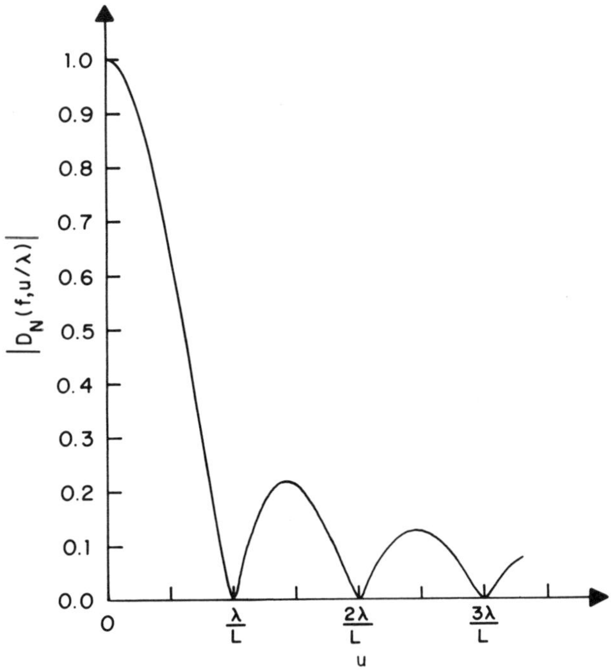

FIG. 3.3-3 Magnitude of the normalized far-field beam pattern of the rectangular window plotted as a function of the direction cosine u.

The directivity function $D(f, f_X) = D(f, u/\lambda)$ given by Eq. (3.3-7) can be evaluated at any value of u, in theory. However, it is only the interval $-1 \leq u \leq 1$ that pertains to the real aperture problem since u is a direction cosine given by

$$u = \sin\theta \cos\psi, \tag{3.2-6}$$

and, as a result, $|u| \leq 1$ for $0 \leq \theta \leq \pi$ and $0 \leq \psi \leq 2\pi$. The interval $-1 \leq u \leq 1$ is called the *visible region*. The value $u = 0$, which implies $\psi = \pi/2$, corresponds to an observation point being at *broadside* or *on-axis* relative to the aperture, and $u = \pm 1$, which implies $\theta = \pi/2, \psi = 0$ for $u = +1$ and $\theta = \pi/2, \psi = \pi$ for $u = -1$, corresponds to an observation point being at the *horizons* or at *end-fire* geometry (see Fig. 3.3-1).

The far-field directivity function of the rectangular window could also have been computed by using the method of *successive differentiation* instead of by direct integration. The method of successive differentiation is most advantageous when the amplitude window is composed of straight line segments, although the method can be applied to arbitrarily shaped windows. This method is based on a spatial Fourier transform pair, which we shall derive, involving the far-field beam pattern and the nth-order derivative of the complex aperture function.

Consider the following analysis: since

$$D(f, f_X) = F_x\{A(f, x)\}, \tag{3.3-9}$$

$$A(f, x) = F_{f_X}^{-1}\{D(f, f_X)\} \tag{3.3-21}$$

or

$$A(f, x) = \int_{-\infty}^{\infty} D(f, f_X) \exp(-j2\pi f_X x)\, df_X. \tag{3.3-22}$$

If we take the derivative with respect to x of both sides of Eq. (3.3-22), we obtain

$$\frac{d}{dx} A(f, x) = \int_{-\infty}^{\infty} (-j2\pi f_X) D(f, f_X) \exp(-j2\pi f_X x)\, df_X, \tag{3.3-23}$$

which can be expressed as

$$\frac{d}{dx} A(f, x) = F_{f_X}^{-1}\{-j2\pi f_X D(f, f_X)\}. \tag{3.3-24}$$

From Eq. (3.3-24), we can write

$$D(f, f_X) = F_x\left\{\frac{d}{dx} A(f, x)\right\} \Big/ (-j2\pi f_X), \tag{3.3-25}$$

or, in general,

$$D(f, f_X) = F_x\left\{\frac{d^n}{dx^n} A(f, x)\right\} \Big/ (-j2\pi f_X)^n, \tag{3.3-26}$$

where d^n/dx^n represents an nth-order derivative with respect to x. The result given by Eq. (3.3-26) is the basis for the method of successive differentiation. We shall demonstrate this method by using it to compute the far-field beam pattern of the triangular amplitude window (also known as the *Bartlett* or *Fejer* window).

EXAMPLE 3.3-2: THE TRIANGULAR WINDOW

The triangular amplitude window is given by

$$a(f, x) = \text{tri}(L), \tag{3.3-27}$$

where (see Fig. 3.3-4)

$$\text{tri}(L) = 1 - \frac{|x|}{L/2}, \qquad |x| \le \frac{L}{2}. \tag{3.3-28}$$

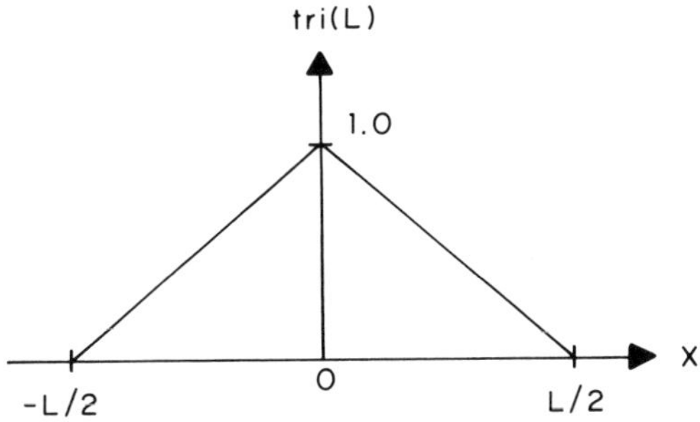

FIG. 3.3-4 Triangular amplitude window.

As mentioned previously, the method of successive differentiation is most advantageous when the amplitude window is piecewise linear since single or multiple differentiations of such a window will ultimately yield impulse functions that have simple spatial Fourier transforms. For example, if we differentiate the triangular window twice, we obtain the impulse functions shown in Fig. 3.3-5b. As a result, we can express the second derivative as follows:

$$\frac{d^2}{dx^2}a(f,x) = \frac{2}{L}\left[\delta\left(x + \frac{L}{2}\right) - 2\delta(x) + \delta\left(x - \frac{L}{2}\right)\right]. \tag{3.3-29}$$

Taking the spatial Fourier transform of both sides of Eq. (3.3-29) yields

$$F_x\left\{\frac{d^2}{dx^2}a(f,x)\right\} = \frac{2}{L}[\exp(-j\pi L f_X) - 2 + \exp(+j\pi L f_X)] \tag{3.3-30}$$

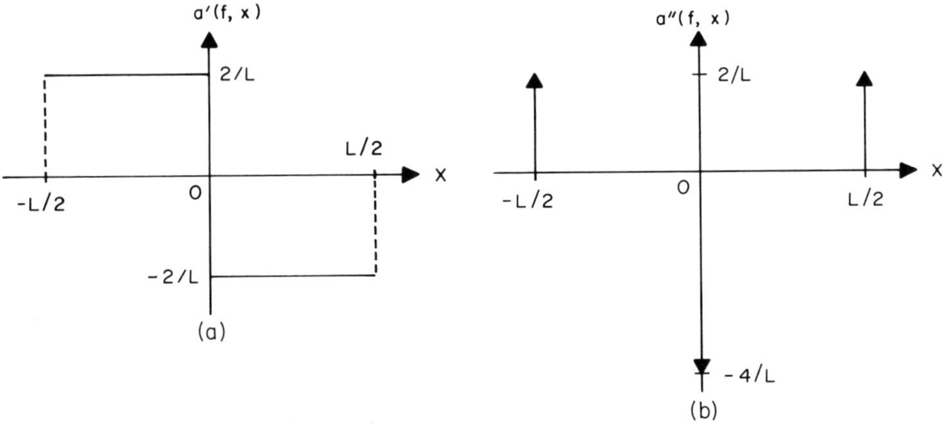

FIG. 3.3-5 Derivatives of the triangular amplitude window. (a) First derivative, where $a'(f,x) \equiv da(f,x)/dx$; (b) second derivative, where $a''(f,x) \equiv d^2a(f,x)/dx^2$.

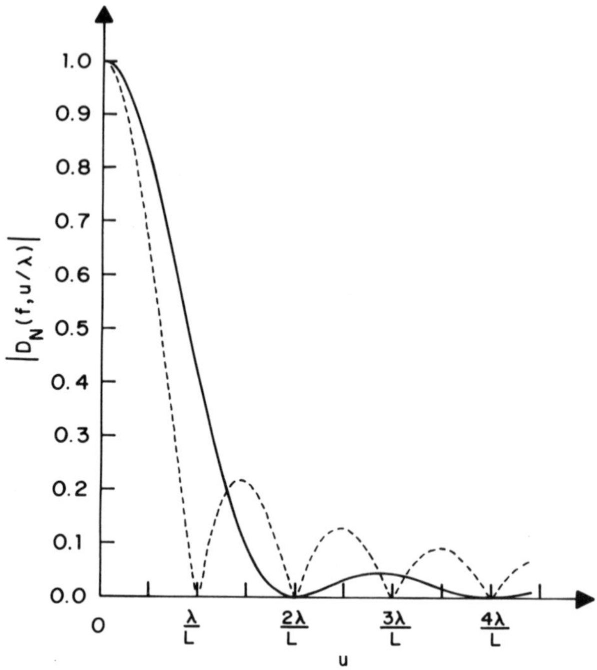

FIG. 3.3-6 Magnitude of the normalized far-field beam patterns of the rectangular window (---) and triangular window (——) plotted as a function of the direction cosine u.

since

$$F_x\{\delta(x \pm d)\} = \exp(\mp j2\pi f_X d). \tag{3.3-31}$$

If we now substitute Eq. (3.3-30) into Eq. (3.3-26) with $n = 2$ and make use of the identity $(a - b)^2 = (a^2 - 2ab + b^2)$, we obtain

$$D(f, f_X) = \frac{2}{(-j2\pi f_X)^2 L}\left[\exp\left(\frac{-j\pi L f_X}{2}\right) - \exp\left(\frac{+j\pi L f_X}{2}\right)\right]^2 \tag{3.3-32}$$

or

$$D(f, f_X) = \frac{L}{2}\left[\frac{\sin(\pi L f_X/2)}{\pi L f_X/2}\right]^2, \tag{3.3-33}$$

which is the unnormalized far-field directivity function of the triangular amplitude window. The normalization factor for the triangular window can be obtained by evaluating Eq. (3.3-33) at $f_X = 0$, that is, at broadside. Doing so yields

$$D(f,0) = L/2, \tag{3.3-34}$$

and upon substituting Eqs. (3.3-33) and (3.3-34) into Eq. (3.3-15), we obtain

$$D_N(f, f_X) = \left[\frac{\sin(\pi L f_X/2)}{\pi L f_X/2}\right]^2, \tag{3.3-35}$$

which is the normalized far-field directivity function of the triangular window (see Fig. 3.3-6). □

The levels of the first sidelobe of the normalized far-field beam patterns of the rectangular and triangular windows are approximately −13 and −27 dB, respectively. However, the main lobe of the triangular window's beam pattern is broader than the rectangular window's main lobe. Note that while the rectangular amplitude window is discontinuous at the end points $x = \pm L/2$ (see Fig. 3.3-2), the triangular amplitude window approaches zero at $x = \pm L/2$ in a comparatively smooth fashion (see Fig. 3.3-4). In general, therefore, the smoother an amplitude window approaches zero at the end points $x = \pm L/2$, the lower the sidelobe levels of its beam pattern will be. However, a reduction in sidelobe level generally results in an increase in the width of the beam pattern's main lobe. This trend will become more apparent as we consider additional amplitude windows.

Another useful method for the computation of far-field beam patterns is the method of *multiplication and convolution*. This method is applicable when the aperture function $A(f,x)$ is a finite portion of a periodic function or the product of two simpler functions. For example, if

$$A(f,x) = A_1(f,x)A_2(f,x), \tag{3.3-36}$$

then the far-field directivity function is given by

$$D(f, f_X) = F_x\{A(f,x)\},$$

$$D(f, f_X) = F_x\{A_1(f,x)\} * F_x\{A_2(f,x)\},$$

or

$$D(f, f_X) = D_1(f, f_X) * D_2(f, f_X), \tag{3.3-37}$$

where the asterisk denotes convolution with respect to the spatial frequency f_X and D_1 and D_2 are the far-field beam patterns of the aperture functions A_1 and A_2, respectively.

EXAMPLE 3.3-3

We shall demonstrate the method of multiplication and convolution by computing the far-field beam pattern of the following aperture function:

$$A(f,x) = \cos(2\pi x/d)\,\mathrm{rect}(L). \tag{3.3-38}$$

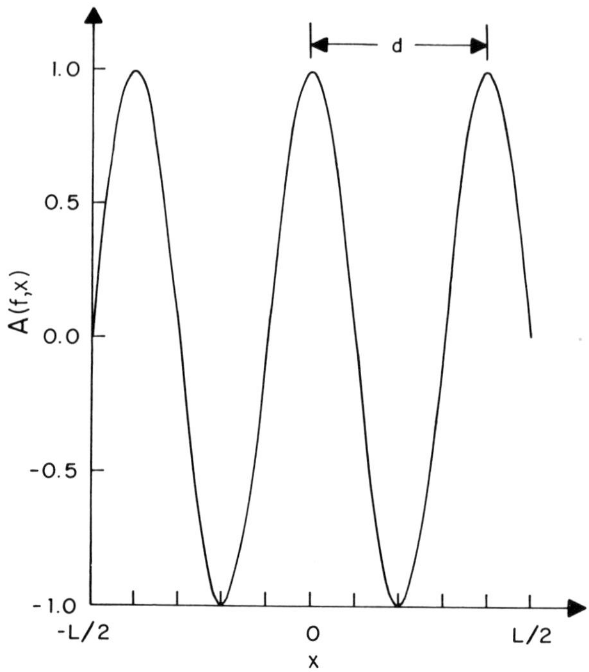

FIG. 3.3-7 Plot of the aperture function given by Eq. (3.3-38) with $L = 5d/2$.

This function is shown in Fig. 3.3-7 for $L = 5d/2$, where d is the *spatial period* in meters. Note that this function is analogous to a continuous-wave (cw) pulse. Comparing Eqs. (3.3-36) and (3.3-38), we let

$$A_1(f, x) = \cos(2\pi x/d) \tag{3.3-39}$$

and

$$A_2(f, x) = \text{rect}(L). \tag{3.3-40}$$

Since

$$\cos(2\pi x/d) = \tfrac{1}{2}[\exp(+j2\pi x/d) + \exp(-j2\pi x/d)] \tag{3.3-41}$$

and

$$F_x\{\exp(\pm j2\pi f'_X x)\} = \delta(f_X \pm f'_X), \tag{3.3-42}$$

$$F_x\{\cos(2\pi x/d)\} = \tfrac{1}{2}[\delta(f_X + 1/d) + \delta(f_X - 1/d)]. \tag{3.3-43}$$

Therefore, since

$$D_1(f, f_X) = F_x\{A_1(f, x)\} = F_x\{\cos(2\pi x/d)\}, \tag{3.3-44}$$

from Eq. (3.3-43) we have

$$D_1(f, f_X) = \tfrac{1}{2}[\delta(f_X + 1/d) + \delta(f_X - 1/d)]; \tag{3.3-45}$$

and since

$$D_2(f, f_X) = F_x\{A_2(f, x)\} = F_x\{\mathrm{rect}(L)\}, \tag{3.3-46}$$

from Eq. (3.3-14) we have

$$D_2(f, f_X) = \sin(\pi L f_X)/(\pi f_X) \tag{3.3-47}$$

or

$$D_2(f, f_X) = L \sin(\pi L f_X)/(\pi L f_X). \tag{3.3-48}$$

By referring back to Eq. (3.3-37), we see that the far-field beam pattern of the aperture function given by Eq. (3.3-38) can be obtained by convolving (with respect to f_X) Eq. (3.3-45) with Eq. (3.3-48). Since Eq. (3.3-45) is a sum of impulse functions, the convolution process is trivial since

$$g(t) * [\delta(t + t_0) + \delta(t - t_0)] = g(t) * \delta(t + t_0) + g(t) * \delta(t - t_0), \tag{3.3-49}$$

$$g(t) * [\delta(t + t_0) + \delta(t - t_0)] = g(t + t_0) + g(t - t_0), \tag{3.3-50}$$

where the asterisk $(*)$ denotes convolution with respect to t in this case and t_0 is an arbitrary constant. Therefore, substituting Eqs. (3.3-45) and (3.3-48) into Eq. (3.3-37) and making use of Eq. (3.3-50) yields

$$D(f, f_X) = F_x\{\cos(2\pi x/d)\,\mathrm{rect}(L)\}, \tag{3.3-51}$$

where

$$F_x\left\{\cos\left(\frac{2\pi x}{d}\right)\mathrm{rect}(L)\right\} = \frac{L}{2}\left\{\frac{\sin[\pi L(f_X + 1/d)]}{\pi L(f_X + 1/d)} + \frac{\sin[\pi L(f_X - 1/d)]}{\pi L(f_X - 1/d)}\right\}. \tag{3.3-52}$$

The magnitude of the unnormalized far-field directivity function given by Eq. (3.3-52) is plotted in Fig. 3.3-8 for $L = 5d/2$. Note that as $d \to \infty$ the aperture function shown in Fig. 3.3-7 approaches rect(L) [see Eq. (3.3-38)] and the beam pattern given by Eq. (3.3-52) approaches

$$L \sin(\pi L f_X)/(\pi L f_X),$$

which is the far-field directivity function of the rectangular window. □

Equation (3.3-52) is a very important result since the far-field directivity functions of the cosine, Hanning, Hamming, and Blackman amplitude windows can be obtained from it, as we shall demonstrate next.

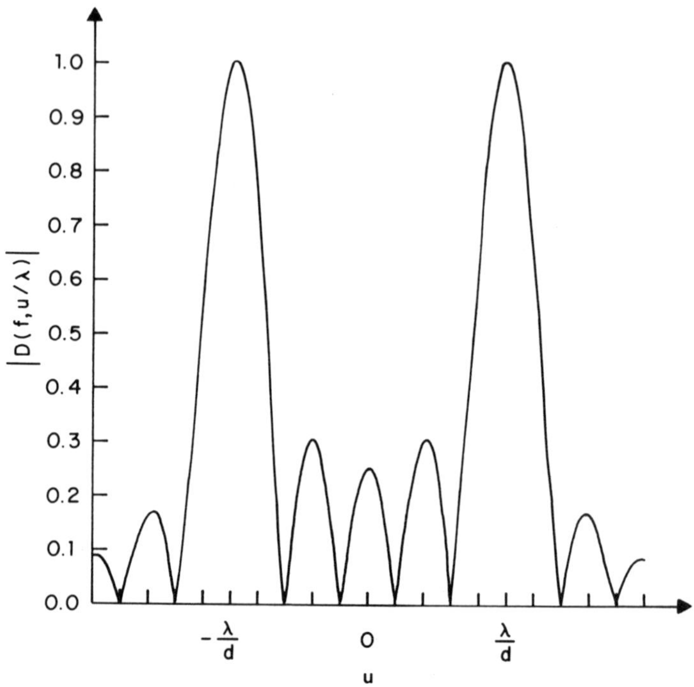

FIG. 3.3-8 Magnitude of the unnormalized far-field beam pattern given by Eq. (3.3-52) plotted as a function of the direction cosine u for $L = 5d/2$.

EXAMPLE 3.3-4: THE COSINE WINDOW

The cosine amplitude window is given by

$$a(f, x) = \cos(\pi x/L)\,\mathrm{rect}(L), \tag{3.3-53}$$

which can be obtained from Eq. (3.3-38) by setting $d = 2L$ (see Fig. 3.3-9). Substituting $d = 2L$ into Eq. (3.3-52) yields

$$F_x\left\{\cos\left(\frac{\pi x}{L}\right)\mathrm{rect}(L)\right\} = \frac{L}{2}\left\{\frac{\sin(\pi L f_X + \pi/2)}{\pi L f_X + \pi/2} + \frac{\sin(\pi L f_X - \pi/2)}{\pi L f_X - \pi/2}\right\}, \tag{3.3-54}$$

or, since $\sin(\pi L f_X \pm \pi/2) = \pm\cos(\pi L f_X)$,

$$F_x\left\{\cos\left(\frac{\pi x}{L}\right)\mathrm{rect}(L)\right\} = \frac{2L}{\pi}\frac{\cos(\pi L f_X)}{[1 - (2L f_X)^2]}, \tag{3.3-55}$$

which is the unnormalized far-field beam pattern of the cosine amplitude

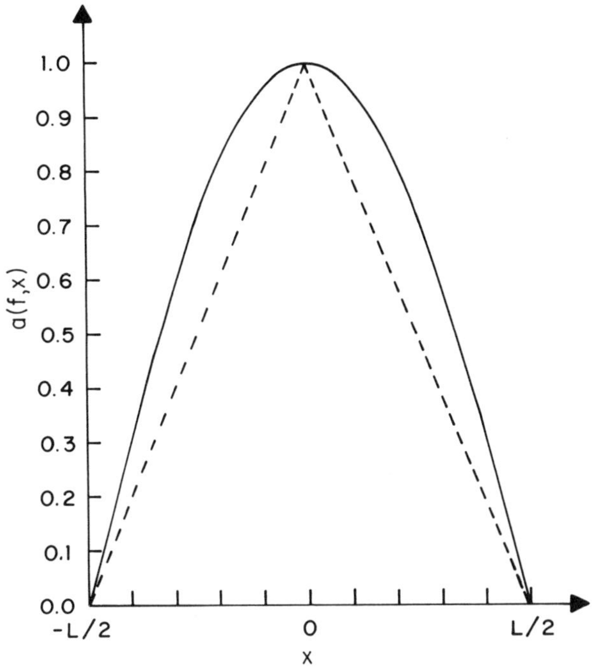

FIG. 3.3-9 Cosine amplitude window (——) and triangular amplitude window (---).

window. Since the maximum value of Eq. (3.3-55) occurs at $f_X = 0$ and is equal to $2L/\pi$, the normalized far-field beam pattern is given by (see Fig. 3.3-10)

$$D_N(f, f_X) = \frac{\cos(\pi L f_X)}{1 - (2Lf_X)^2}. \tag{3.3-56}$$

□

The level of the first sidelobe of the normalized far-field beam pattern of the cosine window is approximately −23 dB, which is in between those of the rectangular and triangular windows. This is not surprising, since the cosine amplitude window approaches zero at the end points $x = \pm L/2$ in a less abrupt manner than the rectangular window but more abruptly than the triangular window (see Fig. 3.3-9).

The next amplitude window to be discussed is the *Hanning* window, which is also known as the *raised cosine* or *cosine squared* window. The Hanning amplitude window is given by (see Fig. 3.3-11)

$$a(f, x) = \cos^2(\pi x/L)\,\text{rect}(L) \tag{3.3-57}$$

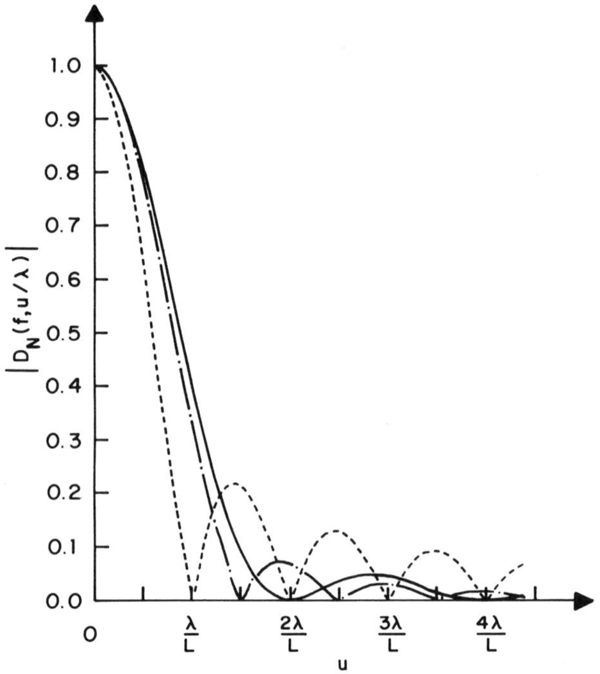

FIG. 3.3-10 Magnitude of the normalized far-field beam patterns of the rectangular window (– – –), triangular window (——), and cosine window (—•—) plotted as a function of the direction cosine u.

or

$$a(f, x) = 0.5[1.0 + \cos(2\pi x/L)] \operatorname{rect}(L) \tag{3.3-58}$$

since

$$\cos^2 \alpha = 0.5[1.0 + \cos(2\alpha)].$$

If Eq. (3.3-58) is rewritten as

$$a(f, x) = 0.5 \operatorname{rect}(L) + 0.5 \cos(2\pi x/L) \operatorname{rect}(L), \tag{3.3-59}$$

then

$$D(f, f_X) = 0.5F_x\{\operatorname{rect}(L)\} + 0.5F_x\{\cos(2\pi x/L) \operatorname{rect}(L)\}, \tag{3.3-60}$$

which is the unnormalized far-field directivity function of the Hanning amplitude window. The first transform in Eq. (3.3-60) is given by Eq. (3.3-14), while the second transform can be obtained from Eq. (3.3-52) by setting $d = L$.

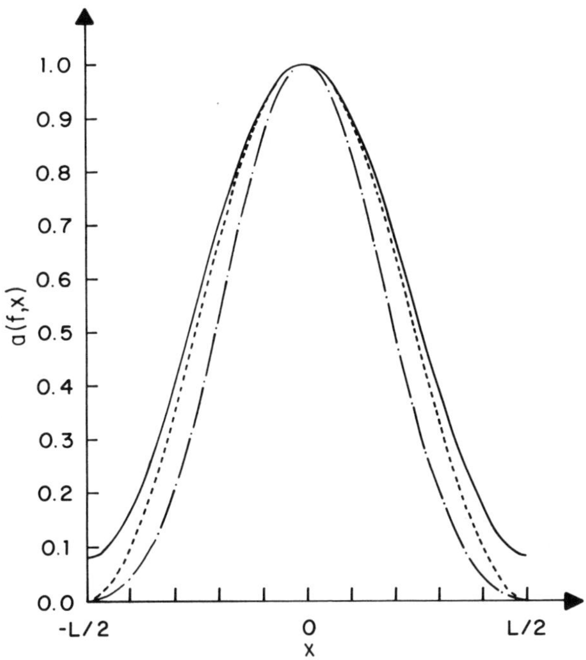

FIG. 3.3-11 Hanning (- - -), Hamming (——), and Blackman (—·—) amplitude windows.

The last two amplitude windows to be discussed are the Hamming and Blackman windows. The *Hamming* amplitude window is given by

$$a(f,x) = [0.54 + 0.46\cos(2\pi x/L)]\,\mathrm{rect}(L), \tag{3.3-61}$$

while the *Blackman* amplitude window is given by (see Fig. 3.3-11)

$$a(f,x) = [0.42 + 0.5\cos(2\pi x/L) + 0.08\cos(4\pi x/L)]\,\mathrm{rect}(L). \tag{3.3-62}$$

The levels of the first sidelobe of the normalized far-field beam patterns of the Hanning, Hamming, and Blackman windows are approximately -32, -44, and -58 dB, respectively (see Fig. 3.3-12). From Fig. 3.3-12 it can be seen that although the level of the first sidelobe of the Hamming beam pattern is less than that of the Hanning, the beamwidth of the Hamming beam pattern is *less* than that of the Hanning, which is an exception to the general rule that beamwidth increases as sidelobe level decreases. Therefore, the Hamming amplitude window is superior to the Hanning window.

The method of multiplication and convolution is also useful when the aperture function $A(f,x)$ can be expressed as the convolution of two simpler

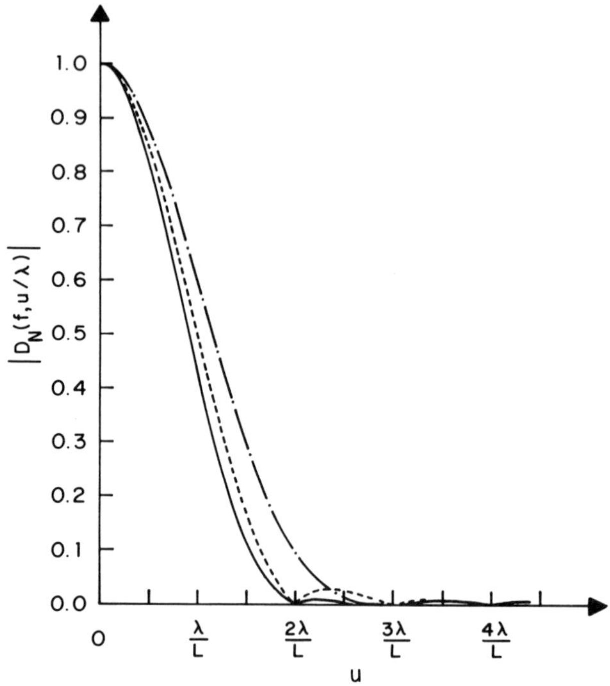

FIG. 3.3-12 Magnitude of the normalized far-field beam patterns of the Hanning window (---), Hamming window (——), and Blackman window (—·—) plotted as a function of the direction cosine u.

functions. For example, if

$$A(f,x) = A_1(f,x) * A_2(f,x), \tag{3.3-63}$$

where the asterisk ($*$) denotes convolution with respect to x, then the far-field directivity function is given by

$$D(f,f_X) = F_x\{A(f,x)\},$$
$$D(f,f_X) = F_x\{A_1(f,x)\}F_x\{A_2(f,x)\},$$

or

$$D(f,f_X) = D_1(f,f_X)D_2(f,f_X), \tag{3.3-64}$$

where D_1 and D_2 are the far-field beam patterns of the aperture functions A_1 and A_2, respectively.

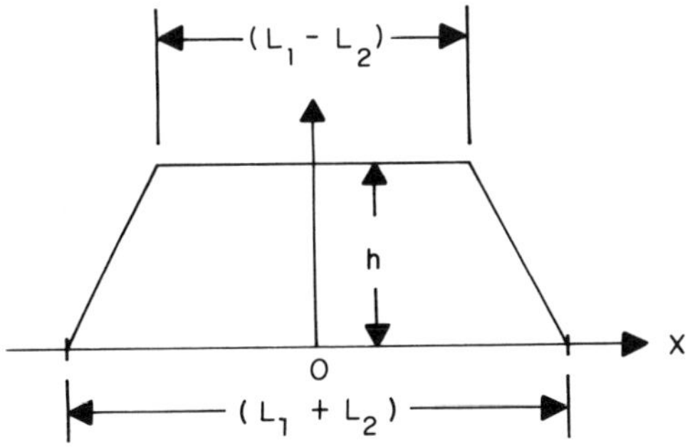

FIG. 3.3-13 Trapezoidal amplitude window.

EXAMPLE 3.3-5

Consider an aperture function $A(f, x) = a(f, x)$ that is a trapezoid, as shown in Fig. 3.3-13. Since a trapezoid can be expressed mathematically as being equal to the convolution between two rectangles, the amplitude window shown in Fig. 3.3-13 is given by

$$a(f, x) = h_1 \operatorname{rect}(L_1) * h_2 \operatorname{rect}(L_2), \qquad L_1 > L_2,$$

where the height h is given by

$$h = h_1 h_2 L_2, \qquad L_1 > L_2.$$

Therefore

$$D(f, f_X) = h_1 h_2 L_1 L_2 \frac{\sin(\pi L_1 f_X)}{\pi L_1 f_X} \frac{\sin(\pi L_2 f_X)}{\pi L_2 f_X}.$$

Note that if $h_2 = h_1 = 1$ and $L_2 = L_1$, then $h = L_1$,

$$a(f, x) = L_1 \operatorname{tri}(2L_1) = \operatorname{rect}(L_1) * \operatorname{rect}(L_1),$$

and

$$D(f, f_X) = L_1^2 \left[\frac{\sin(\pi L_1 f_X)}{\pi L_1 f_X}\right]^2. \quad \square$$

All of the aperture functions discussed in this section were even functions and, as a result, their far-field beam patterns were real and even. As a rule, if the aperture function is *even*, the resulting far-field beam pattern will be *real* and *even*, and if the aperture function is an *odd* function, the beam pattern will be *imaginary* and *odd*.

3.3.2 Beamwidth

The term "beamwidth" corresponds to some measure of the width of the main lobe of the far-field directivity function. The most common measure is the 3-dB beamwidth, which is defined as the width of the main lobe between those two points that are 3 dB below the peak. Figure 3.3-14 shows the magnitude of a typical, normalized far-field beam pattern plotted as a function of the direction cosine u, where the 3-dB beamwidth, designated Δu, is given by

$$\Delta u = u_+ - u_- > 0, \tag{3.3-65}$$

or, since the beam pattern is symmetric in "u" space,

$$\Delta u = 2u_+, \tag{3.3-66}$$

where $u_+ > 0$ and $u_- < 0$, as shown. Note that

$$20\log_{10}(\sqrt{2}/2) = 10\log_{10}(1/2) = -3\text{ dB}.$$

The problem now is to express the 3-dB beamwidth Δu in u space in terms of the 3-dB beamwidths $\Delta\theta$ and $\Delta\psi$ associated with the spherical angles θ and ψ, respectively. Note that Δu is *dimensionless*, while $\Delta\theta$ and $\Delta\psi$ have units of *degrees*. Since $u = \sin\theta\cos\psi$ in general, let (see Fig. 3.3-15)

$$u_+ = \sin\theta_+ \cos\psi_+ > 0, \tag{3.3-67}$$

where

$$\theta_+ = \Delta\theta/2, \tag{3.3-68}$$

$$\psi_+ = (\pi/2) - (\Delta\psi/2), \tag{3.3-69}$$

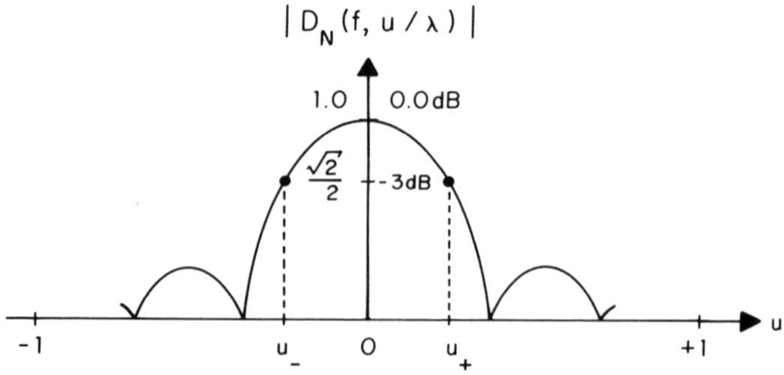

FIG. 3.3-14 Magnitude of a typical, normalized far-field beam pattern plotted as a function of the direction cosine u.

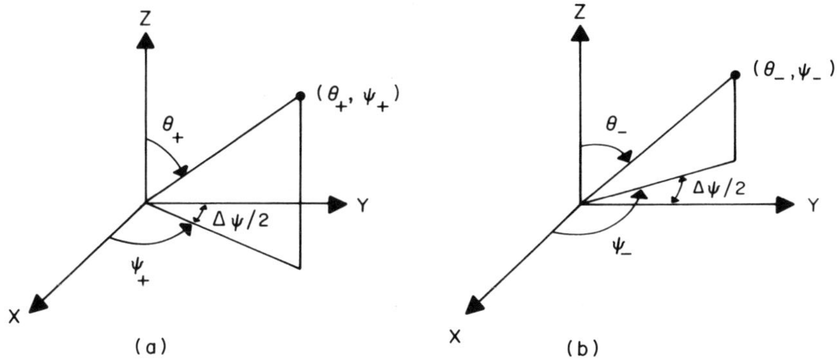

FIG. 3.3-15 Representation of spherical angles; (a) (θ_+, ψ_+); (b) (θ_-, ψ_-).

and

$$u_- = \sin\theta_- \cos\psi_- < 0, \tag{3.3-70}$$

where

$$\theta_- = \theta_+ = \Delta\theta/2 \tag{3.3-71}$$

and

$$\psi_- = (\pi/2) + (\Delta\psi/2). \tag{3.3-72}$$

Note that with ψ_+ and ψ_- given by Eqs. (3.3-69) and (3.3-72), respectively, it is assured that $u_+ > 0$ and $u_- < 0$ since

$$\cos\left(\frac{\pi}{2} - \frac{\Delta\psi}{2}\right) = \sin\left(\frac{\Delta\psi}{2}\right) > 0 \tag{3.3-73}$$

and

$$\cos\left(\frac{\pi}{2} + \frac{\Delta\psi}{2}\right) = -\sin\left(\frac{\Delta\psi}{2}\right) < 0. \tag{3.3-74}$$

Substituting Eqs. (3.3-68) and (3.3-69) into Eq. (3.3-67) yields

$$u_+ = \sin(\Delta\theta/2)\sin(\Delta\psi/2), \tag{3.3-75}$$

where use was made of Eq. (3.3-73), and substituting Eqs. (3.3-71) and (3.3-72) into Eq. (3.3-70) yields

$$u_- = -\sin(\Delta\theta/2)\sin(\Delta\psi/2), \tag{3.3-76}$$

where use was made of Eq. (3.3-74). Finally, upon substituting Eqs. (3.3-75) and (3.3-76) into Eq. (3.3-65), we obtain

$$\Delta u = 2\sin(\Delta\theta/2)\sin(\Delta\psi/2), \tag{3.3-77}$$

which is the desired result. However, we now have only one equation but two unknowns, $\Delta\theta$ and $\Delta\psi$. Therefore, when dealing with a linear aperture, we must restrict ourselves to either the vertical or horizontal beam patterns. Consider the following two examples.

EXAMPLE 3.3-6

Assume we are given a linear aperture lying along the X axis. As a result, the vertical beam pattern occupies the XZ plane. In the XZ plane, $u = \sin\theta$ when $\psi = 0$ or $u = -\sin\theta$ when $\psi = \pi$. Therefore, let (see Fig. 3.3-16)

$$u_+ = \sin\theta_+ > 0, \qquad \psi_+ = 0 \tag{3.3-78}$$

and

$$u_- = -\sin\theta_- < 0, \qquad \psi_- = \pi, \tag{3.3-79}$$

where

$$\theta_+ = \theta_- = \Delta\theta/2. \tag{3.3-80}$$

Thus,

$$\Delta u = u_+ - u_- = 2\sin(\Delta\theta/2) > 0 \tag{3.3-81}$$

or

$$\Delta\theta = 2\sin^{-1}(\Delta u/2), \qquad \Delta u/2 \le 1, \tag{3.3-82}$$

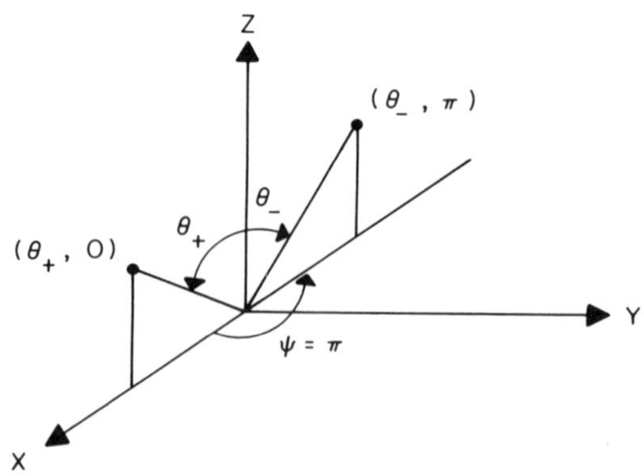

FIG. 3.3-16 Representation of spherical angles $(\theta_+, \psi_+ = 0)$ and $(\theta_-, \psi_- = \pi)$.

where $\Delta\theta$ is the beamwidth (in degrees) of the vertical beam pattern in the XZ plane. □

EXAMPLE 3.3-7

Assume we are given a linear aperture lying along the X axis. As a result, the horizontal beam pattern occupies the XY plane. In the XY plane, $u = \cos\psi$ since $\theta = \pi/2$. Therefore, let (see Fig. 3.3-17)

$$u_+ = \cos\psi_+ > 0, \qquad \theta_+ = \pi/2 \tag{3.3-83}$$

and

$$u_- = \cos\psi_- < 0, \qquad \theta_- = \pi/2, \tag{3.3-84}$$

where

$$\psi_+ = (\pi/2) - (\Delta\psi/2) \tag{3.3-69}$$

and

$$\psi_- = (\pi/2) + (\Delta\psi/2). \tag{3.3-72}$$

Thus, with the use of Eqs. (3.3-73) and (3.3-74),

$$\Delta u = u_+ - u_- = 2\sin(\Delta\psi/2) > 0 \tag{3.3-85}$$

or

$$\Delta\psi = 2\sin^{-1}(\Delta u/2), \qquad \Delta u/2 \leq 1, \tag{3.3-86}$$

where $\Delta\psi$ is the beamwidth (in degrees) of the horizontal beam pattern in the XY plane. Compare Eq. (3.3-86) with Eq. (3.3-82). □

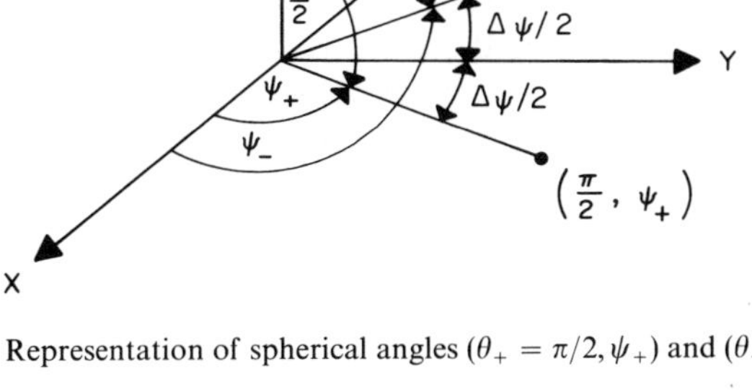

FIG. 3.3-17 Representation of spherical angles $(\theta_+ = \pi/2, \psi_+)$ and $(\theta_- = \pi/2, \psi_-)$.

By examining Eqs. (3.3-82) and (3.3-86), it can be seen that in order to actually compute values for $\Delta\theta$ and $\Delta\psi$, we need to know Δu. Consider the next example.

EXAMPLE 3.3-8

Let us compute the 3-dB beamwidth Δu in direction cosine space of the normalized far-field directivity function of the rectangular amplitude window given by Eq. (3.3-20).

From Eq. (3.3-20), we can write that

$$\frac{\sin(\pi L u_+/\lambda)}{\pi L u_+/\lambda} = \frac{\sqrt{2}}{2},$$

or, rewriting,

$$\sin(\pi x) - (\pi\sqrt{2}/2)x = 0, \tag{3.3-87}$$

where [see Eq. (3.3-66)]

$$x = L u_+/\lambda = L\,\Delta u/(2\lambda),$$

so that

$$\Delta u = 2x\lambda/L. \tag{3.3-88}$$

The solution of Eq. (3.3-87) is $x \approx 0.443$, and upon substituting this result into Eq. (3.3-88), we obtain $\Delta u \approx 0.886\lambda/L$ for the rectangular amplitude window. □

3.3.3 Beam Steering

Recall that, in general, the complex frequency response or complex aperture function $A(f, x)$ can be represented as

$$A(f, x) = a(f, x)\exp[+j\theta(f, x)], \tag{3.3-89}$$

where $a(f, x)$ is known as the amplitude window and $\theta(f, x)$ is the phase function or phase variation across the linear aperture. In Section 3.3.2 we ignored $\theta(f, x)$ and concentrated our efforts on calculating the far-field beam patterns of various amplitude windows. In this section we shall study what happens to a far-field beam pattern when $\theta(f, x)$ is nonzero, in particular when $\theta(f, x)$ is a linear function of x.

Let us begin the analysis by expressing $\theta(f, x)$ in terms of the following polynomial:

$$\theta(f, x) = \theta_0 + \theta_1(f)x + \theta_2(f)x^2 + \cdots + \theta_N(f)x^N, \tag{3.3-90}$$

where θ_0 represents a *constant phase shift* across the aperture, $\theta_1(f)x$ represents a *linear phase shift* (linear in x) across the aperture and is responsible for *beam steering* or *beam tilting*, and $\theta_2(f)x^2$ represents a *quadratic phase shift* (quadratic in x) across the aperture and is responsible for *focusing*. The remaining higher-order terms in Eq. (3.3-90) are not considered. Note that the coefficients $\theta_1(f)$ and $\theta_2(f)$ are functions of the frequency f.

Next, let $D(f, f_X)$ be the far-field beam pattern of the amplitude window $a(f, x)$, that is,

$$D(f, f_X) = F_x\{a(f, x)\} \tag{3.3-91}$$

or

$$D(f, f_X) = \int_{-\infty}^{\infty} a(f, x) \exp(+j2\pi f_X x)\, dx, \tag{3.3-92}$$

and let $D'(f, f_X)$ be the far-field beam pattern of the aperture function $A(f, x)$ given by Eq. (3.3-89), that is,

$$D'(f, f_X) = F_x\{a(f, x) \exp[+j\theta(f, x)]\} \tag{3.3-93}$$

or

$$D'(f, f_X) = \int_{-\infty}^{\infty} a(f, x) \exp[+j\theta(f, x)] \exp(+j2\pi f_X x)\, dx. \tag{3.3-94}$$

Now assume that the phase variation $\theta(f, x)$ across the aperture is a linear function of x, that is,

$$\theta(f, x) = \theta_1(f)x, \tag{3.3-95}$$

where

$$\theta_1(f) = -2\pi f'_X \quad \text{rad/m}, \tag{3.3-96}$$

$$f'_X = u'/\lambda = fu'/c, \tag{3.3-97}$$

and

$$u' = \sin\theta' \cos\psi'. \tag{3.3-98}$$

Note that the linear phase variation given by Eq. (3.3-95) is an *odd* function of x and is a straight line that passes through the origin with slope given by Eq. (3.3-96). If Eqs. (3.3-95) and (3.3-96) are substituted into Eqs. (3.3-93) and (3.3-94), then

$$D'(f, f_X) = F_x\{a(f, x) \exp(-j2\pi f'_X x)\} \tag{3.3-99}$$

and

$$D'(f, f_X) = \int_{-\infty}^{\infty} a(f, x) \exp[+j2\pi(f_X - f'_X)x]\, dx, \tag{3.3-100}$$

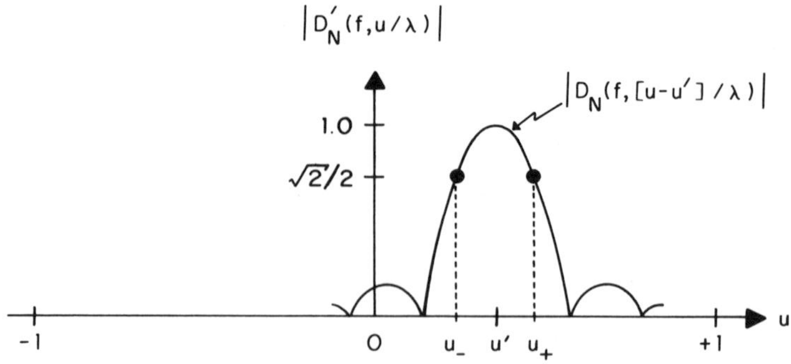

FIG. 3.3-18 Magnitude of a typical normalized far-field beam pattern steered in the direction $u = u'$ in direction cosine space.

respectively. By comparing Eq. (3.3-100) with Eq. (3.3-92),

$$D'(f, f_X) = D(f, f_X - f'_X), \tag{3.3-101}$$

or, using Eq. (3.3-99),

$$F_x\{a(f, x)\exp(-j2\pi f'_X x)\} = D(f, f_X - f'_X), \tag{3.3-102}$$

where $D(f, f_X)$ is given by Eq. (3.3-91) or Eq. (3.3-92). Therefore, a linear phase variation across the length of the aperture will cause the beam pattern $D(f, u/\lambda)$ to be steered or tilted in the direction $u = u'$ in direction cosine space (see Fig. 3.3-18). Note that

$$|D'_N(f, u'/\lambda)| = |D_N(f, 0)| = 1$$

for the beam pattern shown in Fig. 3.3-18.

The relationship [see Eq. (3.3-101)]

$$D'(f, u/\lambda) = D(f, (u - u')/\lambda)$$

indicates that the far-field directivity function D' is simply a translated version of D. Thus, when the beam pattern D is steered or tilted, its shape and beamwidth remain unchanged when plotted as a function of the direction cosine u (see Fig. 3.3-18). However, when the beam pattern D is steered, its shape and beamwidth do change when plotted as a function of the spherical angles θ and ψ.

Beamwidth at an Arbitrary Tilt Angle. When a given beam pattern is steered or tilted from broadside or on-axis toward the horizons or end-fire, its *beamwidth will increase from a minimum value at broadside to a maximum value at end-fire.*

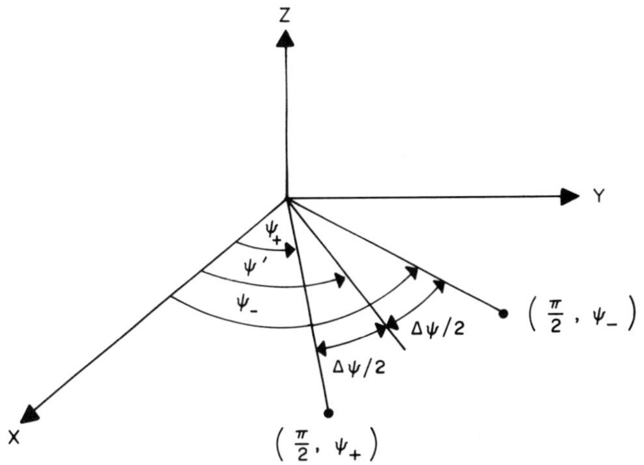

FIG. 3.3-19 Representation of spherical angles $(\theta_+ = \pi/2, \psi_+)$ and $(\theta_- = \pi/2, \psi_-)$ and beam tilt angle ψ'.

In order to compute the beamwidth, first assume we are given a linear aperture lying along the X axis with a linear phase variation across its length. Next, let us restrict ourselves to the horizontal beam pattern, where ψ' is the beam tilt angle in the XY plane (see Fig. 3.3-19). In the XY plane, $u = \cos\psi$ since $\theta = \pi/2$. Therefore, let

$$u_+ = \cos\psi_+ > 0, \qquad \theta_+ = \pi/2 \tag{3.3-103}$$

and

$$u_- = \cos\psi_- > 0, \qquad \theta_- = \pi/2, \tag{3.3-104}$$

where

$$u_+ > u_-, \tag{3.3-105}$$

$$\psi_+ = \psi' - (\Delta\psi/2), \tag{3.3-106}$$

and

$$\psi_- = \psi' + (\Delta\psi/2). \tag{3.3-107}$$

Thus,

$$\Delta u = u_+ - u_- = 2\sin(\psi')\sin(\Delta\psi/2) > 0 \tag{3.3-108}$$

since

$$\cos(\alpha - \beta) - \cos(\alpha + \beta) = 2\sin\alpha\sin\beta$$

or

$$\Delta\psi = 2\sin^{-1}[\Delta u/(2\sin\psi')], \qquad \Delta u/(2\sin\psi') \leq 1, \tag{3.3-109}$$

where $\Delta\psi$ is the beamwidth (in degrees) of the horizontal beam pattern in the XY plane. Note that $\psi' = \pi/2$ corresponds to an untilted horizontal beam pattern (see Fig. 3.3-17). With $\psi' = \pi/2$, Eq. (3.3-109) reduces to Eq. (3.3-86). Also note that $\psi' = 0$ or $\psi' = \pi$ corresponds to steering the horizontal beam pattern to the horizon or end-fire geometry. However, with $\psi' = 0$ or $\psi' = \pi$, $\Delta u/(2\sin\psi') = \infty$ and, as a result, Eq. (3.3-109) is not valid. In fact, Eq. (3.3-109) is only valid when $\Delta u/(2\sin\psi') \leq 1$, as indicated. Therefore, a different approach must be used to determine the beamwidth at end-fire geometry.

Beamwidth at End-Fire Geometry. Restricting ourselves to the horizontal beam pattern, consider the case when the beam tilt angle ψ' in the XY plane is zero (see Fig. 3.3-19). In the XY plane, $u = \cos\psi$ since $\theta = \pi/2$. Therefore, let $u' = \cos\psi'$, and with $\psi' = 0$, $u' = 1$ (see Fig. 3.3-20). Also, let

$$u_- = \cos\psi_- > 0, \qquad \theta_- = \pi/2, \tag{3.3-104}$$

where, upon substituting $\psi' = 0$ into Eq. (3.3-107),

$$\psi_- = \Delta\psi/2. \tag{3.3-110}$$

Referring to Fig. 3.3-20, we see that

$$u_- = 1 - (\Delta u/2), \tag{3.3-111}$$

so that, in conjunction with Eqs. (3.3-104) and (3.3-110),

$$\cos(\Delta\psi/2) = 1 - (\Delta u/2) \tag{3.3-112}$$

or

$$\Delta\psi = 2\cos^{-1}[1 - (\Delta u/2)], \tag{3.3-113}$$

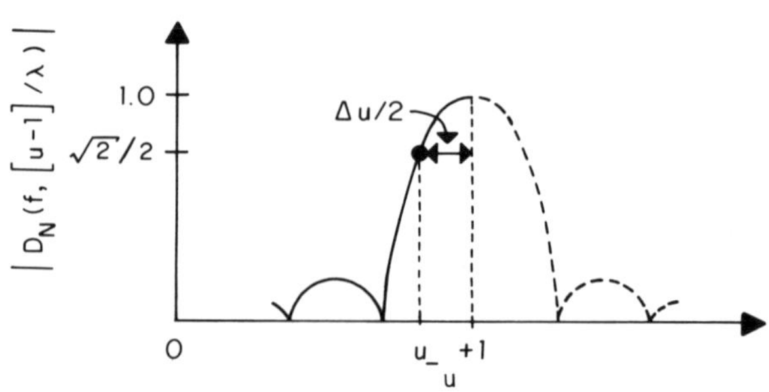

FIG. 3.3-20 Magnitude of a typical normalized far-field beam pattern steered in the direction $u' = 1$ (end-fire).

where $\Delta\psi$ is the beamwidth (in degrees) of the horizontal beam pattern in the XY plane at end-fire geometry.

3.4 LINEAR APERTURES AND NEAR-FIELD DIRECTIVITY FUNCTIONS

3.4.1 Aperture Focusing

Recall that the near-field directivity function of a general, volume aperture is [see Eq. (3.2-35)]

$$D(f,r,\boldsymbol{\alpha}) = \int_{V_0} A(f,\mathbf{r}_0)\exp[-jkr_0^2/(2r)]\exp(+j2\pi\boldsymbol{\alpha}\cdot\mathbf{r}_0)\,dV_0. \tag{3.4-1}$$

Assume we are given a linear aperture lying along the X axis, as shown in Fig. 3.4-1. Also shown is a near-field observation point with spherical coordinates (r,θ,ψ), where $r < \pi(L/2)^2/\lambda$ meters. As a result,

$$\mathbf{r}_0 = (x_0, 0, 0), \tag{3.4-2}$$

$$r_0^2 = |\mathbf{r}_0|^2 = x_0^2, \tag{3.4-3}$$

$$A(f,\mathbf{r}_0) = A(f,x_0), \tag{3.4-4}$$

$$dV_0 \rightarrow dx_0, \tag{3.4-5}$$

and

$$\boldsymbol{\alpha}\cdot\mathbf{r}_0 = (f_X, f_Y, f_Z)\cdot(x_0,0,0) = f_X x_0. \tag{3.4-6}$$

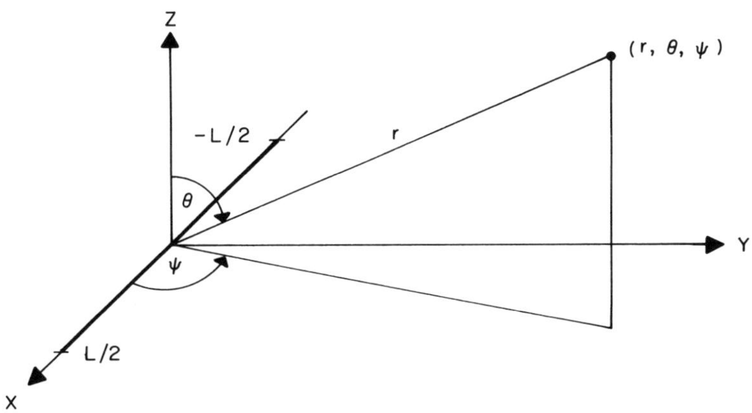

FIG. 3.4-1 Thin, elongated transducer of length L meters lying along the X axis. Also shown is a near-field observation point with spherical coordinates (r,θ,ψ).

Therefore, substituting Eqs. (3.4-3) through (3.4-6) into Eq. (3.4-1) yields

$$D(f, r, f_X) = \int_{-L/2}^{L/2} A(f, x_0) \exp[-jkx_0^2/(2r)] \exp(+j2\pi f_X x_0)\, dx_0, \quad (3.4\text{-}7)$$

which is the near-field directivity function of a linear aperture, where

$$f_X = u/\lambda = \sin\theta \cos\psi/\lambda. \quad (3.4\text{-}8)$$

Equation (3.4-7) can also be written as

$$D(f, r, f_X) = F_{x_0}\{A(f, x_0) \exp[-jkx_0^2/(2r)]\}, \quad (3.4\text{-}9)$$

where, in general, the complex aperture function $A(f, x_0)$ is given by

$$A(f, x_0) = a(f, x_0) \exp[+j\theta(f, x_0)]. \quad (3.4\text{-}10)$$

Let us investigate what happens to Eq. (3.4-7) when the phase variation $\theta(f, x_0)$ is a quadratic function of x_0.

Let us assume that the phase variation $\theta(f, x_0)$ across the length of the aperture is a quadratic function of x_0, that is [see Eq. (3.3-90)],

$$\theta(f, x_0) = \theta_2(f) x_0^2, \quad (3.4\text{-}11)$$

where

$$\theta_2(f) = k/(2r) \quad \text{rad/m}^2 \quad (3.4\text{-}12)$$

and

$$k = 2\pi f/c \quad (3.4\text{-}13)$$

is the wave number. Note that the quadratic phase variation given by Eq. (3.4-11) is an *even* function of x_0 and is a parabola that passes through the origin. If Eqs. (3.4-10) through (3.4-12) are substituted into Eqs. (3.4-7) and (3.4-9), then

$$D(f, r, f_X) = \int_{-L/2}^{L/2} a(f, x_0) \exp(+j2\pi f_X x_0)\, dx_0 \quad (3.4\text{-}14)$$

and

$$D(f, r, f_X) = F_{x_0}\{a(f, x_0)\} = D(f, f_X), \quad (3.4\text{-}15)$$

respectively, where $D(f, f_X)$ is the far-field beam pattern of the amplitude window $a(f, x_0)$. Thus, Eq. (3.4-15) indicates that a far-field beam pattern exists at the near-field distance r from the aperture when the phase variation across the aperture is given by Eqs. (3.4-11) through (3.4-13). The quadratic phase variation compensates for the wavefront curvature at near-field distances from the aperture. The physical situation represented by Eq. (3.4-15) is known as *aperture focusing*.

A good way to think about aperture focusing is to compare it with taking a photograph. If the subject is far enough away (in the far-field), we do not concern ourselves with the actual range to the subject—we simply set the focus on the camera at infinity and take the picture. However, if the subject is nearby (in the near-field), we do concern ourselves with the actual range to the subject since we must focus the camera everytime the subject moves.

3.4.2 Beam Steering and Aperture Focusing

Now assume that the phase variation $\theta(f, x_0)$ across the length of the aperture is equal to the sum of a linear and a quadratic phase shift, that is,

$$\theta(f, x_0) = \theta_1(f)x_0 + \theta_2(f)x_0^2, \tag{3.4-16}$$

where

$$\theta_1(f) = -2\pi f'_X \quad \text{rad/m}, \tag{3.3-96}$$

$$\theta_2(f) = k/(2r) \quad \text{rad/m}^2 \tag{3.4-12}$$

$$f'_X = u'/\lambda = \sin\theta' \cos\psi'/\lambda, \tag{3.3-97}$$

and

$$k = 2\pi f/c. \tag{3.4-13}$$

If Eqs. (3.4-10) and (3.4-16) are substituted into Eqs. (3.4-7) and (3.4-9), then

$$D(f, r, f_X) = \int_{-L/2}^{L/2} a(f, x_0)\exp[+j2\pi(f_X - f'_X)x_0]\,dx_0 \tag{3.4-17}$$

and

$$D(f, r, f_X) = F_{x_0}\{a(f, x_0)\exp(-j2\pi f'_X x_0)\} = D(f, f_X - f'_X), \tag{3.4-18}$$

respectively, where $D(f, f_X - f'_X)$ is the far-field beam pattern of the amplitude window $a(f, x_0)$ steered in the direction $u = u'$ in direction cosine space.

3.5 PLANAR APERTURES AND FAR-FIELD DIRECTIVITY FUNCTIONS

Consider the case of a thin, planar transducer of arbitrary shape lying in the XY plane, as shown in Fig. 3.5-1. This physical situation is known as a *planar aperture*. Also shown in Fig. 3.5-1 is an observation point with spherical coordinates (ρ, θ, ψ). The observation point is assumed to be in the far-field

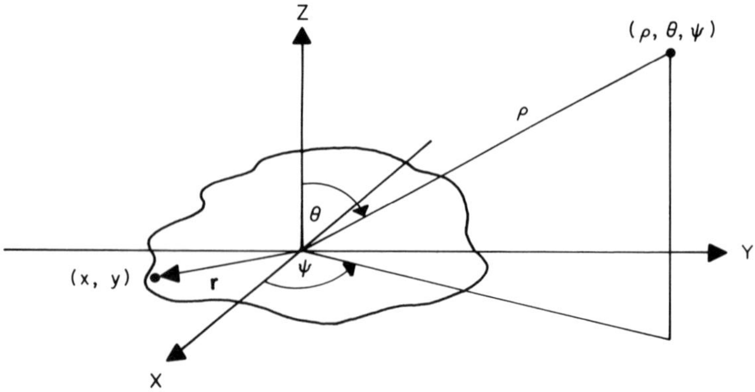

FIG. 3.5-1 Thin, planar transducer of arbitrary shape lying in the XY plane. Also shown is a far-field observation point with spherical coordinates (ρ, θ, ψ).

region of the aperture, that is, at a range $\rho > \pi R^2/\lambda$ meters, where R is the maximum radial extent of the transducer.

Recall that the far-field directivity function of a general, volume aperture is [see Eq. (3.2-42)]

$$D(f, \boldsymbol{\alpha}) = \int_V A(f, \mathbf{r}) \exp(+j2\pi\boldsymbol{\alpha} \cdot \mathbf{r})\, dV. \tag{3.5-1}$$

Since the planar aperture illustrated in Fig. 3.5-1 is shown lying in the XY plane,

$$\mathbf{r} = (x, y, 0), \tag{3.5-2}$$

$$A(f, \mathbf{r}) = A(f, x, y), \tag{3.5-3}$$

$$dV \to dx\, dy, \tag{3.5-4}$$

and

$$\boldsymbol{\alpha} \cdot \mathbf{r} = (f_X, f_Y, f_Z) \cdot (x, y, 0) = f_X x + f_Y y. \tag{3.5-5}$$

Therefore, substituting Eqs. (3.5-3) through (3.5-5) into Eq. (3.5-1) yields

$$D(f, f_X, f_Y) = \int_{-\infty}^{\infty} \int_{-\infty}^{\infty} A(f, x, y) \exp[+j2\pi(f_X x + f_Y y)]\, dx\, dy, \tag{3.5-6}$$

where

$$f_X = u/\lambda = \sin\theta \cos\psi/\lambda \tag{3.5-7}$$

and

$$f_Y = v/\lambda = \sin\theta \sin\psi/\lambda. \tag{3.5-8}$$

Equation (3.5-6) can also be written as

$$D(f, f_X, f_Y) = F_x F_y\{A(f, x, y)\}, \tag{3.5-9}$$

where, in general, the complex aperture function $A(f, x, y)$ is given by

$$A(f, x, y) = a(f, x, y)\exp[+j\theta(f, x, y)]. \tag{3.5-10}$$

Since the spatial frequencies f_X and f_Y can be expressed in terms of θ and ψ, Eq. (3.5-6) specifies the far-field beam pattern of a planar aperture as a function of the frequency f and the spherical angles θ and ψ, that is,

$$D(f, f_X, f_Y) \equiv D(f, \theta, \psi). \tag{3.5-11}$$

Figure 3.5-2 shows the magnitude of a typical normalized far-field beam pattern plotted as a function of the direction cosines u and v.

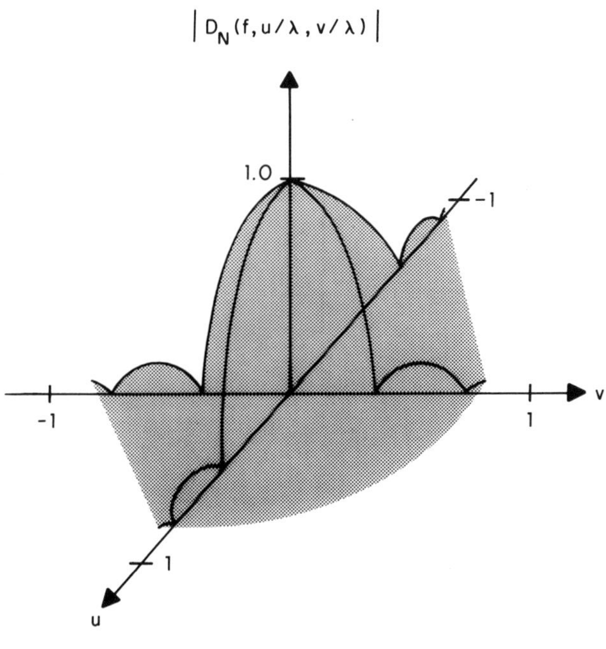

FIG. 3.5-2 Magnitude of a typical normalized far-field beam pattern plotted as a function of the direction cosines u and v.

3.5.1 Beam Steering

Let $D(f, f_X, f_Y)$ be the far-field beam pattern of the amplitude window $a(f, x, y)$, that is,

$$D(f, f_X, f_Y) = F_x F_y\{a(f, x, y)\} \tag{3.5-12}$$

or

$$D(f, f_X, f_Y) = \int_{-\infty}^{\infty}\int_{-\infty}^{\infty} a(f, x, y)\exp[+j2\pi(f_X x + f_Y y)]\,dx\,dy, \tag{3.5-13}$$

and let $D'(f, f_X, f_Y)$ be the far-field beam pattern of the aperture function $A(f, x, y)$ given by Eq. (3.5-10), that is,

$$D'(f, f_X, f_Y) = F_x F_y\{a(f, x, y)\exp[+j\theta(f, x, y)]\} \tag{3.5-14}$$

or

$$D'(f, f_X, f_Y) = \int_{-\infty}^{\infty}\int_{-\infty}^{\infty} a(f, x, y)\exp[+j\theta(f, x, y)] \times \exp[+j2\pi(f_X x + f_Y y)]\,dx\,dy. \tag{3.5-15}$$

Now assume that the phase variation $\theta(f, x, y)$ across the aperture is a linear function of both x and y, that is,

$$\theta(f, x, y) = -2\pi f'_X x - 2\pi f'_Y y, \tag{3.5-16}$$

where

$$f'_X = u'/\lambda = \sin\theta' \cos\psi'/\lambda \tag{3.5-17}$$

and

$$f'_Y = v'/\lambda = \sin\theta' \sin\psi'/\lambda. \tag{3.5-18}$$

If Eq. (3.5-16) is substituted into Eqs. (3.5-14) and (3.5-15), then

$$D'(f, f_X, f_Y) = F_x F_y\{a(f, x, y)\exp[-j2\pi(f'_X x + f'_Y y)]\} \tag{3.5-19}$$

and

$$D'(f, f_X, f_Y) = \int_{-\infty}^{\infty}\int_{-\infty}^{\infty} a(f, x, y)\exp\{+j2\pi[(f_X - f'_X)x + (f_Y - f'_Y)y]\}\,dx\,dy, \tag{3.5-20}$$

respectively. By comparing Eq. (3.5-20) with Eq. (3.5-13),

$$D'(f, f_X, f_Y) = D(f, f_X - f'_X, f_Y - f'_Y), \tag{3.5-21}$$

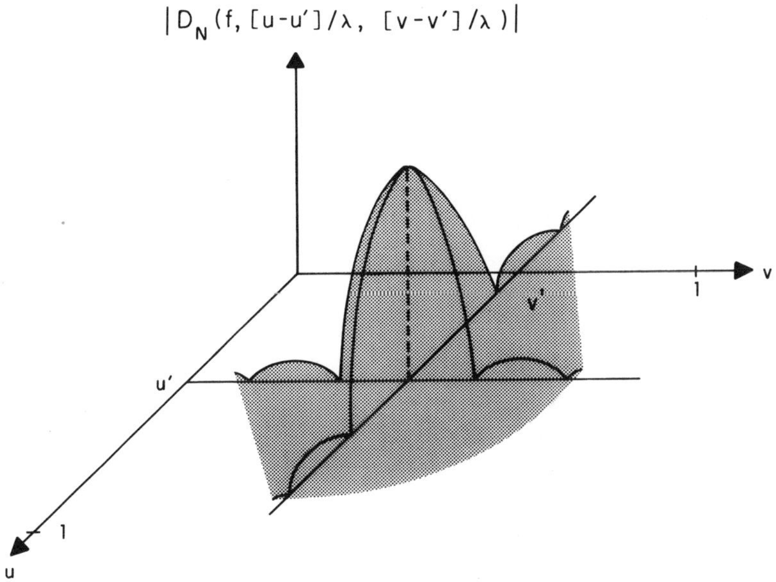

FIG. 3.5-3 Magnitude of a typical normalized far-field beam pattern steered in the direction $u = u'$ and $v = v'$ in direction cosine space.

or, using Eq. (3.5-19),

$$F_x F_y\{a(f, x, y)\exp[-j2\pi(f'_X x + f'_Y y)]\} = D(f, f_X - f'_X, f_Y - f'_Y), \quad (3.5\text{-}22)$$

where $D(f, f_X, f_Y)$ is given by Eq. (3.5-12) or Eq. (3.5-13). Figure 3.5-3 shows the magnitude of a typical, normalized far-field beam pattern steered in the direction $u = u'$ and $v = v'$ in direction cosine space.

3.5.2 Separable Functions—Rectangular Coordinates

A function of two independent variables is called *separable* if it can be written as a product of two functions, each of which depends on only one of the independent variables. Thus, a function $g(x, y)$ is separable in the rectangular coordinates x and y if

$$g(x, y) = g_X(x)g_Y(y). \quad (3.5\text{-}23)$$

Let us assume that the complex aperture function $A(f, x, y)$ is separable in x and y, that is,

$$A(f, x, y) = A_X(f, x)A_Y(f, y), \quad (3.5\text{-}24)$$

where

$$A_X(f,x) = a_X(f,x)\exp[+j\theta_X(f,x)] \tag{3.5-25}$$

and

$$A_Y(f,y) = a_Y(f,y)\exp[+j\theta_Y(f,y)]. \tag{3.5-26}$$

If Eq. (3.5-24) is substituted into Eq. (3.5-6), then

$$D(f,f_X,f_Y) = \int_{-\infty}^{\infty} A_X(f,x)\exp(+j2\pi f_X x)\,dx \int_{-\infty}^{\infty} A_Y(f,y)\exp(+j2\pi f_Y y)\,dy \tag{3.5-27}$$

or

$$D(f,f_X,f_Y) = D_X(f,f_X)D_Y(f,f_Y), \tag{3.5-28}$$

where

$$D_X(f,f_X) = F_x\{A_X(f,x)\} \tag{3.5-29}$$

and

$$D_Y(f,f_Y) = F_y\{A_Y(f,y)\}. \tag{3.5-30}$$

Therefore, Eq. (3.5-28) indicates that if the aperture function $A(f,x,y)$ is separable in x and y, then the corresponding far-field beam pattern $D(f,f_X,f_Y)$ is equal to the product of the individual beam patterns $D_X(f,f_X)$ and $D_Y(f,f_Y)$.

EXAMPLE 3.5-1: THE RECTANGULAR PISTON

One of the most common planar apertures separable in the rectangular coordinates x and y is the *rectangular piston*, shown in Fig. 3.5-4. The aperture function for the rectangular piston is given by

$$A(f,x,y) = \text{rect}(L_X)\,\text{rect}(L_Y), \tag{3.5-31}$$

and the corresponding normalized far-field beam pattern is

$$D_N(f,f_X,f_Y) = \frac{\sin(\pi L_X f_X)}{\pi L_X f_X}\,\frac{\sin(\pi L_Y f_Y)}{\pi L_Y f_Y}, \tag{3.5-32}$$

where use was made of Eq. (3.3-18) and Eqs. (3.5-28) through (3.5-30). Note that the range $\rho > \pi R^2/\lambda$ meters, where

$$R = [(L_X/2)^2 + (L_Y/2)^2]^{1/2} \quad \text{meters}$$

is the maximum radial extent of the transducer. □

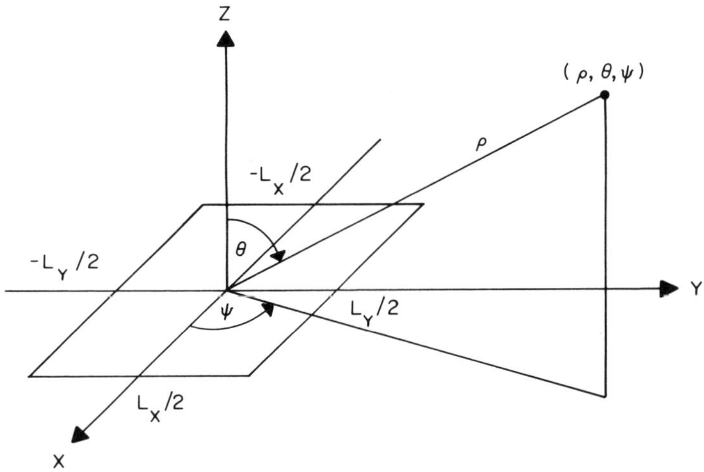

FIG. 3.5-4 The rectangular piston. Also shown is a far-field observation point with spherical coordinates (ρ, θ, ψ).

3.5.3 Separable Functions—Polar Coordinates

Given a complex aperture function $A(f, x, y)$ expressed in the rectangular coordinates x and y, we know that its far-field directivity function is given by the following two-dimensional spatial Fourier transform:

$$D(f, f_X, f_Y) = \int_{-\infty}^{\infty} \int_{-\infty}^{\infty} A(f, x, y) \exp[+j2\pi(f_X x + f_Y y)]\, dx\, dy. \quad (3.5\text{-}6)$$

Using Eq. (3.5-6) as a starting point, consider the problem of computing the far-field beam pattern of a planar aperture, circular in shape, with radius a meters and with a complex aperture function $A(f, r, \phi)$ expressed in the polar coordinates r and ϕ (see Fig. 3.5-5). By referring to Fig. 3.5-5, we can write that

$$x = r \cos \phi, \quad (3.5\text{-}33)$$

$$y = r \sin \phi, \quad (3.5\text{-}34)$$

$$dx\, dy \rightarrow r\, dr\, d\phi, \quad (3.5\text{-}35)$$

and

$$A(f, x, y) = A(f, r \cos \phi, r \sin \phi) \equiv A(f, r, \phi). \quad (3.5\text{-}36)$$

Since

$$f_X = u/\lambda = \sin \theta \cos \psi/\lambda \quad (3.5\text{-}37)$$

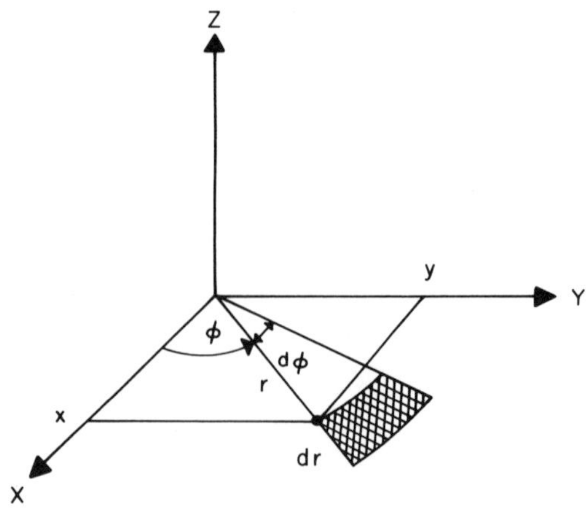

FIG. 3.5-5 Relationship between the rectangular coordinates (x, y) and polar coordinates (r, ϕ). Also shown is the infinitesimal area $r\,dr\,d\phi$.

and

$$f_Y = v/\lambda = \sin\theta \sin\psi/\lambda, \tag{3.5-38}$$

substituting Eqs. (3.5-33) through (3.5-38) into Eq. (3.5-6) yields

$$D(f,\theta,\psi) = \int_0^{2\pi}\int_0^a A(f,r,\phi) \times \exp\left[+j\left(\frac{2\pi r}{\lambda}\right)\sin\theta\,(\cos\psi\cos\phi + \sin\psi\sin\phi)\right] r\,dr\,d\phi, \tag{3.5-39}$$

or, since

$$\cos(\alpha - \beta) = \cos\alpha\cos\beta + \sin\alpha\sin\beta,$$

we obtain the desired far-field directivity function

$$D(f,\theta,\psi) = \int_0^{2\pi}\int_0^a A(f,r,\phi)\exp\left[+j\left(\frac{2\pi r}{\lambda}\right)\sin\theta\cos(\psi - \phi)\right] r\,dr\,d\phi, \tag{3.5-40}$$

where

$$A(f,r,\phi) = a(f,r,\phi)\exp[+j\theta(f,r,\phi)]. \tag{3.5-41}$$

Note that Eq. (3.5-40) will be used in Chapter 4 to obtain an expression for the far-field beam pattern of a planar array of concentric circular arrays.

Equation (3.5-40) is the far-field beam pattern of the general, complex aperture function $A(f, r, \phi)$ as given by Eq. (3.5-41). Now let us assume that $A(f, r, \phi)$ is separable in the polar coordinates r and ϕ, that is,

$$A(f, r, \phi) = A_r(f, r) A_\phi(f, \phi), \tag{3.5-42}$$

where

$$A_r(f, r) = a_r(f, r) \exp[+j\theta_r(f, r)] \tag{3.5-43}$$

and

$$A_\phi(f, \phi) = a_\phi(f, \phi) \exp[+j\theta_\phi(f, \phi)]. \tag{3.5-44}$$

If Eq. (3.5-42) is substituted into Eq. (3.5-40), then

$$D(f, \theta, \psi) = \int_0^a r A_r(f, r) \int_0^{2\pi} A_\phi(f, \phi) \exp\left[+j\left(\frac{2\pi r}{\lambda}\right) \sin\theta \cos(\psi - \phi)\right] d\phi \, dr. \tag{3.5-45}$$

Equation (3.5-45) can be rewritten as

$$D(f, \theta, \psi) = \int_0^a r A_r(f, r) \int_0^{2\pi} A_\phi(f, \phi) \exp\left[+jb \sin\left(\frac{\pi}{2} + \psi - \phi\right)\right] d\phi \, dr, \tag{3.5-46}$$

where

$$b = (2\pi r/\lambda) \sin\theta \tag{3.5-47}$$

and

$$\cos(\psi - \phi) = \sin[(\pi/2) + \psi - \phi]. \tag{3.5-48}$$

Note that

$$\exp(+jb \sin\alpha) = \sum_{n=-\infty}^{\infty} J_n(b) \exp(+jn\alpha), \tag{3.5-49}$$

where b is given by Eq. (3.5-47),

$$\alpha = (\pi/2) + \psi - \phi, \tag{3.5-50}$$

and $J_n(\cdot)$ is the *nth-order Bessel function of the first kind*. If Eqs. (3.5-49) and (3.5-50) are substituted into Eq. (3.5-46), and by noting that

$$\exp(+jn\pi/2) = j^n, \tag{3.5-51}$$

then

$$D(f, \theta, \psi) = \sum_{n=-\infty}^{\infty} j^n c_n(f) A_r(f, \theta; n) \exp(+jn\psi), \tag{3.5-52}$$

where

$$c_n(f) = \frac{1}{2\pi} \int_0^{2\pi} A_\phi(f, \phi) \exp(-jn\phi)\, d\phi, \tag{3.5-53}$$

$$A_r(f, \theta; n) = H_n\{A_r(f, r)\}, \tag{3.5-54}$$

and

$$H_n\{A_r(f, r)\} = 2\pi \int_0^a r A_r(f, r) J_n\left(\frac{2\pi r}{\lambda} \sin\theta\right) dr \tag{3.5-55}$$

is the *nth-order Hankel transform* of $A_r(f, r)$. Recall that since the planar aperture was assumed to be circular in shape with radius a meters,

$$A_r(f, r) = 0, \qquad r > a. \tag{3.5-56}$$

Therefore, when the complex frequency response or aperture function $A(f, r, \phi)$ is separable in polar coordinates [see Eq. (3.5-42)], its far-field directivity function can be computed using Eqs. (3.5-52) through (3.5-55).

EXAMPLE 3.5-2

Suppose we are given an aperture function $A(f, r, \phi)$ separable in polar coordinates, that is,

$$A(f, r, \phi) = A_r(f, r) A_\phi(f, \phi), \tag{3.5-42}$$

where

$$A_\phi(f, \phi) = \exp(+jm\phi) \tag{3.5-57}$$

and m is an integer. If Eq. (3.5-57) is substituted into Eq. (3.5-53), then

$$c_n(f) = \frac{1}{2\pi} \int_0^{2\pi} \exp[+j(m - n)\phi]\, d\phi = \delta_{mn}, \tag{3.5-58}$$

where

$$\delta_{mn} = \begin{cases} 0, & m \neq n, \\ 1, & m = n, \end{cases} \tag{3.5-59}$$

is the *Kronecker delta*. Substituting Eq. (3.5-58) into Eq. (3.5-52) yields the far-field directivity function

$$D(f, \theta, \psi) = j^m A_r(f, \theta; m) \exp(+jm\psi), \tag{3.5-60}$$

where

$$A_r(f, \theta; m) = H_m\{A_r(f, r)\}. \tag{3.5-61}$$

If $m = 0$, then

$$A(f, r, \phi) = A_r(f, r) \tag{3.5-62}$$

since $A_\phi(f, \phi) = 1$. The aperture function given by Eq. (3.5-62) is said to be *circularly symmetric* since it can be written as a function of the radius r alone. With $m = 0$, the directivity function given by Eq. (3.5-60) reduces to

$$D(f, \theta, \psi) = A_r(f, \theta; 0) = H_0\{A_r(f, r)\}, \tag{3.5-63}$$

where

$$H_0\{A_r(f, r)\} = 2\pi \int_0^a r A_r(f, r) J_0\left(\frac{2\pi r}{\lambda} \sin\theta\right) dr \tag{3.5-64}$$

is the zero-order Hankel transform of $A_r(f, r)$ and $J_0(\cdot)$ is the zero-order Bessel function of the first kind. Equation (3.5-64) is also known as the *Fourier–Bessel transform* of $A_r(f, r)$. □

EXAMPLE 3.5-3: THE CIRCULAR PISTON

The most frequently encountered aperture function with circular symmetry is given by

$$A(f, r, \phi) = A_r(f, r), \tag{3.5-62}$$

where

$$A_r(f, r) = \text{circ}(a) = \begin{cases} 1, & r \leq a, \\ 0, & r > a. \end{cases} \tag{3.5-65}$$

The aperture function described by Eqs. (3.5-62) and (3.5-65) is known as the *circular piston* (see Fig. 3.5-6). The far-field directivity function of the circular piston can be obtained by substituting Eq. (3.5-65) into Eqs. (3.5-63) and (3.5-64). Doing so yields

$$D(f, \theta, \psi) = 2\pi \int_0^a r J_0\left(\frac{2\pi r}{\lambda} \sin\theta\right) dr, \tag{3.5-66}$$

and by making use of the identity

$$\int_0^x \alpha J_0(\alpha)\, d\alpha = x J_1(x), \tag{3.5-67}$$

$$D(f, \theta, \psi) = a \frac{J_1[(2\pi a/\lambda)\sin\theta]}{(\sin\theta)/\lambda}, \tag{3.5-68}$$

where $J_1(\cdot)$ is the first-order Bessel function of the first kind. Equation (3.5-68) is the unnormalized far-field beam pattern. It can be shown that the

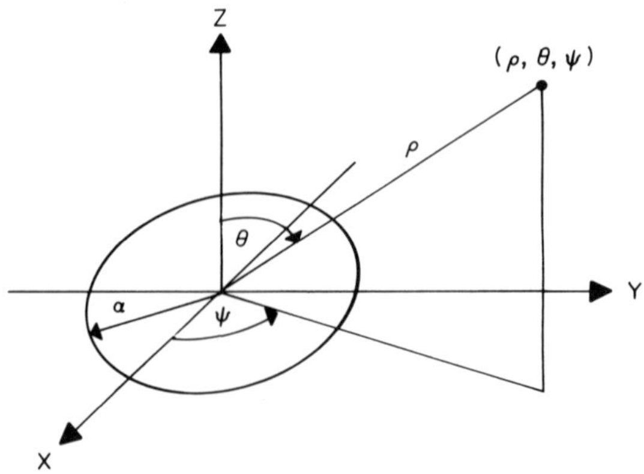

FIG. 3.5-6 The circular piston. Also shown is a far-field observation point with spherical coordinates (ρ, θ, ψ).

normalization factor is given by

$$D(f, \theta, \psi)\Big|_{\theta=0} = \pi a^2, \tag{3.5-69}$$

so that the normalized far-field directivity function of the circular piston is equal to

$$D_N(f, \theta, \psi) = 2\frac{J_1[(2\pi a/\lambda)\sin\theta]}{(2\pi a/\lambda)\sin\theta}. \tag{3.5-70}$$

Note that Eq. (3.5-70) is independent of the azimuthal angle ψ. Figure 3.5-7 is a plot of the magnitude of Eq. (3.5-70), where the level of the first sidelobe is approximately -18 dB. Also note that the range $\rho > \pi a^2/\lambda$ meters. □

3.6 PLANAR APERTURES AND NEAR-FIELD DIRECTIVITY FUNCTIONS

3.6.1 Beam Steering and Aperture Focusing

Recall that the near-field directivity function of a general, volume aperture is given by [see Eq. (3.2-35)]

$$D(f, r, \boldsymbol{\alpha}) = \int_{V_0} A(f, \mathbf{r}_0)\exp[-jkr_0^2/(2r)]\exp(+j2\pi\boldsymbol{\alpha}\cdot\mathbf{r}_0)\,dV_0. \tag{3.6-1}$$

Assume we are given a planar aperture lying in the XY plane, as shown in Fig. 3.6-1. Also shown is a near-field observation point with spherical

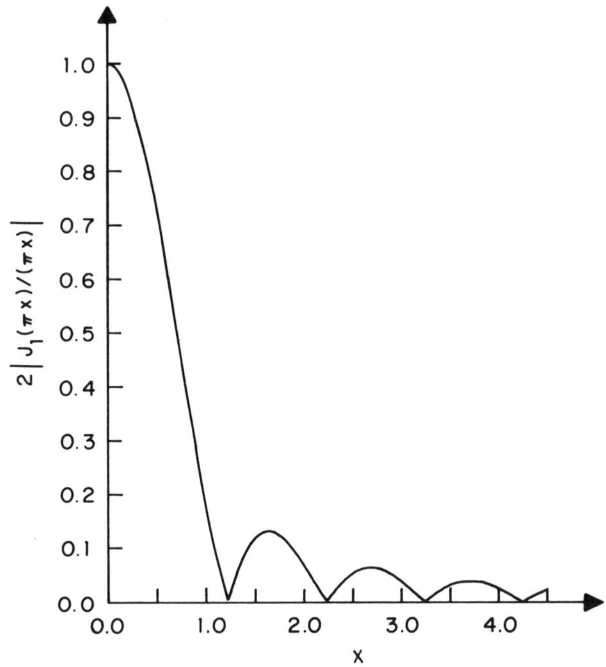

FIG. 3.5-7 Magnitude of the normalized far-field beam pattern of the circular piston, where $x = (2a/\lambda)\sin\theta$.

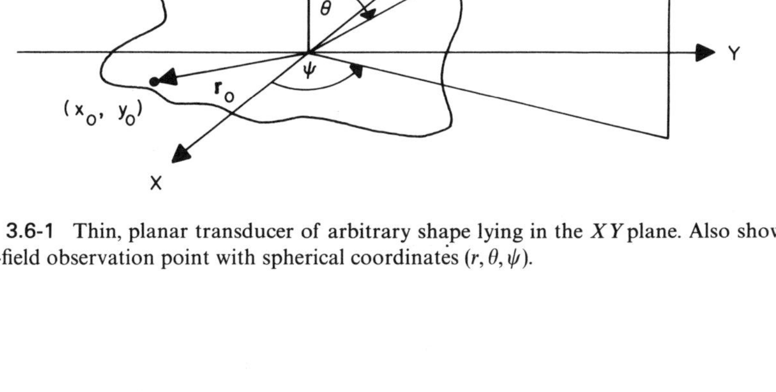

FIG. 3.6-1 Thin, planar transducer of arbitrary shape lying in the XY plane. Also shown is a near-field observation point with spherical coordinates (r, θ, ψ).

coordinates (r, θ, ψ), that is, the range $r < \pi R^2/\lambda$ where R is the maximum radial extent of the transducer. For the planar aperture shown,

$$\mathbf{r}_0 = (x_0, y_0, 0), \tag{3.6-2}$$

$$r_0^2 = |\mathbf{r}_0|^2 = x_0^2 + y_0^2, \tag{3.6-3}$$

$$A(f, \mathbf{r}_0) = A(f, x_0, y_0), \tag{3.6-4}$$

$$dV_0 \rightarrow dx_0\, dy_0, \tag{3.6-5}$$

and

$$\boldsymbol{\alpha} \cdot \mathbf{r}_0 = (f_X, f_Y, f_Z) \cdot (x_0, y_0, 0) = f_X x_0 + f_Y y_0. \tag{3.6-6}$$

Therefore, substituting Eqs. (3.6-3) through (3.6-6) into Eq. (3.6-1) yields

$$D(f, r, f_X, f_Y) = \int_{-\infty}^{\infty} \int_{-\infty}^{\infty} A(f, x_0, y_0) \exp[-jk(x_0^2 + y_0^2)/(2r)] \times \exp[+j2\pi(f_X x_0 + f_Y y_0)]\, dx_0\, dy_0, \tag{3.6-7}$$

which is the near-field directivity function of a planar aperture where

$$f_X = u/\lambda = \sin\theta \cos\psi/\lambda \tag{3.6-8}$$

and

$$f_Y = v/\lambda = \sin\theta \sin\psi/\lambda. \tag{3.6-9}$$

Equation (3.6-7) can also be written as

$$D(f, r, f_X, f_Y) = F_{x_0} F_{y_0}\{A(f, x_0, y_0) \exp[-jk(x_0^2 + y_0^2)/(2r)]\}, \tag{3.6-10}$$

where, in general, the complex frequency response or aperture function $A(f, x_0, y_0)$ is given by

$$A(f, x_0, y_0) = a(f, x_0, y_0) \exp[+j\theta(f, x_0, y_0)]. \tag{3.6-11}$$

Now assume that the phase variation $\theta(f, x_0, y_0)$ across the aperture is equal to the sum of a linear and a quadratic phase shift, that is,

$$\theta(f, x_0, y_0) = -2\pi f_X' x_0 - 2\pi f_Y' y_0 + k(x_0^2 + y_0^2)/(2r), \tag{3.6-12}$$

where

$$f_X' = u'/\lambda = \sin\theta' \cos\psi'/\lambda, \tag{3.6-13}$$

$$f_Y' = v'/\lambda = \sin\theta' \sin\psi'/\lambda, \tag{3.6-14}$$

and $k = 2\pi f/c$. If Eqs. (3.6-11) and (3.6-12) are substituted into Eqs. (3.6-7) and

(3.6-10), then

$$D(f, r, f_X, f_Y) = \int_{-\infty}^{\infty} \int_{-\infty}^{\infty} a(f, x_0, y_0) \times \exp\{+j2\pi[(f_X - f_X')x_0 + (f_Y - f_Y')y_0]\} \, dx_0 \, dy_0 \tag{3.6-15}$$

and

$$D(f, r, f_X, f_Y) = F_{x_0} F_{y_0}\{a(f, x_0, y_0)\exp[-j2\pi(f_X' x_0 + f_Y' y_0)]\}, \tag{3.6-16}$$

respectively, or

$$D(f, r, f_X, f_Y) = D(f, f_X - f_X', f_Y - f_Y'), \tag{3.6-17}$$

where $D(f, f_X - f_X', f_Y - f_Y')$ is the far-field beam pattern of the amplitude window $a(f, x_0, y_0)$ steered in the direction $u = u'$ and $v = v'$ in direction cosine space.

3.7 DIRECTIVITY INDEX

In this section we shall derive an equation for the *directivity* of a transducer (aperture), where the transducer is characterized by its far-field beam pattern. The *directivity index* is simply the decibel equivalent of the directivity.

The directivity of a transducer, when used as a transmitter or *projector*, is a measure of its ability to concentrate the available acoustic power into a preferred direction. When used as a receiver or *hydrophone*, the directivity of a transducer is a measure of its ability to distinguish between several sound sources located at different spatial locations within the medium. Hence, directivity is basically a measure of the beamwidth and sidelobe levels of a given beam pattern. In the analysis that follows, it is assumed that the transducer is being used as a projector. Note that if the transducer is *reversible*, that is, if it can be used either as a projector or as a hydrophone, then the transmit and receive far-field beam patterns of the transducer will be identical.

In order to compute the directivity, we must calculate the time-average intensity vector of the radiated field and then the time-average radiated power. But, in order to compute the intensity and power, we must first calculate the acoustic pressure and acoustic fluid velocity vector due to the source. We begin the analysis by recalling that the velocity potential at a far-field observation point (r, θ, ψ) due to an aperture (source) is given by

$$\varphi(t, r, \theta, \psi) \approx -D(f, \theta, \psi)\frac{\exp\{+j2\pi f[t - (r/c)]\}}{4\pi r} \tag{3.7-1}$$

when an identical, time-harmonic input electrical signal is applied to all spatial locations of the transducer, that is, when

$$x(t, \mathbf{r}) \equiv x(t) = \exp(+j2\pi f t),$$

and where $D(f, \theta, \psi)$ is the unnormalized far-field beam pattern of the aperture (see Example 3.2-2). Also, recall that the acoustic pressure $p(t, r, \theta, \psi)$ and the acoustic fluid (particle) velocity vector $\mathbf{u}(t, r, \theta, \psi)$ can be obtained from the velocity potential $\varphi(t, r, \theta, \psi)$ as follows [see Eqs. (1.1-4) and (1.1-5)]:

$$p(t, r, \theta, \psi) = -\rho_0 \frac{\partial}{\partial t} \varphi(t, r, \theta, \psi) \tag{3.7-2}$$

and

$$\mathbf{u}(t, r, \theta, \psi) = \nabla \varphi(t, r, \theta, \psi). \tag{3.7-3}$$

Therefore, substituting Eq. (3.7-1) into Eq. (3.7-2) yields

$$p(t, r, \theta, \psi) \approx jk\rho_0 c D(f, \theta, \psi) \frac{\exp\{+j2\pi f[t - (r/c)]\}}{4\pi r}, \tag{3.7-4}$$

where $\rho_0 c$ is known as the *characteristic impedance* of the medium in rayls (1 rayl = 1 Pa · sec/m). Computing the acoustic fluid velocity vector is a bit more involved. Substituting Eq. (3.7-1) into Eq. (3.7-3) yields

$$\begin{aligned} \mathbf{u}(t, r, \theta, \psi) \approx -\{&D(f, \theta, \psi) \nabla[\exp(-jkr)/r] \\ &+ [\exp(-jkr)/r] \nabla D(f, \theta, \psi)\} \frac{\exp(+j2\pi f t)}{4\pi} \end{aligned} \tag{3.7-5}$$

since

$$\nabla(gh) = g \nabla h + h \nabla g,$$

where ∇ is the gradient in spherical coordinates, that is,

$$\nabla g = \frac{\partial}{\partial r} g \hat{a}_r + \frac{1}{r} \frac{\partial}{\partial \theta} g \hat{a}_\theta + \frac{1}{r \sin \theta} \frac{\partial}{\partial \psi} g \hat{a}_\psi. \tag{3.7-6}$$

From Eq. (3.7-6), we can write that

$$\nabla[\exp(-jkr)/r] = \frac{\partial}{\partial r}[\exp(-jkr)/r] \hat{a}_r \tag{3.7-7}$$

or

$$\nabla[\exp(-jkr)/r] = -k^2 \left[\frac{1}{(kr)^2} + j \frac{1}{kr} \right] \exp(-jkr) \hat{a}_r. \tag{3.7-8}$$

Since it was assumed that the observation point is in the far-field, that is, $kr \gg 1$, Eq. (3.7-8) reduces to

$$\nabla[\exp(-jkr)/r] \approx -jk[\exp(-jkr)/r]\hat{a}_r. \tag{3.7-9}$$

Next, consider the term $\nabla D(f, \theta, \psi)$. Since the far-field beam pattern does not depend on the range r to the observation point, from Eq. (3.7-6) we can write that

$$\nabla D(f, \theta, \psi) = \frac{1}{r}\left[\frac{\partial}{\partial \theta} D(f, \theta, \psi)\hat{a}_\theta + \frac{1}{\sin\theta}\frac{\partial}{\partial \psi} D(f, \theta, \psi)\hat{a}_\psi\right]. \tag{3.7-10}$$

Substituting Eqs. (3.7-9) and (3.7-10) into Eq. (3.7-5) yields

$$\mathbf{u}(t, r, \theta, \psi) \approx \left\{ jkD(f, \theta, \psi)\hat{a}_r - \frac{1}{r}\left[\frac{\partial}{\partial \theta} D(f, \theta, \psi)\hat{a}_\theta + \frac{1}{\sin\theta}\frac{\partial}{\partial \psi} D(f, \theta, \psi)\hat{a}_\psi\right]\right\} \times \frac{\exp\{+j2\pi f[t - (r/c)]\}}{4\pi r}, \tag{3.7-11}$$

and since the observation point is in the far-field, the first term in Eq. (3.7-11) is dominant, and as a result,

$$\mathbf{u}(t, r, \theta, \psi) \approx u(t, r, \theta, \psi)\hat{a}_r = jkD(f, \theta, \psi)\frac{\exp\{+j2\pi f[t - (r/c)]\}}{4\pi r}\hat{a}_r. \tag{3.7-12}$$

Note that Eqs. (3.7-4) and (3.7-12) are related by

$$p(t, r, \theta, \psi) = \rho_0 c u(t, r, \theta, \psi),$$

which is a plane wave relationship between acoustic pressure and acoustic fluid velocity. Thus, in the far-field the radiated field behaves like plane waves traveling in different directions, where the amplitude of any one plane wave is determined by the value of the beam pattern in the direction the plane wave is traveling.

Also, if the size of our *arbitrarily* shaped source is small compared to a wavelength, then the beamwidth of the far-field beam pattern is large and, as a result, the beam pattern appears omnidirectional and Eq. (3.7-4), for example, reduces to the form of a pressure wave from a simple point source.

Now that we have expressions for both the radiated acoustic pressure and the acoustic fluid velocity vector, we can calculate the time-average intensity vector. Since the pressure and velocity expressions given by Eqs. (3.7-4) and (3.7-12), respectively, have time-harmonic dependence [i.e., $\exp(+j2\pi ft)$], the time-average intensity vector $\mathbf{I}_{avg}$ of the field radiated by the source can be computed from the following equation:

$$\mathbf{I}_{avg} = \tfrac{1}{2}\mathrm{Re}\{p(r, \theta, \psi)\mathbf{u}^*(r, \theta, \psi)\}, \tag{3.7-13}$$

where

$$p(t, r, \theta, \psi) = p(r, \theta, \psi)\exp(+j2\pi ft) \tag{3.7-14}$$

and

$$\mathbf{u}(t, r, \theta, \psi) = \mathbf{u}(r, \theta, \psi)\exp(+j2\pi ft). \tag{3.7-15}$$

Re means "take the real part," and the asterisk ($*$) denotes complex conjugate. Substituting Eqs. (3.7-4) and (3.7-12) into Eq. (3.7-13) yields

$$\mathbf{I}_{\text{avg}} \approx \frac{k^2 \rho_0 c}{32\pi^2 r^2} |D(f, \theta, \psi)|^2 \hat{a}_r, \quad kr \gg 1. \tag{3.7-16}$$

The time-average acoustic power radiated by the source (in watts) is given by

$$P_{\text{avg}} = \iint_S \mathbf{I}_{\text{avg}} \cdot d\mathbf{S}, \tag{3.7-17}$$

where S is *any* surface enclosing the source,

$$d\mathbf{S} = dS\,\hat{n}, \tag{3.7-18}$$

and $\hat{n}$ is a unit vector normal to the surface S pointing in the conventional *outward* direction away from the enclosed volume, and hence, the source. For convenience, choose S as a spherical surface of radius r. As a result, Eq. (3.7-18) becomes

$$d\mathbf{S} = r^2 \sin\theta\, d\theta\, d\psi\, \hat{a}_r, \tag{3.7-19}$$

where $\hat{a}_r$ is the unit vector in the radial direction, normal to the surface of the sphere. Therefore, substituting Eqs. (3.7-16) and (3.7-19) into Eq. (3.7-17) yields

$$P_{\text{avg}} \approx \frac{k^2 \rho_0 c}{32\pi^2} \int_0^{2\pi} \int_0^{\pi} |D(f, \theta, \psi)|^2 \sin\theta\, d\theta\, d\psi, \qquad kr \gg 1 \tag{3.7-20}$$

or

$$P_{\text{avg}} \approx \frac{k^2 \rho_0 c}{32\pi^2} \int_0^{4\pi} |D(f, \theta, \psi)|^2\, d\Omega, \qquad kr \gg 1 \tag{3.7-21}$$

since the infinitesimal element of solid angle $d\Omega = \sin\theta\, d\theta\, d\psi$ for a spherical surface.

The *directivity* D of an aperture (source) is defined by

$$D \triangleq I_0 / I_{\text{ref}}, \tag{3.7-22}$$

where I_0 is the on-axis ($\theta = 0°$) far-field time-average intensity measured at some distance r from the source and I_{ref} is a reference intensity. To be more specific, the on-axis intensity is given by

$$I_0 = |\mathbf{I}_{avg}|\Big|_{\theta=0^\circ} = \frac{k^2\rho_0 c}{32\pi^2 r^2}|D(f,0,\psi)|^2 \quad \text{W/m}^2 \tag{3.7-23}$$

and the reference intensity, defined by

$$I_{ref} \triangleq P_{avg}/(4\pi r^2) \quad \text{W/m}^2, \tag{3.7-24}$$

becomes

$$I_{ref} = \left(\frac{k^2\rho_0 c}{32\pi^2 r^2}\right)\frac{|D(f,0,\psi)|^2}{4\pi}\int_0^{4\pi}|D_N(f,\theta,\psi)|^2\,d\Omega, \tag{3.7-25}$$

where

$$D_N(f,\theta,\psi) = D(f,\theta,\psi)/D(f,0,\psi) \tag{3.7-26}$$

is the normalized far-field directivity function of the aperture. Note that when $\theta = 0^\circ$, the direction cosines reduce to $u = 0$, $v = 0$, and $w = 1$ so that the quantity $D(f,0,\psi)$ is really *not* a function of ψ. From Eq. (3.7-24), it can be seen that I_{ref} is defined as the intensity from a simple, omnidirectional point source that radiates the same power P_{avg} as the general directional source. Therefore, upon substituting Eqs. (3.7-23) and (3.7-25) into Eq. (3.7-22), we finally obtain the following expressions for the directivity of an aperture (source):

$$D = \frac{4\pi}{\int_0^{4\pi}|D_N(f,\theta,\psi)|^2\,d\Omega} \tag{3.7-27}$$

or

$$D = \frac{4\pi}{\int_0^{2\pi}\int_0^{\pi}|D_N(f,\theta,\psi)|^2\sin\theta\,d\theta\,d\psi}. \tag{3.7-28}$$

The *directivity index* DI is defined as

$$\text{DI} \triangleq 10\log_{10} D \quad \text{dB}, \tag{3.7-29}$$

which is the decibel equivalent of the directivity D. Note that the directivity index is one of the terms in the sonar equation.

Values of the directivity D range from a minimum of unity (DI $= 0$ dB) for an omnidirectional source ($|D_N(f,\theta,\psi)| = 1$) such as a radially pulsating sphere to large numbers for highly directional sources. For example, for an omnidirectional *hemispherical* source mounted on an infinite baffle, the directivity $D = 2$ (DI $= 3$ dB) since $0 \le \psi \le 2\pi$ *but* $0 \le \theta \le \pi/2$ [see Eq. (3.7-28)]. In general, the directivity of an aperture (source) increases with increasing frequency or physical size since the beamwidth of the corresponding far-field beam pattern decreases with increasing frequency or size.

Finally, it should be noted that both the directivity and the directivity index of an *array* of electroacoustic transducers can also be computed from Eqs. (3.7-28) and (3.7-29), respectively, as long as the normalized far-field beam pattern of the array is known. Arrays are discussed next in Chapter 4.

PROBLEMS

3-1 The direction of the position vector **r** is specified by two sets of angles, as shown in Fig. P3-1. The first set of angles is (α, β, γ), and the second set is the spherical angles (θ, ψ). The numbers $\cos\alpha$, $\cos\beta$, and $\cos\gamma$ are called the direction cosines of **r** since

$$\mathbf{r} = |\mathbf{r}|\hat{a}_r = r(\cos\alpha\,\hat{x} + \cos\beta\,\hat{y} + \cos\gamma\,\hat{z}),$$

where $\hat{a}_r$ is a unit vector. If we denote $\cos\alpha$, $\cos\beta$, and $\cos\gamma$ by u, v, and w, respectively, derive expressions for u, v, and w in terms of the spherical angles θ and ψ, that is, verify Eqs. (3.2-6) through (3.2-8).

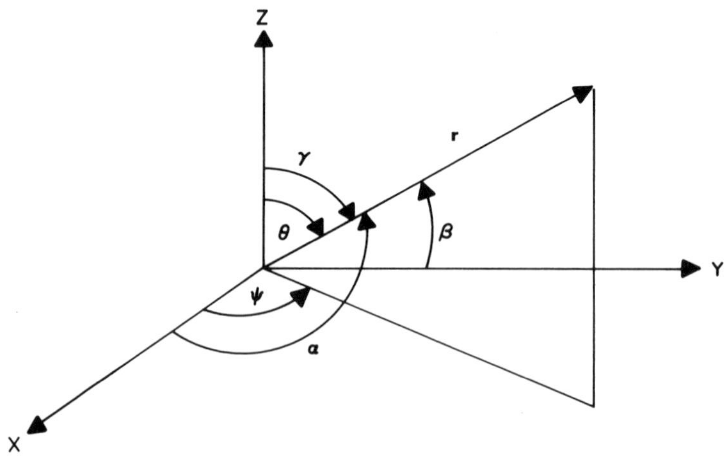

FIG. P3-1

3-2 Verify Eqs. (3.2-12) and (3.2-13).

3-3 Verify Eq. (3.2-17).

3-4 Show that Eq. (3.3-13) is equal to Eq. (3.3-14).

3-5 Verify Eq. (3.3-17).

3-6 Compute the normalized far-field directivity functions of the following aperture functions:

(a) $A(f, y) = \text{rect}(L)$,

(b) $A(f, z) = \text{rect}(L)$.

Express your answers in terms of the spherical angles θ and ψ.

3-7 Using Eq. (3.3-22) as the starting point, show that the third derivative of the aperture function $A(f, x)$ with respect to x is related to its far-field beam pattern $D(f, f_X)$ by the following expression:

$$D(f, f_X) = F_x\left\{\frac{d^3}{dx^3} A(f, x)\right\} \bigg/ (-j2\pi f_X)^3.$$

3-8 Verify Eq. (3.3-31).

3-9 Show that

$$F_x\{\text{tri}(L)\} = \frac{L}{2}\left[\frac{\sin(\pi L f_X/2)}{\pi L f_X/2}\right]^2$$

by direct integration.

3-10 Using the method of successive differentiation, compute the *unnormalized* far-field directivity function of the rectangular amplitude window shown in Fig. 3.3-2. Compare your result with Eq. (3.3-14).

3-11 Using the method of successive differentiation, compute the *unnormalized* far-field directivity function of the aperture function shown in Fig. P3-11. Note that since this aperture function is an *odd* function of x, its beam pattern will be *imaginary* and *odd*.

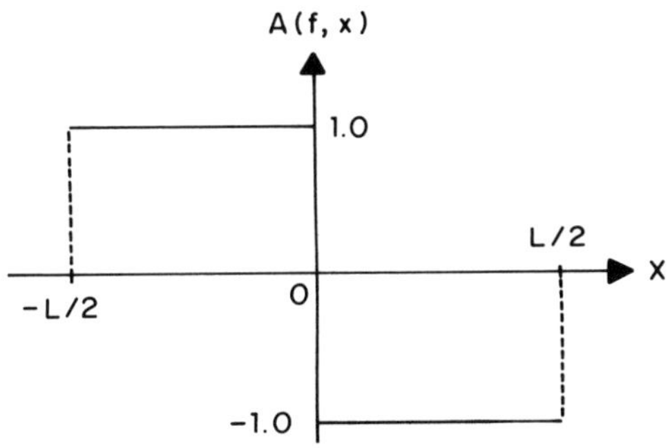

FIG. P3-11

3-12 (a) Using the method of successive differentiation, compute the *unnormalized* far-field directivity function of the trapezoidal amplitude

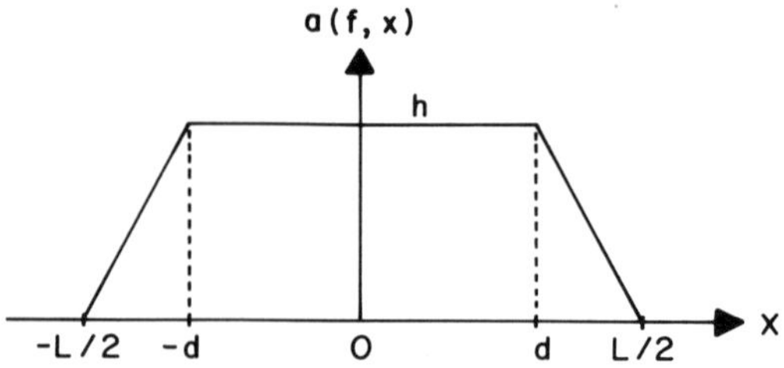

FIG. P3-12

window shown in Fig. P3-12. Use the following trigonometric identity:

$$\cos\alpha - \cos\beta = -2\sin[(\alpha+\beta)/2]\sin[(\alpha-\beta)/2].$$

(b) Using your result from part (a), what is the equation for the beam pattern when $d = 0$ and $h = 1$?

(c) Repeat part (b) using $d = L/2$ and $h = 1$.

3-13 Show that

$$F_{f_X}^{-1}\{\delta(f_X \pm f'_X)\} = \exp(\pm j2\pi f'_X x).$$

From this result, we can write that [see Eq. (3.3-42)]

$$F_x\{\exp(\pm j2\pi f'_X x)\} = \delta(f_X \pm f'_X),$$

and, if $f'_X = 0$, then

$$F_x\{1\} = \delta(f_X) \qquad \text{and} \qquad F_{f_X}^{-1}\{\delta(f_X)\} = 1.$$

3-14 Show that

$$g(t) * \delta(t \pm t_0) = g(t \pm t_0),$$

where the asterisk ($*$) denotes convolution with respect to t and t_0 is an arbitrary constant.

3-15 Obtain expressions for the *normalized* far-field directivity functions of the Hanning, Hamming, and Blackman amplitude windows. Use the following trigonometric identity: $\sin(\alpha \pm \pi) = -\sin\alpha$.

3-16 Compute the beamwidths of the horizontal beam patterns of a rectangular amplitude window for the following combination of values for the length L and frequency f (use $c = 1500$ m/sec):

(a) $L = 0.1$ m, $f = 10$ kHz, 15 kHz, 25 kHz;
(b) $L = 0.5$ m, $f = 10$ kHz, 15 kHz, 25 kHz;
(c) $L = 1.0$ m, $f = 10$ kHz, 15 kHz, 25 kHz.

What is your conclusion about the relationship between the beamwidth, length of the linear aperture, and frequency?

3-17 Using your results from Problem 3-15 and the procedure outlined in Example 3.3-8, compute the 3-dB beamwidths Δu in direction cosine space of the triangular, cosine, Hanning, Hamming, and Blackman amplitude windows.

3-18 Using your results from Problem 3-17, repeat Problem 3-16 for the triangular, cosine, Hanning, Hamming, and Blackman amplitude windows.

3-19 Compute the beamwidths of the horizontal beam patterns for the rectangular, triangular, cosine, Hanning, Hamming, and Blackman amplitude windows at the following beam tilt angles: $\psi' = 90°$ (broadside), 60°, 45°, 30°, and 0° (end-fire). Use $L = 1$ m, $c = 1500$ m/sec, and $f = 25$ kHz.

3-20 Assume we are given a linear aperture lying along the X axis with a linear phase variation across its length.

(a) Show that the beamwidth $\Delta\theta$ (in degrees) of the vertical beam pattern in the XZ plane is given by

$$\Delta\theta = 2\sin^{-1}[\Delta u/(2\cos\theta')], \qquad \Delta u/(2\cos\theta') \leq 1,$$

where θ' is the beam tilt angle in the XZ plane. Note that $\theta' = 0$ corresponds to an untilted vertical beam pattern.

Hint: follow a procedure analogous to the one used to derive Eq. (3.3-109).

(b) Show that the beamwidth $\Delta\theta$ (in degrees) of the vertical beam pattern in the XZ plane at end-fire geometry is given by

$$\Delta\theta = 2\cos^{-1}[1 - (\Delta u/2)].$$

Hint: follow a procedure analogous to the one used to derive Eq. (3.3-113).

3-21 The spherical coordinates of a target's location with respect to a linear aperture $A(f, x)$ lying along the X axis has been determined to be $r_T = 50.0$ m, $\theta_T = 22°$, and $\psi_T = 63°$.

(a) Is the target in the aperture's near-field or far-field? Use $L = 2.0$ m, $c = 1500$ m/sec, and $f = 25$ kHz.

(b) What must be the phase variation $\theta(f, x)$ across the length of the aperture if the main lobe of the far-field beam pattern $F_x\{a(f, x)\}$ is to exist at coordinates (r_T, θ_T, ψ_T)?

(c) The target changes its coordinates to $r_T = 70.0$ m, $\theta_T = 33°$, and $\psi_T = 47°$. Repeat parts (a) and (b).

(d) Repeat parts (a) through (c) for a linear aperture $A(f, y)$ lying along the Y axis.

3-22 What is the normalized far-field beam pattern of the complex aperture function

$$A(f, x, y) = A_X(f, x) A_Y(f, y)$$

when

(a) $A_X(f, x) = \text{rect}(L_X)$,
$A_Y(f, y) = \text{rect}(L_Y)$;

(b) $A_X(f, x) = \text{tri}(L_X)$,
$A_Y(f, y) = \text{tri}(L_Y)$;

(c) $A_X(f, x) = \text{rect}(L_X)\exp(-j2\pi f'_X x)$,
$A_Y(f, y) = \text{tri}(L_Y)\exp(-j2\pi f'_Y y)$;

(d) $A_X(f, x) = \text{tri}(L_X)$,
$A_Y(f, y) = \cos(\pi y/L_Y)\,\text{rect}(L_Y)$;

(e) $A_X(f, x) = \cos(\pi x/L_X)\,\text{rect}(L_X)\exp(-j2\pi f'_X x)$,
$A_Y(f, y) = \cos(\pi y/L_Y)\,\text{rect}(L_Y)\exp(-j2\pi f'_Y y)$.

3-23 Verify Eq. (3.5-69). Note that

$$J_0(0) = 1, \qquad J_n(0) = 0, \qquad n \neq 0,$$

and

$$2J'_n(x) = J_{n-1}(x) - J_{n+1}(x),$$

where

$$J'_n(x) = \frac{d}{dx} J_n(x).$$

3-24 The spherical coordinates of a target's location with respect to a planar aperture $A(f, x, y)$ lying in the XY plane has been determined to be $r_T = 72.0$ m, $\theta_T = 26^\circ$, and $\psi_T = 103^\circ$.

(a) Is the target in the aperture's near-field or far-field? Use $L_X = 2.0$ m, $L_Y = 3.0$ m, $c = 1500$ m/sec, and $f = 15$ kHz.

(b) What must be the phase variation $\theta(f, x, y)$ across the face of the aperture if the main lobe of the far-field beam pattern $F_x F_y\{a(f, x, y)\}$ is to exist at coordinates (r_T, θ_T, ψ_T)?

(c) The target changes its coordinates to $r_T = 120.0$ m, $\theta_T = 38^\circ$, and $\psi_T = 53^\circ$. Repeat parts (a) and (b).

BIBLIOGRAPHY

C. S. Clay and H. Medwin, *Acoustical Oceanography*. Wiley, New York, 1977.
J. D. Gaskill, *Linear Systems, Fourier Transforms, and Optics*. Wiley, New York, 1978.
A. W. Goodman, *Analytic Geometry and The Calculus*. MacMillan, New York, 1963.

J. W. Goodman, *Introduction to Fourier Optics.* McGraw-Hill, New York, 1968.

F. J. Harris, On the use of windows for harmonic analysis with the discrete Fourier transform. *Proc. IEEE* **66**, 51–83 (1978).

F. B. Hildebrand, *Advanced Calculus for Applications.* Prentice-Hall, Englewood Cliffs, New Jersey, 1962.

A. Ishimaru, *Wave Propagation and Scattering in Random Media,* Vol. I. Academic Press, New York, 1978.

L. E. Kinsler and A. R. Frey, *Fundamentals of Acoustics,* 2nd edition. Wiley, New York, 1962.

L. E. Kinsler, A. R. Frey, A. B. Coppens, and J. V. Sanders, *Fundamentals of Acoustics,* 3rd edition. Wiley, New York, 1982.

W. C. Knight, R. G. Pridham, and S. M. Kay, Digital signal processing for sonar. *Proc. IEEE* **69**, 1451–1506 (1981).

D. Middleton, A statistical theory of reverberation and similar first-order scattered fields. Part I: Waveforms and the general process. *IEEE Trans. Inf. Theory* **IT-13**, 372–392 (1967).

P. M. Morse and K. U. Ingard, *Theoretical Acoustics.* McGraw-Hill, New York, 1968.

A. V. Oppenheim and R. W. Schafer, *Digital Signal Processing.* Prentice-Hall, Englewood Cliffs, New Jersey, 1975.

A. Papoulis, *Systems and Transforms with Applications in Optics.* McGraw-Hill, New York, 1968.

A. Papoulis, *Signal Analysis.* McGraw-Hill, New York, 1977.

A. D. Pierce, *Acoustics: An Introduction to Its Physical Principles and Applications.* McGraw-Hill, New York, 1981.

B. D. Steinberg, *Principles of Aperture and Array System Design.* Wiley, New York, 1976.

S. Temkin, *Elements of Acoustics.* Wiley, New York, 1981.

R. E. Ziemer and W. H. Tranter, *Principles of Communications.* Houghton Mifflin Co., Boston, 1976.

4

Arrays

4.1 LINEAR ARRAYS AND FAR-FIELD DIRECTIVITY FUNCTIONS

4.1.1 Product Theorem

An array can be thought of as a *sampled aperture*, that is, an aperture that is excited only at points or in localized areas. An array consists of individual electroacoustic transducers called *elements*.

Consider a linear array composed of an *even* number of *identical, equally spaced* elements, as shown in Fig. 4.1-1. The complex frequency response or complex aperture function $A(f, x)$ of this array can be expressed as

$$A(f,x) = \sum_{n=-N/2}^{-1} c_n e\left(f, x - \left(\frac{2n+1}{2}\right)d\right) + \sum_{n=1}^{N/2} c_n e\left(f, x - \left(\frac{2n-1}{2}\right)d\right), \tag{4.1-1}$$

where N is the total, even number of elements, d is the *interelement spacing* in meters, c_n is the *complex weight* associated with the nth element, and $e(f, x)$ is the *complex frequency response* of the identical elements, also known as the *element function.* The complex weights are used for amplitude and phase weighting, as will be discussed later. In general, the element function $e(f, x)$ represents a linear aperture L meters in length, as shown in Fig. 4.1-1. For example, $e(f, x)$ may be a rectangular or triangular amplitude window. The simplest representation for $e(f, x)$ is $e(f, x) = \delta(x)$, that is, the element function is equal to a delta function that represents a simple point source, such as a radially pulsating sphere.

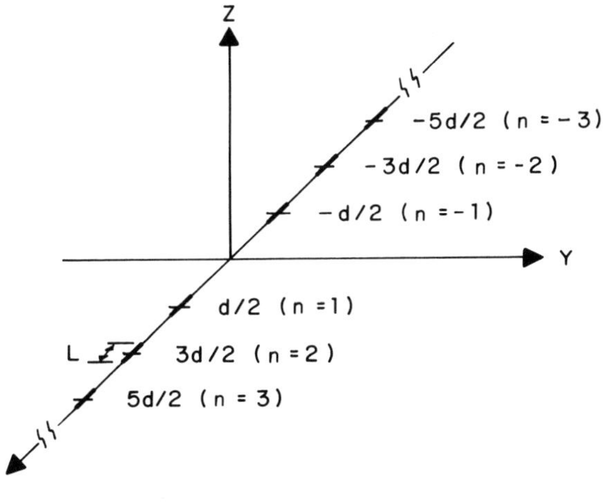

FIG. 4.1-1 A linear array composed of an even number of identical, equally spaced elements lying along the X axis.

The aperture function given by Eq. (4.1-1) can be rewritten by noting that

$$e\left(f, x - \left(\frac{2n+1}{2}\right)d\right) = e(f, x) * \delta\left(x - \left(\frac{2n+1}{2}\right)d\right) \tag{4.1-2}$$

and

$$e\left(f, x - \left(\frac{2n-1}{2}\right)d\right) = e(f, x) * \delta\left(x - \left(\frac{2n-1}{2}\right)d\right), \tag{4.1-3}$$

where the asterisk ($*$) denotes convolution with respect to x [see Eq. (3.3-50)]. In this representation the element function is no longer dependent on the index n. In addition, we can write that

$$\sum_{n=-N/2}^{-1} c_n \delta\left(x - \left(\frac{2n+1}{2}\right)d\right) = \sum_{n=1}^{N/2} c_{-n} \delta\left(x + \left(\frac{2n-1}{2}\right)d\right), \tag{4.1-4}$$

so that

$$A(f, x) = s(x) * e(f, x), \tag{4.1-5}$$

where, for N even,

$$s(x) = \sum_{n=1}^{N/2} \left\{c_{-n} \delta\left(x + \left(\frac{2n-1}{2}\right)d\right) + c_n \delta\left(x - \left(\frac{2n-1}{2}\right)d\right)\right\}. \tag{4.1-6}$$

The function $s(x)$ can be thought of as the complex frequency response or complex aperture function of an equivalent linear array of N even complex-weighted *point sources.* With the exception of the element function, $s(x)$ contains all the remaining information concerning the array such as the total number of elements, the complex weights, and the geometry of the array, that is, the positioning of the elements along the X axis. However, if the elements are not identical and not equally spaced, then the aperture function is given by the following general expression:

$$A(f,x) = \sum_{n=-N/2}^{-1} c_n e_n(f, x + d_n) + \sum_{n=1}^{N/2} c_n e_n(f, x - d_n), \tag{4.1-7}$$

where d_n is a *positive* distance measured from the origin to the center of the nth element.

Next consider a linear array composed of an *odd* number of *identical, equally spaced* elements, as shown in Fig. 4.1-2. The complex frequency response or complex aperture function $A(f, x)$ of this array can be expressed as

$$A(f,x) = \sum_{n=-(N-1)/2}^{(N-1)/2} c_n e(f, x - nd), \tag{4.1-8}$$

where N is the total, odd number of elements. Equation (4.1-8) can also be rewritten as

$$A(f,x) = s(x) * e(f,x), \tag{4.1-9}$$

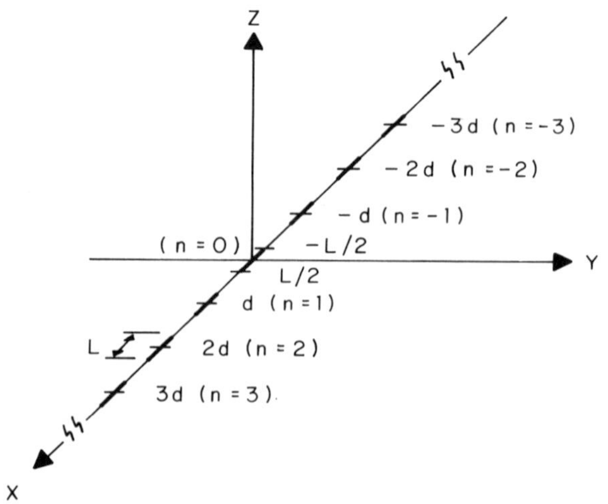

FIG. 4.1-2 A linear array composed of an odd number of identical, equally spaced elements lying along the X axis.

where, for N odd,

$$s(x) = \sum_{n=-(N-1)/2}^{(N-1)/2} c_n \,\delta(x - nd); \tag{4.1-10}$$

or, noting that

$$\sum_{n=-(N-1)/2}^{-1} c_n \,\delta(x - nd) = \sum_{n=1}^{(N-1)/2} c_{-n} \,\delta(x + nd), \tag{4.1-11}$$

$s(x)$, which can be thought of as the complex frequency response or complex aperture function of an equivalent linear array of N odd complex-weighted *point sources*, can also be expressed as

$$s(x) = c_0 \,\delta(x) + \sum_{n=1}^{(N-1)/2} \left\{ c_{-n} \,\delta(x + nd) + c_n \,\delta(x - nd) \right\}. \tag{4.1-12}$$

However, if the elements are not identical and not equally spaced, then the aperture function is given by the following general expression:

$$A(f, x) = c_0 e_0(f, x) + \sum_{n=-(N-1)/2}^{-1} c_n e_n(f, x + d_n) + \sum_{n=1}^{(N-1)/2} c_n e_n(f, x - d_n), \tag{4.1-13}$$

where, as before, d_n is a *positive* distance measured from the origin to the center of the nth element.

In the analysis that follows, it is assumed that all elements in the array (whether N is even or odd) are identical and equally spaced.

The far-field directivity function of a linear array that can be described by an aperture function of the form

$$A(f, x) = s(x) * e(f, x)$$

is given by

$$D(f, f_X) = S(f_X)E(f, f_X), \tag{4.1-14}$$

where

$$E(f, f_X) = F_x\{e(f, x)\} \tag{4.1-15}$$

and

$$S(f_X) = F_x\{s(x)\}; \tag{4.1-16}$$

for N even [see Eq. (3.3-31)],

$$S(f_X) = \sum_{n=1}^{N/2} \left\{ c_{-n} \exp\left[-j2\pi f_X \left(\frac{2n-1}{2} \right) d \right] + c_n \exp\left[+j2\pi f_X \left(\frac{2n-1}{2} \right) d \right] \right\}, \tag{4.1-17}$$

while, for N odd,

$$S(f_X) = \sum_{n=-(N-1)/2}^{(N-1)/2} c_n \exp(+j2\pi f_X nd) \tag{4.1-18}$$

or

$$S(f_X) = c_0 + \sum_{n=1}^{(N-1)/2} \left\{ c_{-n} \exp(-j2\pi f_X nd) + c_n \exp(+j2\pi f_X nd) \right\}. \tag{4.1-19}$$

Equation (4.1-14) is referred to as the *product theorem* for linear arrays. The product theorem states that the far-field directivity function of a linear array of *identical*, complex-weighted elements is equal to the product of the far-field directivity function $E(f, f_X)$ of *one* of the elements and the far-field directivity function $S(f_X)$ of an equivalent linear array of complex-weighted *point sources*. For example, if

$$e(f, x) = \text{rect}(L), \tag{4.1-20}$$

then [see Eq. (3.3-14)]

$$E(f, f_X) = \sin(\pi L f_X)/(\pi f_X), \tag{4.1-21}$$

and, as a result,

$$D(f, f_X) = S(f_X) \sin(\pi L f_X)/(\pi f_X). \tag{4.1-22}$$

If the linear array is composed of point sources, that is, if

$$e(f, x) = \delta(x), \tag{4.1-23}$$

then

$$E(f, f_X) = 1, \tag{4.1-24}$$

and, as a result,

$$D(f, f_X) = S(f_X). \tag{4.1-25}$$

Note that while the product theorem requires identical elements, equal spacing of the elements is *not* required (see Problem 4-1).

Before proceeding further, let us make two observations. First, one can simulate the case in which one or more elements in the array are broken (i.e., not radiating or receiving) by setting $c_n = 0$ for those elements. One can then investigate the effect this situation has on the far-field beam pattern. Second, if the length L of the elements is small compared to a wavelength, then the far-field beam pattern of the array $D(f, f_X)$ is, in effect, dependent on $S(f_X)$ alone since $E(f, f_X)$ is omnidirectional for all practical purposes. For example, consider the element function given by Eq. (4.1-20) and its far-field beam pattern given by Eq. (4.1-21). Since $\sin\alpha \approx \alpha$ for $\alpha \leq 0.0873$ rad,

$$\sin(\pi L f_X) \approx \pi L f_X \tag{4.1-26}$$

whenever

$$\pi L f_X = \pi L u/\lambda \leq 0.0873 \tag{4.1-27}$$

or

$$L/\lambda \leq 0.0278, \tag{4.1-28}$$

where the direction cosine u was set equal to one. Therefore, if either Eq. (4.1-27) or (4.1-28) is satisfied, then $E(f, f_X)$ as given by Eq. (4.1-21) is omnidirectional since

$$E(f, f_X) = \sin(\pi L f_X)/(\pi f_X) \approx L. \tag{4.1-29}$$

Let us next examine the complex weights c_n. In general,

$$c_n = a_n \exp(+j\theta_n), \tag{4.1-30}$$

where a_n and θ_n are *real* amplitude and phase weighting terms, respectively. Now, let us assume that

$$c_{-n} = c_n^* \quad \text{for} \quad \begin{cases} n = 1, 2, \ldots, N/2, & N \text{ even}, \\ n = 1, 2, \ldots, (N-1)/2, & N \text{ odd}, \end{cases} \tag{4.1-31}$$

and investigate the implications of such an assumption, where the asterisk (*) denotes complex conjugate. Since

$$c_{-n} = a_{-n} \exp(+j\theta_{-n}) \tag{4.1-32}$$

and

$$c_n^* = a_n \exp(-j\theta_n), \tag{4.1-33}$$

if $c_{-n} = c_n^*$, then

$$a_{-n} = a_n, \tag{4.1-34}$$

that is, a_n is an *even* function of the index n and

$$\theta_{-n} = -\theta_n, \tag{4.1-35}$$

that is, θ_n is an *odd* function of the index n. Note that since θ_n is an odd function,

$$\theta_0 = 0 \tag{4.1-36}$$

since an odd function passes through the origin ($n = 0$ in this case) and, as a result,

$$c_0 = a_0. \tag{4.1-37}$$

If Eq. (4.1-31) is substituted into Eq. (4.1-17), then for N even,

$$S(f_X) = \sum_{n=1}^{N/2} \left\{ \left[c_n \exp\left(+j2\pi f_X \left(\frac{2n-1}{2} \right) d \right) \right]^* + c_n \exp\left(+j2\pi f_X \left(\frac{2n-1}{2} \right) d \right) \right\}, \tag{4.1-38}$$

and, since

$$Z^* + Z = 2\,\mathrm{Re}\, Z,$$

$$S(f_X) = 2\,\mathrm{Re} \sum_{n=1}^{N/2} c_n \exp\left[+j2\pi f_X \left(\frac{2n-1}{2} \right) d \right] \tag{4.1-39}$$

or

$$S(f_X) = 2 \sum_{n=1}^{N/2} a_n \cos\left[2\pi f_X \left(\frac{2n-1}{2} \right) d + \theta_n \right]. \tag{4.1-40}$$

Similarly, if Eq. (4.1-31) is substituted into Eq. (4.1-19), then for N odd,

$$S(f_X) = a_0 + 2\,\mathrm{Re} \sum_{n=1}^{(N-1)/2} c_n \exp(+j2\pi f_X n d) \tag{4.1-41}$$

or

$$S(f_X) = a_0 + 2 \sum_{n=1}^{(N-1)/2} a_n \cos(2\pi f_X n d + \theta_n). \tag{4.1-42}$$

Therefore, it can be seen from Eqs. (4.1-40) and (4.1-42) that the beam patterns $S(f_X)$ for N even and N odd, respectively, are *real* and *even* functions of f_X when the complex weights obey Eq. (4.1-31). Also, recall from linear-aperture theory that all of the amplitude windows we discussed were even functions of x and that the linear phase variation responsible for beam steering was an odd function of x.

Amplitude and phase weighting will be discussed further in Sections 4.1.2 and 4.1.3, respectively. However, we shall restrict our analysis in these sections to linear arrays of N odd elements only. In addition, unless indicated otherwise, when we refer to a linear array *we shall assume it is lying along the X axis*. This is not a severe restriction, since if we have a linear array lying along either the Y or Z axis, we simply need to replace f_X with f_Y or f_Z, respectively, in the appropriate equations.

4.1.2 Amplitude Weighting

Consider a linear array of N(odd) identical, equally spaced, *amplitude-weighted* elements. The far-field directivity function for such an array is given by

$$D(f, f_X) = S(f_X)E(f, f_X), \qquad (4.1\text{-}14)$$

where, in general,

$$S(f_X) = \sum_{n=-(N-1)/2}^{(N-1)/2} a_n \exp(+j2\pi f_X nd) \qquad (4.1\text{-}43)$$

since $\theta_n = 0$.

The most common amplitude weights are simply sampled values of the continuous amplitude windows already discussed in Section 3.3.1. For example, consider the following set of equations for a_n, where $n = -(N-1)/2, \ldots, 0, \ldots, (N-1)/2$ and N is odd:

(1) *rectangular window*,

$$a_n = 1.0; \qquad (4.1\text{-}44)$$

(2) *triangular window*,

$$a_n = 1.0 - \frac{|n|}{(N-1)/2}; \qquad (4.1\text{-}45)$$

(3) *cosine window*,

$$a_n = \cos[\pi n/(N-1)]; \qquad (4.1\text{-}46)$$

(4) *Hanning window*,

$$a_n = 0.5 + 0.5\cos[2\pi n/(N-1)]; \qquad (4.1\text{-}47)$$

(5) *Hamming window*,

$$a_n = 0.54 + 0.46\cos[2\pi n/(N-1)]; \qquad (4.1\text{-}48)$$

(6) *Blackman window*,

$$a_n = 0.42 + 0.50\cos[2\pi n/(N-1)] + 0.08\cos[4\pi n/(N-1)]. \qquad (4.1\text{-}49)$$

The process of amplitude weighting an array of elements is also known as amplitude shading or as applying an amplitude window to the array. Amplitude weighting can be implemented by connecting each element in the array in series with an electronic amplifier. The gain of each amplifier is determined by which amplitude window is being used. Note that the above-mentioned amplitude windows are also used in digital filter design and spectrum analysis problems as well.

Since the amplitude weights given by Eqs. (4.1-44) through (4.1-49) are even functions of the index n, Eq. (4.1-42) can be used instead of Eq. (4.1-43) to evaluate the far-field beam pattern $S(f_X)$ by setting $\theta_n = 0$. Note that Eq. (4.1-42) only requires $(N - 1)/2 + 1 = (N + 1)/2$ values of a_n to compute $S(f_X)$, while Eq. (4.1-43) requires all N values. However, it should be mentioned that by using Eq. (4.1-43), it is possible to obtain closed-form expressions for all the amplitude windows given by Eqs. (4.1-44) through (4.1-49). For example, for the rectangular amplitude window [see Eq. (4.1-44)], Eq. (4.1-43) becomes

$$S(f_X) = \frac{\sin(\pi f_X N d)}{\sin(\pi f_X d)}, \tag{4.1-50}$$

and for the triangular amplitude window [see Eq. (4.1-45)], Eq. (4.1-43) becomes

$$S(f_X) = \frac{2}{(N-1)} \frac{\sin^2[\pi f_X \frac{1}{2}(N-1)d]}{\sin^2(\pi f_X d)}. \tag{4.1-51}$$

Compare Eqs. (4.1-50) and (4.1-51) with Eqs. (3.3-14) and (3.3-33), respectively.

EXAMPLE 4.1-1

Given a linear array of N(odd) identical, equally spaced elements amplitude weighted by a rectangular window, if

$$e(f, x) = \mathrm{rect}(L),$$

then from the product theorem, the far-field directivity function of the linear array is given by [see Eqs. (4.1-22) and (4.1-50)]

$$D(f, f_X) = \frac{\sin(\pi f_X N d)}{\sin(\pi f_X d)} \frac{\sin(\pi L f_X)}{\pi f_X}.$$

If the linear array is composed of point sources, that is, if

$$e(f, x) = \delta(x),$$

then [see Eqs. (4.1-25) and (4.1-50)]

$$D(f, f_X) = \frac{\sin(\pi f_X N d)}{\sin(\pi f_X d)}. \quad \square$$

EXAMPLE 4.1-2

Given a linear array of N (odd) identical, equally spaced elements amplitude weighted by a triangular window, if

$$e(f, x) = \mathrm{rect}(L),$$

then from the product theorem, the far-field directivity function of this linear array is given by [see Eqs. (4.1-22) and (4.1-51)]

$$D(f, f_X) = \frac{2}{(N-1)} \frac{\sin^2[\pi f_X \frac{1}{2}(N-1)d]}{\sin^2(\pi f_X d)} \frac{\sin(\pi L f_X)}{\pi f_X}.$$

If the linear array is composed of point sources, that is, if

$$e(f, x) = \delta(x),$$

then [see Eqs. (4.1-25) and (4.1-51)]

$$D(f, f_X) = \frac{2}{(N-1)} \frac{\sin^2[\pi f_X \frac{1}{2}(N-1)d]}{\sin^2(\pi f_X d)}. \quad \square$$

EXAMPLE 4.1-3

Let us evaluate Eq. (4.1-43) for the following amplitude window:

$$a_n = \cos[b\pi n/(N-1)],$$

where b is an arbitrary constant, N is odd, and $n = -(N-1)/2, \ldots, 0, \ldots, (N-1)/2$. Note that the amplitude windows given by Eqs. (4.1-46) through (4.1-49) all involve one or more terms of this form where b takes on the values of 1, 2, and 4. Since

$$\begin{aligned} a_n &= \cos[b\pi n/(N-1)] \\ &= 0.5 \exp[+jb\pi n/(N-1)] + 0.5 \exp[-jb\pi n/(N-1)] \end{aligned}$$

and

$$\sum_{k=-K}^{K} \exp(+jk\theta) = \frac{\sin\{[(2K+1)/2]\theta\}}{\sin(\theta/2)},$$

Eq. (4.1-43) becomes [compare with Eq. (3.3-52)]

$$S(f_X) = \frac{\sin\{\pi f_X N d + b\pi N/[2(N-1)]\}}{2\sin\{\pi f_X d + b\pi/[2(N-1)]\}} + \frac{\sin\{\pi f_X N d - b\pi N/[2(N-1)]\}}{2\sin\{\pi f_X d - b\pi/[2(N-1)]\}}.$$

Therefore, by using this result and Eq. (4.1-50) we can obtain closed-form expressions for $S(f_X)$ for the cosine, Hanning, Hamming, and Blackman amplitude windows given by Eqs. (4.1-46) through (4.1-49), respectively. Note that if we let $b = 0$, then

$$a_n = \cos 0 = 1$$

and

$$S(f_X) = \frac{\sin(\pi f_X N d)}{\sin(\pi f_X d)},$$

which is the result for the rectangular amplitude window given by Eq. (4.1-50). □

If a closed-form expression is also available for $E(f, f_X)$, then a closed-form expression can be obtained for the far-field directivity function $D(f, f_X) = S(f_X)E(f, f_X)$ of the linear array. Therefore, the 3-dB beamwidth Δu in direction cosine space can be obtained separately either for the normalized function $S_N(f_X)$ or for the normalized function $D_N(f, f_X)$ using the procedure outlined in Example 3.3-8. Once Δu has been determined, the 3-dB beamwidths $\Delta\theta$ and $\Delta\psi$, in degrees, of the vertical and horizontal beam patterns, respectively, can be computed using Eqs. (3.3-82) and (3.3-86), respectively (see Problems 4-2 and 4-3).

In the process of obtaining solutions to Problems 4-2 and 4-3, one should note that the expressions for the 3-dB beamwidths depend upon the total number of elements in the array N and the ratio λ/d. The interelement spacing d is usually chosen to be a fraction of a wavelength, with $d = \lambda/2$ the most common choice (see Section 4.3). Therefore, with $d = \lambda/2$, the beamwidth no longer depends upon frequency but depends, instead, only on the length of the array, which is directly proportional to N. As N increases, the beamwidth decreases and vice versa.

Dolph–Chebyshev method of amplitude weighting. We know from our discussion of linear apertures (see Section 3.3.1) that different amplitude windows yield far-field beam patterns with different sidelobe levels and beamwidths, and that as sidelobe levels decrease, beamwidth generally increases. This principle also applies to the far-field directivity functions of amplitude-weighted linear arrays. However, the Dolph–Chebyshev method of amplitude shading makes it possible to *optimize* the far-field beam patterns of arrays so that for any specified sidelobe level relative to the level of the main lobe, the *narrowest* possible main-lobe beamwidth is achieved or, for any specified main-lobe beamwidth, the *lowest* possible sidelobe level is achieved. These relationships are subject to physical limitations such as the number and spacing of the elements, the operating frequency, and the directional properties of the elements themselves as characterized by $E(f, f_X)$. In addition, although the Dolph–Chebyshev method produces the lowest possible sidelobe level for a specified main-lobe beamwidth, *all* of the sidelobes are at this same, minimum level. Let us demonstrate this method by the following example.

EXAMPLE 4.1-4: THE DOLPH–CHEBYSHEV METHOD

Suppose we are given a linear array of $N = 7$ identical, equally spaced elements and it is desired to amplitude weight this array using the Dolph–Chebyshev method (note that this method applies to both N even and N odd).

Since the amplitude weights a_n produced by the Dolph–Chebyshev method are even functions of the index n, the far-field beam pattern of this array is given by

$$D(f, f_X) = S(f_X)E(f, f_X), \tag{4.1-14}$$

where [see Eq. (4.1-42)]

$$S(f_X) = a_0 + 2 \sum_{n=1}^{3} a_n \cos(2\pi f_X nd) \tag{4.1-52}$$

since $N = 7$ and $\theta_n = 0$. If we let

$$\alpha = \pi f_X d, \tag{4.1-53}$$

then Eq. (4.1-52) can be rewritten as

$$S(f_X) = a_0 + 2a_1 \cos(2\alpha) + 2a_2 \cos(4\alpha) + 2a_3 \cos(6\alpha). \tag{4.1-54}$$

Since the Dolph–Chebyshev method depends on certain properties of *Chebyshev polynomials*, we must digress a moment to discuss these properties before continuing further.

The expression

$$T_n(x) = \cos[n(\cos^{-1} x)] \tag{4.1-55}$$

defines an nth-degree polynomial known as a Chebyshev polynomial. If we also let

$$\alpha = \cos^{-1} x \qquad \text{or} \qquad x = \cos \alpha, \tag{4.1-56}$$

then Eq. (4.1-55) can be written as

$$T_n(x) = \cos(n\alpha), \tag{4.1-57}$$

and upon comparing Eq. (4.1-57) with Eq. (4.1-54), it can be seen that the beam pattern $S(f_X)$ given by Eq. (4.1-54) can be expressed as a sum of Chebyshev polynomials and that the sum of these polynomials is itself a polynomial. The degree of this polynomial, which is called the *array polynomial*, is $n = (N - 1)$, or one less than the total number of elements in the array.

All Chebyshev polynomials possess the following important properties.

(1) For $n \geq 2$,

$$T_n(x) = 2xT_{n-1}(x) - T_{n-2}(x), \tag{4.1-58}$$

where

$$T_0(x) = 1 \qquad \text{and} \qquad T_1(x) = x.$$

(2) $T_n(x)$ has n real roots and they lie between $-1 < x < 1$.

(3) $T_n(x)$ has maxima and minima that occur alternately at

$$x_k = \cos(k\pi/n), \qquad k = 1, 2, \ldots, n-1,$$

where $-1 < x_k < 1$ and the absolute value of these maxima and minima equals unity, that is, $|T_n(x_k)| = 1$. Thus, the polynomials $T_n(x)$ have "equal ripple" characteristics in the interval $-1 < x < 1$.

(4) At the end points of the interval $-1 \le x \le 1$, $|T_n(\pm 1)| = 1$; however, $|T_n(x)| > 1$ for $|x| > 1$ (see Fig. 4.1-3) and

$$\left|\frac{d}{dx} T_n(x)\right|\Bigg|_{x=\pm 1} = n^2.$$

The Chebyshev polynomials possess some additional properties, but they are not pertinent to the present problem.

In the Dolph–Chebyshev method, the magnitude of the main-lobe level corresponds to the value of $T_n(x_0)$, where $x_0 > 1$, and the magnitude of the

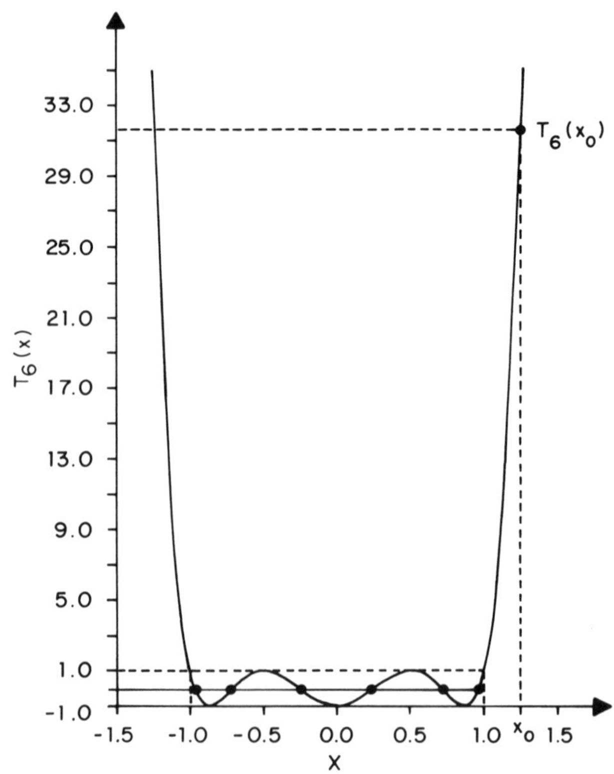

FIG. 4.1-3 Sixth-degree Chebyshev polynomial $T_6(x)$ as given by Eq. (4.1-66).

sidelobe level is numerically equal to the absolute value of the interior maxima and minima, that is, unity. Therefore, the ratio

$$\frac{\text{main-lobe level}}{\text{sidelobe level}} = r > 1, \tag{4.1-59}$$

which can be made as large as desired, is used to determine the value of x_0, that is, we solve for x_0 in $T_n(x_0) = r$. Note that it is the "equal ripple" characteristic of Chebyshev polynomials as discussed in property (3) that is responsible for generating sidelobes all at the same level in the Dolph–Chebyshev method of amplitude weighting.

Returning now to our example, suppose we require that the main-lobe level be 30 dB above the sidelobe level, that is,

$$20 \log_{10} r = 30 \quad \text{dB} \tag{4.1-60}$$

or

$$r = 31.623. \tag{4.1-61}$$

Since we have $N = 7$ elements, we let $n = N - 1 = 6$ and next solve

$$T_6(x_0) = r = 31.623 \tag{4.1-62}$$

for x_0. However, it is not necessary to solve Eq. (4.1-62) for x_0, since x_0 can be calculated from the expression

$$x_0 = 0.5\left[[r + (r^2 - 1)^{1/2}]^{1/n} + \left(\frac{1}{r + (r^2 - 1)^{1/2}} \right)^{1/n} \right], \tag{4.1-63}$$

or, since $r^2 \gg 1$,

$$x_0 \approx 0.5\left[(2r)^{1/n} + \frac{1}{(2r)^{1/n}} \right]. \tag{4.1-64}$$

Substituting $r = 31.623$ and $n = N - 1 = 6$ into Eq. (4.1-64) yields

$$x_0 = 1.249. \tag{4.1-65}$$

From the recurrence relationship given by Eq. (4.1-58), we obtain the following expression for the sixth-degree Chebyshev polynomial $T_6(x)$ (see Fig. 4.1-3):

$$T_6(x) = 32x^6 - 48x^4 + 18x^2 - 1. \tag{4.1-66}$$

Next, replace x with $x_0 x$ in Eq. (4.1-66) so that

$$T_6(x_0 x) = 32x_0^6 x^6 - 48x_0^4 x^4 + 18x_0^2 x^2 - 1. \tag{4.1-67}$$

Since

$$x = \cos\alpha \tag{4.1-56}$$

and

$$\cos(2\alpha) = 2\cos^2\alpha - 1 = 2x^2 - 1, \tag{4.1-68}$$

$$\cos(4\alpha) = 8\cos^4\alpha - 8\cos^2\alpha + 1 \tag{4.1-69}$$

or

$$\cos(4\alpha) = 8x^4 - 8x^2 + 1, \tag{4.1-70}$$

and

$$\cos(6\alpha) = 32\cos^6\alpha - 48\cos^4\alpha + 18\cos^2\alpha - 1 \tag{4.1-71}$$

or

$$\cos(6\alpha) = 32x^6 - 48x^4 + 18x^2 - 1, \tag{4.1-72}$$

$S(f_X)$ as given by Eq. (4.1-54) can be rewritten as

$$\begin{aligned} S(f_X) = {} & 64a_3x^6 + 16(a_2 - 6a_3)x^4 + 4(a_1 - 4a_2 + 9a_3)x^2 \\ & + [a_0 - 2(a_1 - a_2 + a_3)]. \end{aligned} \tag{4.1-73}$$

The next step is to calculate the amplitude weights. By comparing Eqs. (4.1-67) and (4.1-73) and equating coefficients of like powers of x, we obtain the following set of four equations in four unknowns:

$$64a_3 = 32x_0^6, \tag{4.1-74}$$

$$16(a_2 - 6a_3) = -48x_0^4, \tag{4.1-75}$$

$$4(a_1 - 4a_2 + 9a_3) = 18x_0^2, \tag{4.1-76}$$

and

$$a_0 - 2(a_1 - a_2 + a_3) = -1. \tag{4.1-77}$$

Solving Eqs. (4.1-74) through (4.1-77) for the amplitude weights yields

$$a_0 = 2x_0^2(5x_0^4 - 9x_0^2 + 4.5) - 1, \tag{4.1-78}$$

$$a_1 = x_0^2(7.5x_0^4 - 12x_0^2 + 4.5), \tag{4.1-79}$$

$$a_2 = 3x_0^4(x_0^2 - 1), \tag{4.1-80}$$

and

$$a_3 = 0.5x_0^6, \tag{4.1-81}$$

and upon substituting $x_0 = 1.249$ into Eqs. (4.1-78) through (4.1-81), we obtain

$$a_0 = 7.199, \qquad a_1 = 6.29, \qquad a_2 = 4.089, \qquad a_3 = 1.898,$$

or, dividing by $a_0 = 7.199$, we obtain the *normalized* values

$$a_0 = 1.0, \tag{4.1-82}$$

$$a_1 = 0.8737, \tag{4.1-83}$$

$$a_2 = 0.5680, \tag{4.1-84}$$

and

$$a_3 = 0.2636. \tag{4.1-85}$$

Since the amplitude weights are even functions of the index n,

$$a_{-1} = a_1 = 0.8737, \tag{4.1-86}$$

$$a_{-2} = a_2 = 0.5680, \tag{4.1-87}$$

and

$$a_{-3} = a_3 = 0.2636. \tag{4.1-88}$$

Let us next compute the resulting beamwidth of the main lobe of $S(f_X)$ for a specified value of the ratio r. Since

$$\alpha = \cos^{-1} x \tag{4.1-56}$$

and

$$\alpha = \pi f_X d, \tag{4.1-53}$$

$$u = \sin\theta \cos\psi = [\lambda/(\pi d)] \cos^{-1} x \tag{4.1-89}$$

since

$$f_X = u/\lambda.$$

Let us restrict ourselves to the vertical beam pattern of $S(f_X)$, which occupies the XZ plane since the linear array lies along the X axis (see Example 3.3-6 for a linear aperture). In the XZ plane, $u = \sin\theta$ when $\psi = 0$. If we let (see Fig. 3.3-16)

$$u_+ = \sin\theta_+ > 0, \qquad \psi_+ = 0, \tag{3.3-78}$$

then, from Eq. (4.1-89),

$$u_+ = \sin\theta_+ = [\lambda/(\pi d)] \cos^{-1} x_+ \tag{4.1-90}$$

or

$$\theta_+ = \sin^{-1}\{[\lambda/(\pi d)] \cos^{-1} x_+\}, \tag{4.1-91}$$

where, from Eq. (3.3-80),

$$\Delta\theta = 2\theta_+ = 2\sin^{-1}\{[\lambda/(\pi d)] \cos^{-1} x_+\} \tag{4.1-92}$$

is the beamwidth (in degrees) of the vertical beam pattern in the XZ plane. The x_+ term is the value of x that corresponds to the 3-dB down point θ_+, and it is related to the ratio r by the following expression:

$$x_+ = \frac{(1.414r)^{1/n} + 1/(1.414r)^{1/n}}{(2r)^{1/n} + 1/(2r)^{1/n}}. \tag{4.1-93}$$

Thus, Eqs. (4.1-92) and (4.1-93) are used together to determine the beamwidth $\Delta\theta$.

Continuing with our example, if we substitute $r = 31.623$ and $n = N - 1 = 6$ into Eq. (4.1-93), we obtain

$$x_+ = 0.967, \tag{4.1-94}$$

and substituting this value of x_+ into Eq. (4.1-92) yields

$$\Delta\theta = 2\sin^{-1}[0.258\lambda/(\pi d)]. \tag{4.1-95}$$

In order to evaluate Eq. (4.1-95), let the interelement spacing $d = \lambda/2$. In Section 4.3 we shall demonstrate that $d < \lambda/2$ is a necessary criterion for avoiding grating lobes under all conditions of beam steering for linear arrays when N is odd. Therefore, substituting $d = \lambda/2$ into Eq. (4.1-95) yields

$$\Delta\theta = 18.907° \tag{4.1-96}$$

as the beamwidth of the vertical beam pattern of $S(f_X)$, whose main-lobe level is 30 dB above the sidelobe level. □

Let us finish our discussion of amplitude weighting by noting that if the amplitude weights are an *even* function of the index n, $S(f_X)$ will be a *real* and *even* function, and if the amplitude weights are an *odd* function of the index n, $S(f_X)$ will be *imaginary* and *odd*.

4.1.3 The Phased Array (Beam Steering)

Consider a linear array of N (odd) identical, equally spaced point sources. From the product theorem, the far-field directivity function of this array is given by

$$D(f, f_X) = S(f_X), \tag{4.1-25}$$

where, in general,

$$S(f_X) = \sum_{n=-(N-1)/2}^{(N-1)/2} c_n \exp(+j2\pi f_X nd) \tag{4.1-18}$$

and

$$c_n = a_n \exp(+j\theta_n). \tag{4.1-30}$$

Let $S'(f_X)$ represent the far-field beam pattern of the array when it is complex weighted and let $S(f_X)$ represent the far-field beam pattern of the array when it is amplitude weighted, that is,

$$S'(f_X) = \sum_{n=-(N-1)/2}^{(N-1)/2} a_n \exp[+j(2\pi f_X nd + \theta_n)] \tag{4.1-97}$$

and

$$S(f_X) = \sum_{n=-(N-1)/2}^{(N-1)/2} a_n \exp(+j2\pi f_X nd). \tag{4.1-98}$$

If the phase weights θ_n are given by [compare with Eqs. (3.3-95) through (3.3-98)]

$$\theta_n = -2\pi f'_X nd, \qquad n = -(N-1)/2, \ldots, 0, \ldots, (N-1)/2, \tag{4.1-99}$$

where

$$f'_X = u'/\lambda = \sin\theta' \cos\psi'/\lambda, \tag{4.1-100}$$

then substituting Eq. (4.1-99) into Eq. (4.1-97) yields

$$S'(f_X) = \sum_{n=-(N-1)/2}^{(N-1)/2} a_n \exp[+j2\pi(f_X - f'_X)\, nd] \tag{4.1-101}$$

or

$$S'(f_X) = S(f_X - f'_X). \tag{4.1-102}$$

Equation (4.1-102) indicates that the far-field beam pattern $S(f_X)$ can be steered in the direction $u = u'$ in direction cosine space when the array of point sources is phase weighted according to Eq. (4.1-99). Note that θ_n as given by Eq. (4.1-99) is a *linear* phase variation (linear in nd) applied across the array and that θ_n is an *odd* function of the index n, that is,

$$\theta_0 = 0$$

and

$$\theta_{-n} = -\theta_n, \qquad n = 1, 2, \ldots, (N-1)/2.$$

Therefore, whether or not the amplitude weights a_n possess even symmetry or any symmetry, in order to do beam steering, the phase weights θ_n *must* have *odd* symmetry.

Finally, note that a phase shift θ_n is equivalent to a *time delay* τ_n, that is,

$$\theta_n = 2\pi f \tau_n$$

or

$$\tau_n = \theta_n/(2\pi f) \quad \text{sec.} \tag{4.1-103}$$

Substituting Eqs. (4.1-99) and (4.1-100) into Eq. (4.1-103) yields

$$\tau_n = -u'nd/c, \qquad n = -(N-1)/2,\ldots,0,\ldots,(N-1)/2. \tag{4.1-104}$$

According to Eq. (4.1-104), half of the time delays will be positive and half will be negative. In order to insure that all the time delays τ_n are *positive* for practical applications, the constant $|u'|[(N-1)/2](d/c)$ should be added to all values of τ_n computed from Eq. (4.1-104) so that $\tau_{(N-1)/2}$ equals zero when u' is positive and $\tau_{-(N-1)/2}$ equals zero when u' is negative. Assuming that the 3-dB beamwidth Δu in direction cosine space can be obtained for the normalized far-field beam pattern $D_N(f, f_X) = S_N(f_X)$ where the unnormalized beam pattern $S(f_X)$ is given by Eq. (4.1-98) [refer to Example 4.1-3 and the discussion that follows it], the resulting 3-dB beamwidths $\Delta\theta$ and $\Delta\psi$, in degrees, of the vertical and horizontal beam patterns, respectively, of the steered far-field beam pattern of the array can be computed. For example, $\Delta\theta$ can be computed using the results of Problem 3-20(a) or (b) at end-fire, and $\Delta\psi$ can be computed using Eqs. (3.3-109) or (3.3-113) at end-fire.

4.1.4 Far-field Beam Patterns and the Discrete Fourier Transform

The far-field directivity function of a linear array of N (odd) identical, equally spaced, complex-weighted point sources is given by

$$D(f, f_X) = S(f_X), \tag{4.1-25}$$

where, in general,

$$S(f_X) = \sum_{n=-(N-1)/2}^{(N-1)/2} c_n \exp(+j2\pi f_X nd) \tag{4.1-18}$$

and

$$c_n = a_n \exp(+j\theta_n). \tag{4.1-30}$$

Since in most practical applications $c_{-n} = c_n^*$, that is, the amplitude weights a_n are an even function of the index n and, as we have discussed in Section 4.1.3, the phase weights θ_n must be an odd function of n in order to do beam steering, Eq. (4.1-18) can be written as

$$S(f_X) = a_0 + 2\,\mathrm{Re} \sum_{n=1}^{(N-1)/2} c_n \exp(+j2\pi f_X nd) \tag{4.1-41}$$

or

$$S(f_X) = 2\,\mathrm{Re} \sum_{n=0}^{M-1} c_n \exp(+j2\pi f_X nd), \tag{4.1-105}$$

where

$$c_0 \triangleq a_0/2 \tag{4.1-106}$$

and

$$M - 1 = (N - 1)/2 \tag{4.1-107}$$

or

$$M = (N - 1)/2 + 1 \tag{4.1-108}$$

is the total number of complex numbers c_n. The summation

$$\sum_{n=0}^{M-1} c_n \exp(+j2\pi f_X nd) \tag{4.1-109}$$

is in the form of a *discrete spatial Fourier transform* (DSFT) of the sequence of complex numbers c_n, $n = 0, 1, \ldots, M - 1$. Note that this summation is evaluated at continuous values of the spatial frequency f_X.

A more efficient way to compute $S(f_X)$ is to evaluate it only at the discrete spatial frequencies

$$f_X = mf_{X_S}, \qquad m = 0, 1, \ldots, M - 1, \tag{4.1-110}$$

where

$$f_{X_S} = 1/(Md) \tag{4.1-111}$$

represents the maximum allowable spatial-frequency spacing such that the continuous curve $S(f_X)$ can be uniquely interpolated from the finite number of values $S(mf_{X_S})$, $m = 0, 1, \ldots, M - 1$. Therefore, substituting Eqs. (4.1-110) and (4.1-111) into Eq. (4.1-105) yields

$$S(mf_{X_S}) = 2\,\mathrm{Re} \sum_{n=0}^{M-1} c_n W_M^{mn}, \qquad m = 0, 1, \ldots, M - 1, \tag{4.1-112}$$

where

$$W_M = \exp(+j2\pi/M). \tag{4.1-113}$$

The summation

$$\sum_{n=0}^{M-1} c_n W_M^{mn}, \qquad m = 0, 1, \ldots, M - 1, \tag{4.1-114}$$

where f_{X_S} and W_M are given by Eqs. (4.1-111) and (4.1-113), respectively, is called the *discrete Fourier transform* (DFT) of the sequence of complex numbers c_n, $n = 0, 1, \ldots, M - 1$. One can then use a *fast-Fourier-transform* (FFT) computer algorithm to compute the discrete Fourier transform. Thus, the far-field directivity function of a linear array of N (odd) identical, equally spaced, complex-weighted point sources can be computed by using a FFT computer algorithm.

The beam-pattern expression given by Eq. (4.1-112) is periodic with period M, that is,

$$S(mf_{X_S}) = S[(m + M)f_{X_S}]. \tag{4.1-115}$$

In addition, $S(mf_{X_S})$ is real and is an even function about the origin $m = 0$ when no beam steering is being done, that is, when $\theta_n = 0$ for all n. However, when beam steering is being done, then $S(mf_{X_S})$ is an even function about $m = m'$, where m' corresponds to the FFT bin number where the main lobe is located. Also, since

$$f_X = u/\lambda = mf_{X_S}, \qquad m = 0, 1, \ldots, M - 1, \tag{4.1-116}$$

$$u = u_m = \lambda mf_{X_S}, \qquad m = 0, 1, \ldots, M - 1, \tag{4.1-117}$$

or, from Eq. (4.1-111),

$$u_m = m\lambda/(Md), \qquad m = 0, 1, \ldots, M - 1. \tag{4.1-118}$$

Equation (4.1-118) indicates that we can relate the FFT bin number m to a corresponding value of the direction cosine $u = u_m$ and to values of the spherical angles $\theta = \theta_m$ and $\psi = \psi_m$ since, in general, $u_m = \sin\theta_m \cos\psi_m$. If we restrict ourselves to the vertical beam pattern, then $u_m = \pm\sin\theta_m$ since $\psi_m = 0$ or π. If we restrict ourselves to the horizontal beam pattern, then $u_m = \cos\psi_m$ since $\theta_m = \pi/2$. For example, if $d = \lambda$ and $m = \pm M$, then according to Eq. (4.1-118)

$$u_{\pm M} = \pm 1,$$

which are the end points of the visible region where ($\theta_{+M} = \pi/2$, $\psi_{+M} = 0$) for $u_{+M} = +1$ and ($\theta_{-M} = \pi/2$, $\psi_{-M} = \pi$) for $u_{-M} = -1$. Therefore, if no beam steering is being done and if $d = \lambda$, then the main lobe of $S(mf_{X_S})$ given by Eq. (4.1-112) will appear not only at $m = 0$ (which corresponds to $u = u_0 = 0$) but also at $m = \pm M$ (which corresponds to $u = u_{\pm M} = \pm 1$) since $S(mf_{X_S})$ is periodic with period M [see Eq. (4.1-115)]. These additional main lobes within the visible region $|u| \leq 1$ are referred to as *grating lobes*, which is the subject of Section 4.3.

Most commonly available FFT algorithms assume that the total number M of complex data points c_n, $n = 0, 1, \ldots, M - 1$, is equal to an integer power of two, that is,

$$M = 2^b, \qquad b = 1, 2, 3, \ldots. \tag{4.1-119}$$

If M does not satisfy Eq. (4.1-119), one need only add enough zeros Z to the original data sequence until the total number of data points $M + Z$ is equal to an integer power of two. This is known as "padding with zeros." Therefore,

Eq. (4.1-112) can be rewritten as

$$S(mf_{X_S}) = 2\,\text{Re}\sum_{n=0}^{M-1} c_n W_{M+Z}^{mn}, \qquad m = 0, 1, \ldots, M-1, \ldots, M-1+Z, \tag{4.1-120}$$

where

$$c_n = 0, \qquad n = M, M+1, \ldots, M-1+Z, \tag{4.1-121}$$

$$W_{M+Z} = \exp[+j2\pi/(M+Z)], \tag{4.1-122}$$

and

$$f_{X_S} = 1/[(M+Z)d]. \tag{4.1-123}$$

For the *same* value of the argument mf_{X_S}, the value of $S(mf_{X_S})$ computed by Eq. (4.1-120) is identical to that for $S(mf_{X_S})$ computed by Eq. (4.1-112) since, according to Eq. (4.1-121), the summation given by Eq. (4.1-120) was obtained by simply adding zeros to the original summation given by Eq. (4.1-112). However, padding with zeros does *increase the resolution* since the spatial-frequency spacing is now given by Eq. (4.1-123), which is equal to ($Z = 0$) or less than ($Z > 0$) the maximum allowable spatial-frequency spacing given by Eq. (4.1-111). Note that Eq. (4.1-120) is periodic with period $M + Z$, that is,

$$S(mf_{X_S}) = S([m + (M+Z)]f_{X_S}), \tag{4.1-124}$$

where f_{X_S} is given by Eq. (4.1-123). Also note that Eq. (4.1-120) reduces to Eq. (4.1-112) when $Z = 0$.

Finally, the *normalized* form of Eq. (4.1-120) is given by

$$S_N(mf_{X_S}) = 2\,\text{Re}\left\{\sum_{n=0}^{M-1} c_n W_{M+Z}^{mn}\right\}\Bigg/\left(2\sum_{n=0}^{M-1} a_n\right),$$
$$m = 0, 1, \ldots, M-1, \ldots, M-1+Z, \tag{4.1-125}$$

or, cancelling the factor 2,

$$S_N(mf_{X_S}) = \text{Re}\left\{\sum_{n=0}^{M-1} c_n W_{M+Z}^{mn}\right\}\Bigg/\sum_{n=0}^{M-1} a_n,$$
$$m = 0, 1, \ldots, M-1, \ldots, M-1+Z, \tag{4.1-126}$$

where the relationship between the FFT bin number m and the direction cosine $u = u_m$ is now given by

$$u_m = m\lambda/[(M+Z)d], \qquad m = 0, 1, \ldots, M-1, \ldots, M-1+Z, \tag{4.1-127}$$

which was obtained by substituting Eq. (4.1-123) into Eq. (4.1-117). Keep in mind that the complex weight c_0 is defined by Eq. (4.1-106). In addition, when

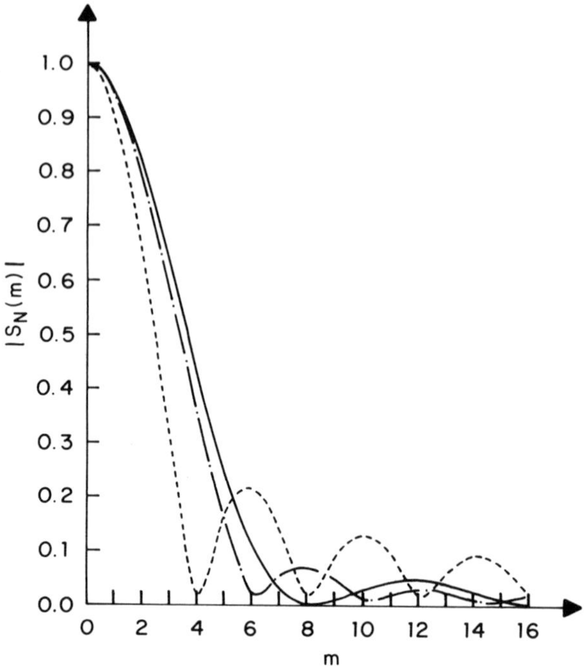

FIG. 4.1-4 Magnitude of the normalized far-field beam pattern given by Eq. (4.1-126) for a rectangular window (– – –), triangular window (——), and cosine window (—·—) plotted as a function of the FFT bin number m.

evaluating Eq. (4.1-126), *one must replace* a_0 *with* $a_0/2$ *in the denominator as well.* Also, recall that M and the total number of elements in the array N are related by Eq. (4.1-108).

Figures 4.1-4 and 4.1-5 are plots of the magnitude of Eq. (4.1-126) evaluated using the amplitude windows given by Eqs. (4.1-44) through (4.1-46) and Eqs. (4.1-47) through (4.1-49), respectively. The values of the parameters used in Eq. (4.1-126) were $N = 63$, which implies $M = 32$, $Z = 224$, and $\theta_n = 0$ for all n. Compare Figs. 4.1-4 and 4.1-5 with Figs. 3.3-10 and 3.3-12, respectively.

4.2 LINEAR ARRAYS AND NEAR-FIELD DIRECTIVITY FUNCTIONS

Recall that the near-field directivity function of a complex linear aperture function $A(f, x_0)$ is given by

$$D(f, r, f_X) = \int_{-L/2}^{L/2} A(f, x_0) \exp[-jkx_0^2/(2r)] \exp(+j2\pi f_X x_0)\, dx_0. \quad (3.4\text{-}7)$$

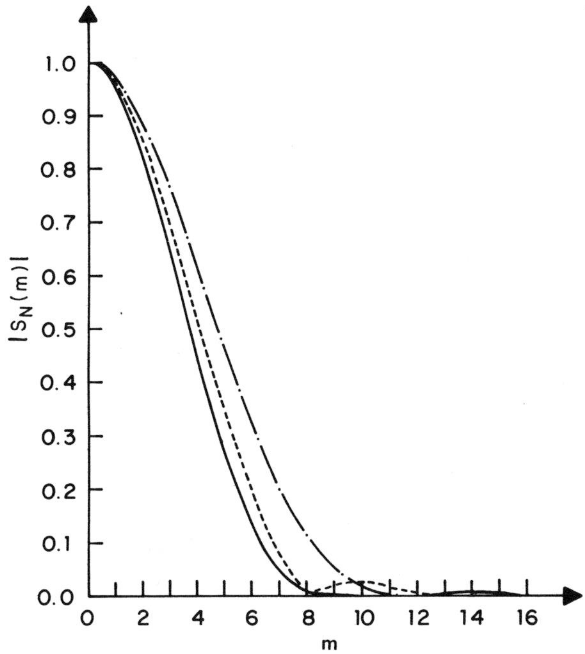

FIG. 4.1-5 Magnitude of the normalized far-field beam pattern given by Eq. (4.1-126) for a Hanning window (---), Hamming window (——), and Blackman window (—•—) plotted as a function of the FFT bin number *m*.

Now consider a linear array of N (odd) identical, equally spaced, complex-weighted point sources. Since the element function for this array is a delta function, that is [see Eq. (4.1-23)],

$$e(f, x_0) = \delta(x_0), \tag{4.2-1}$$

the aperture function can be expressed as [see Eqs. (4.1-9) and (4.1-10)]

$$A(f, x_0) = s(x_0) * e(f, x_0) \tag{4.2-2}$$

$$= s(x_0) * \delta(x_0) = s(x_0), \tag{4.2-3}$$

or

$$A(f, x_0) = \sum_{n=-(N-1)/2}^{(N-1)/2} c_n \delta(x_0 - nd) \tag{4.2-4}$$

for N odd. Substituting Eq. (4.2-4) into Eq. (3.4-7) and making use of the sifting property of the delta function yields

$$D(f, r, f_X) = \sum_{n=-(N-1)/2}^{(N-1)/2} c_n \exp[-jk(nd)^2/(2r)] \exp(+j2\pi f_X nd), \tag{4.2-5}$$

where

$$c_n = a_n \exp(+j\theta_n). \tag{4.1-30}$$

Equation (4.2-5) is the near-field directivity function of the linear array under consideration.

4.2.1 Beam Steering and Array Focusing

Consider phase weighting the point sources in the array according to the following rule [compare with Eq. (3.4–16)]:

$$\theta_n = -2\pi f'_X nd + k(nd)^2/(2r), \qquad n = -(N-1)/2, \ldots, 0, \ldots, (N-1)/2, \tag{4.2-6}$$

where

$$f'_X = u'/\lambda = \sin\theta' \cos\psi'/\lambda \tag{3.3-97}$$

and

$$k = 2\pi f/c. \tag{3.4-13}$$

The first term in Eq. (4.2-6) represents a linear phase variation (linear in nd) and is responsible for beam steering, while the second term represents a quadratic phase variation (quadratic in nd) and is responsible for array focusing. This can be seen by substituting Eqs. (4.1-30) and (4.2-6) into Eq. (4.2-5). Doing so yields

$$D(f, r, f_X) = S(f_X - f'_X), \tag{4.2-7}$$

where

$$S(f_X) = \sum_{n=-(N-1)/2}^{(N-1)/2} a_n \exp(+j2\pi f_X nd) \tag{4.1-98}$$

is the far-field beam pattern of the array when it is only amplitude weighted. Thus, Eq. (4.2-7) indicates that if we phase weight the array according to Eq. (4.2-6), then we can steer and focus the main lobe of the far-field beam pattern $S(f_X)$, given by Eq. (4.1-98), to some location $(r, u = u')$ in the near-field of the array.

4.3 GRATING LOBES

The normalized far-field directivity function of a linearly phased array of N (odd) identical, equally spaced point sources is given by [see Eq. (4.1-101)]

$$D_N(f, f_X) = \sum_{n=-(N-1)/2}^{(N-1)/2} a_n \exp[+j2\pi(f_X - f'_X)nd]/D(f, f'_X), \tag{4.3-1}$$

where the normalization factor $D(f, f_X')$ is given by

$$D(f, f_X') = \sum_{n=-(N-1)/2}^{(N-1)/2} a_n. \tag{4.3-2}$$

The maximum value of $|D_N(f, f_X)|$ occurs at $f_X = f_X'$, that is,

$$|D_N(f, f_X')| = \left| \sum_{n=-(N-1)/2}^{(N-1)/2} a_n \right| \Big/ |D(f, f_X')| = 1. \tag{4.3-3}$$

However, the maximum value of unity can also occur even if $f_X \neq f_X'$ so long as

$$(f_X - f_X')d = \pm m, \qquad m = 1, 2, 3, \ldots \tag{4.3-4}$$

or

$$f_X = f_X' \pm (m/d), \qquad m = 1, 2, 3, \ldots. \tag{4.3-5}$$

For example, substituting Eq. (4.3-4) into Eq. (4.3-1) yields

$$D_N(f, f_X) = \sum_{n=-(N-1)/2}^{(N-1)/2} a_n \exp(\pm j2mn\pi)/D(f, f_X'), \tag{4.3-6}$$

which is equal to [see Eq. (4.3-2)]

$$\sum_{n=-(N-1)/2}^{(N-1)/2} a_n/D(f, f_X') = 1. \tag{4.3-7}$$

Therefore, if we express Eq. (4.3-5) in terms of the direction cosines u and u', that is,

$$u = u' \pm (m\lambda/d), \qquad m = 1, 2, 3, \ldots, \tag{4.3-8}$$

then we can state that *grating lobes*, which are main lobes occurring at positions other than $u = u'$, will exist at

$$u = u_g = u' \pm (m\lambda/d), \qquad m = 1, 2, 3, \ldots \tag{4.3-9}$$

if u_g is in the visible region $|u_g| \leq 1$ (see Fig. 4.3-1). Generally speaking, grating lobes are undesirable.

It can be seen from Fig. 4.3-1 that the first grating lobes ($m = 1$) appear at

$$u_g = u' \pm (\lambda/d) \tag{4.3-10}$$

if $|u_g| \leq 1$. If no beam steering is being done, then $u' = 0$ and Eq. (4.3-10) reduces to

$$u_g = \pm\lambda/d, \tag{4.3-11}$$

and if we choose $d \leq \lambda$, then $|u_g| \geq 1$. Therefore, when no beam steering is

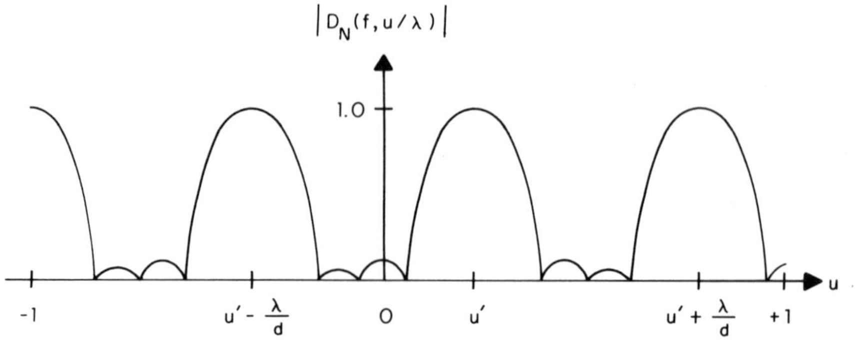

FIG. 4.3-1 Desired main lobe steered to $u = u'$ and extraneous grating lobes.

being done, grating lobes can be avoided if

$$d < \lambda. \tag{4.3-12}$$

Let us next consider the case of steering the main lobe to end-fire geometry ($u' = \pm 1$). The end-fire case represents the extreme amount of beam steering. The first grating lobe ($m = 1$) for $u' = 1$ appears at (see Fig. 4.3-1)

$$u_g = 1 - (\lambda/d), \tag{4.3-13}$$

and if $d \leq \lambda/2$, then $|u_g| \geq 1$. Similarly, the first grating lobe ($m = 1$) for $u' = -1$ appears at

$$u_g = -1 + (\lambda/d), \tag{4.3-14}$$

and if $d \leq \lambda/2$, then $|u_g| \geq 1$. Therefore, when $u' = \pm 1$ (end-fire), grating lobes can be avoided if

$$d < \lambda/2. \tag{4.3-15}$$

As a result, Eq. (4.3-15) is the criterion for the avoidance of *all* grating lobes under *all* conditions of beam steering.

4.4 ARRAY GAIN

Consider a linear array of N (odd) complex-weighted elements being used in the *passive mode*, that is, as a receive array. Let $s_i(t)$ and $n_i(t)$ represent the *random output signal* and *noise components* from element i, respectively. Note that

$$s_i(t) = s(t, x_i) = s(t, id) \tag{4.4-1}$$

and

$$n_i(t) = n(t, x_i) = n(t, id), \tag{4.4-2}$$

where d is the interelement spacing. Next, let $s_T(t)$ and $n_T(t)$ represent the *total output signal* and *noise* from the array, that is,

$$s_T(t) = \sum_{i=-(N-1)/2}^{(N-1)/2} s_i(t) \tag{4.4-3}$$

and

$$n_T(t) = \sum_{i=-(N-1)/2}^{(N-1)/2} n_i(t). \tag{4.4-4}$$

Therefore, the *average output signal-to-noise power ratio* for the array is defined as

$$\mathrm{SNR_A} \triangleq \mathrm{E}\{|s_T(t)|^2\}/\mathrm{E}\{|n_T(t)|^2\}, \tag{4.4-5}$$

where

$$\mathrm{E}\{|s_T(t)|^2\} = \mathrm{USP} + \sum_{i=-(N-1)/2}^{(N-1)/2} \sum_{\substack{j=-(N-1)/2 \\ i \neq j}}^{(N-1)/2} \mathrm{E}\{s_i(t)s_j^*(t)\} \tag{4.4-6}$$

is the *average total output signal power* and

$$\mathrm{USP} = \sum_{i=-(N-1)/2}^{(N-1)/2} \mathrm{E}\{|s_i(t)|^2\} \tag{4.4-7}$$

is the *uncorrelated output signal power*, and, similarly,

$$\mathrm{E}\{|n_T(t)|^2\} = \mathrm{UNP} + \sum_{i=-(N-1)/2}^{(N-1)/2} \sum_{\substack{j=-(N-1)/2 \\ i \neq j}}^{(N-1)/2} \mathrm{E}\{n_i(t)n_j^*(t)\} \tag{4.4-8}$$

is the *average total output noise power* and

$$\mathrm{UNP} = \sum_{i=-(N-1)/2}^{(N-1)/2} \mathrm{E}\{|n_i(t)|^2\} \tag{4.4-9}$$

is the *uncorrelated output noise power*. In Eqs. (4.4-5) through (4.4-9), $\mathrm{E}\{\cdot\}$ is the *expectation* or *ensemble average operator* and the asterisk $(*)$ denotes complex conjugate.

The *array gain* (AG) is defined as

$$\mathrm{AG} \triangleq 10\log_{10}\left\{\frac{\mathrm{SNR_A}}{\sigma_s^2/\sigma_n^2}\right\} \text{ dB} \tag{4.4-10}$$

or

$$\mathrm{AG} \triangleq 10\log_{10}\left\{\frac{\sigma_n^2[\mathrm{USP} + \sum_{i=-(N-1)/2}^{(N-1)/2}\sum_{j=-(N-1)/2,\, i\neq j}^{(N-1)/2} \mathrm{E}\{s_i(t)s_j^*(t)\}]}{\sigma_s^2[\mathrm{UNP} + \sum_{i=-(N-1)/2}^{(N-1)/2}\sum_{j=-(N-1)/2,\, i\neq j}^{(N-1)/2} \mathrm{E}\{n_i(t)n_j^*(t)\}]}\right\} \text{ dB}, \tag{4.4-11}$$

where the ratio σ_s^2/σ_n^2 is the *input* signal-to-noise power ratio at a *single element* in the array and is assumed to be the *same* at all elements. The quantities σ_s^2 and σ_n^2 are the variances of the zero-mean input signal and noise, respectively. The uncorrelated signal- and noise-power terms USP and UNP, respectively, σ_s^2, σ_n^2, and the signal and noise *correlation functions* $E\{s_i(t)s_j^*(t)\}$ and $E\{n_i(t)n_j^*(t)\}$, respectively, are dependent on the basic properties of the signal and noise sound fields in which the array is placed. Thus, in general, the same array will have different AG values when placed in different signal and noise sound fields. Also, as can be seen from Eq. (4.4-11), the AG will decrease as the *signal coherence* measured by $E\{s_i(t)s_j^*(t)\}$ decreases and as the *noise coherence* measured by $E\{n_i(t)n_j^*(t)\}$ increases.

EXAMPLE 4.4-1

Assume that both the output signal and output noise components are uncorrelated across the array, that is, assume that

$$\operatorname{cov}\{s_i(t), s_j(t)\} = E\{s_i(t)s_j^*(t)\} - E\{s_i(t)\}\,E\{s_j^*(t)\} = 0, \qquad i \neq j, \quad (4.4\text{-}12)$$

and

$$\operatorname{cov}\{n_i(t), n_j(t)\} = E\{n_i(t)n_j^*(t)\} - E\{n_i(t)\}\,E\{n_j^*(t)\} = 0, \qquad i \neq j, \quad (4.4\text{-}13)$$

where $\operatorname{cov}(X, Y)$ is the covariance between the random quantities X and Y. If we further assume that both the output signal and output noise components are zero mean, that is,

$$E\{s_i(t)\} = 0 \qquad \text{and} \qquad E\{n_i(t)\} = 0$$

for all i, then [see Eqs. (4.4-12) and (4.4-13)]

$$E\{s_i(t)s_j^*(t)\} = 0, \qquad i \neq j, \quad (4.4\text{-}14)$$

and

$$E\{n_i(t)n_j^*(t)\} = 0, \qquad i \neq j. \quad (4.4\text{-}15)$$

Substituting Eqs. (4.4-14) and (4.4-15) into Eq. (4.4-11) yields

$$\mathrm{AG} = 10\log_{10}\left\{\frac{\sigma_n^2}{\sigma_s^2}\,\frac{\mathrm{USP}}{\mathrm{UNP}}\right\} \quad \mathrm{dB}.$$

If $E\{|s_i(t)|^2\} = \sigma_s^2$ and $E\{|n_i(t)|^2\} = \sigma_n^2$ for all i, then from Eq. (4.4-7), USP $= N\sigma_s^2$, and from Eq. (4.4-9), UNP $= N\sigma_n^2$. Therefore,

$$\mathrm{AG} = 10\log_{10} 1 = 0.0 \quad \mathrm{dB}.$$

An AG value of 0.0 dB means that the array is unable to distinguish signal from noise. □

EXAMPLE 4.4-2

Assume that the output noise components are zero mean and uncorrelated across the array. Therefore, from Eq. (4.4-13),

$$\operatorname{cov}\{n_i(t), n_j(t)\} = \mathrm{E}\{n_i(t)n_j^*(t)\} = 0, \qquad i \neq j. \tag{4.4-16}$$

In addition, assume that the output signal components are zero mean and identical across the array, that is,

$$s_i(t) = s_j(t) = s(t), \qquad i \neq j. \tag{4.4-17}$$

As a result,

$$\mathrm{E}\{s_i(t)s_j^*(t)\} = \mathrm{E}\{|s(t)|^2\} = \sigma_\mathrm{s}^2, \tag{4.4-18}$$

and, from Eq. (4.4-7),

$$\mathrm{USP} = \mathrm{N}\sigma_\mathrm{s}^2, \tag{4.4-19}$$

where σ_s^2 is the variance of $s(t)$. Substituting Eqs. (4.4-16), (4.4-18), and (4.4-19) into Eq. (4.4-11) yields

$$\mathrm{AG} = 10\log_{10}\left\{\frac{\sigma_\mathrm{n}^2}{\mathrm{UNP}}\left[N + \sum_{i=-(N-1)/2}^{(N-1)/2}\ \sum_{\substack{j=-(N-1)/2\\ i\neq j}}^{(N-1)/2} 1\right]\right\} \quad \mathrm{dB}. \tag{4.4-20}$$

Since

$$\sum_{i=-(N-1)/2}^{(N-1)/2}\ \sum_{\substack{j=-(N-1)/2\\ i\neq j}}^{(N-1)/2} 1 = \sum_{i=1}^{N}\ \sum_{\substack{j=1\\ i\neq j}}^{N} 1 = N(N-1), \tag{4.4-21}$$

substituting Eq. (4.4-21) into Eq. (4.4-20) yields

$$\mathrm{AG} = 10\log_{10}\left\{\frac{N^2\sigma_\mathrm{n}^2}{\mathrm{UNP}}\right\} \quad \mathrm{dB}.$$

If $\mathrm{E}\{|n_i(t)|^2\} = \sigma_\mathrm{n}^2$ for all i, then from Eq. (4.4-9), $\mathrm{UNP} = N\sigma_\mathrm{n}^2$. Therefore,

$$\mathrm{AG} = 10\log_{10} N \quad \mathrm{dB}, \tag{4.4-22}$$

where N is the total number of elements in the array.

The assumption that the output signal components are identical across the array [see Eq. (4.4-17)] implies that a rectangular amplitude window is applied to the array and that beam steering is done (via phase weighting or time delays) in order to *cophase* all output signals. As a result, each output signal will have the same amplitude and phase if the input sound field to the array is a plane wave. If the input sound field exhibits *wavefront curvature*, then beam steering and focusing must be done in order to cophase all of the output signals. □

EXAMPLE 4.4-3

Assume that both the output signal and output noise components are zero mean and identical across the array, that is,

$$s_i(t) = s_j(t) = s(t), \qquad i \neq j, \tag{4.4-17}$$

and

$$n_i(t) = n_j(t) = n(t), \qquad i \neq j. \tag{4.4-23}$$

As a result,

$$E\{s_i(t)s_j^*(t)\} = E\{|s(t)|^2\} = \sigma_{\mathrm{s}}^2, \tag{4.4-18}$$

$$\mathrm{USP} = N\sigma_{\mathrm{s}}^2, \tag{4.4-19}$$

$$E\{n_i(t)n_j^*(t)\} = E\{|n(t)|^2\} = \sigma_{\mathrm{n}}^2, \tag{4.4-24}$$

and

$$\mathrm{UNP} = N\sigma_{\mathrm{n}}^2, \tag{4.4-25}$$

where σ_{s}^2 and σ_{n}^2 are the variances of $s(t)$ and $n(t)$, respectively. Substituting Eqs. (4.4-18), (4.4-19), (4.4-21), (4.4-24), and (4.4-25) into Eq. (4.4-11) yields

$$\mathrm{AG} = 10 \log_{10} 1 = 0.0 \quad \mathrm{dB}. \ \square$$

4.5 PLANAR ARRAYS AND FAR-FIELD DIRECTIVITY FUNCTIONS

4.5.1 Product Theorem

Consider a planar array composed of an *odd* number $(M \times N)$ of *identical* elements, as shown in Fig. 4.5-1. The complex frequency response or complex aperture function $A(f, x, y)$ of this array can be expressed as

$$A(f, x, y) = \sum_{m=-(M-1)/2}^{(M-1)/2} \sum_{n=-(N-1)/2}^{(N-1)/2} c_{mn} e(f, x - md_X, y - nd_Y), \tag{4.5-1}$$

where M (odd) and N (odd) are the total number of elements in the X and Y directions, respectively; d_X and d_Y are the constant interelement spacings in the X and Y directions, respectively;

$$c_{mn} = a_{mn} \exp(+j\theta_{mn}) \tag{4.5-2}$$

is the complex weight associated with element (m, n), where a_{mn} and θ_{mn} are real

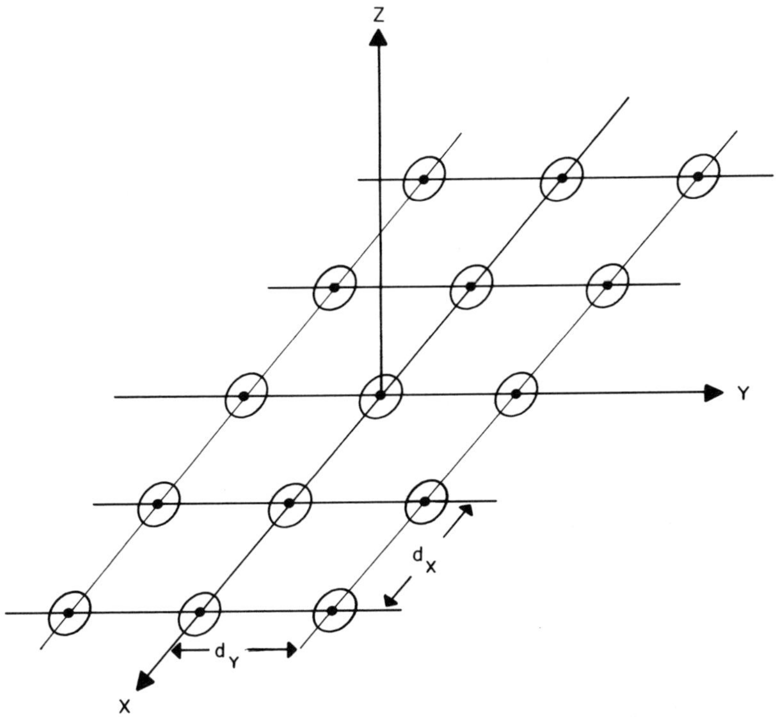

FIG. 4.5-1 A planar array composed of an odd number of identical elements lying in the XY plane.

amplitude- and phase-weighting terms, respectively; and $e(f, x, y)$ is the complex frequency response of the identical elements, also known as the *element function*. Note that, in general, $d_X \neq d_Y$ and $M \neq N$.

Since

$$e(f, x - md_X, y - nd_Y) = e(f, x, y) ** \delta(x - md_X)\,\delta(y - nd_Y), \tag{4.5-3}$$

where the double asterisk (**) denotes a two-dimensional convolution with respect to x and y, the aperture function given by Eq. (4.5-1) can be rewritten as

$$A(f, x, y) = s(x, y) ** e(f, x, y), \tag{4.5-4}$$

where

$$s(x, y) = \sum_{m=-(M-1)/2}^{(M-1)/2} \sum_{n=-(N-1)/2}^{(N-1)/2} c_{mn}\,\delta(x - md_X)\,\delta(y - nd_Y). \tag{4.5-5}$$

The function $s(x, y)$ can be thought of as the complex frequency response or complex aperture function of an equivalent planar array of $M \times N$ odd complex-weighted *point sources.* With the exception of the element function, $s(x, y)$ contains all the remaining information concerning the array such as the total number of elements, the complex weights, and the geometry of the array, that is, the positioning of the elements in the XY plane. Therefore, the far-field directivity function of a planar array whose aperture function is given by Eqs. (4.5-4) and (4.5-5) is equal to

$$D(f, f_X, f_Y) = S(f_X, f_Y)E(f, f_X, f_Y), \tag{4.5-6}$$

where

$$E(f, f_X, f_Y) = F_x F_y\{e(f, x, y)\} \tag{4.5-7}$$

and

$$S(f_X, f_Y) = F_x F_y\{s(x, y)\}, \tag{4.5-8}$$

or, from Eq. (4.5-5),

$$S(f_X, f_Y) = \sum_{m=-(M-1)/2}^{(M-1)/2} \sum_{n=-(N-1)/2}^{(N-1)/2} c_{mn} \exp[+j2\pi(f_X m d_X + f_Y n d_Y)], \tag{4.5-9}$$

where

$$f_X = u/\lambda = \sin\theta\cos\psi/\lambda \tag{4.5-10}$$

and

$$f_Y = v/\lambda = \sin\theta\sin\psi/\lambda. \tag{4.5-11}$$

Equation (4.5-6) is referred to as the *product theorem* for planar arrays. The product theorem states that the far-field directivity function of a planar array of *identical*, complex-weighted elements is equal to the product of the far-field directivity function $E(f, f_X, f_Y)$ of *one* of the elements and the far-field directivity function $S(f_X, f_Y)$ of an equivalent planar array of complex-weighted *point sources.* While the product theorem requires identical elements, equal spacing of the elements is *not* required, as was discussed in the case of linear arrays (see Section 4.1.1 and Problem 4-1). Note that $S(f_X, f_Y)$ as given by Eq. (4.5-9) is in the form of a *two-dimensional discrete spatial Fourier transform* of the complex numbers c_{mn}.

EXAMPLE 4.5-1

Consider a planar array of $M \times N$ complex-weighted point sources, where both M and N are odd. Therefore,

$$e(f, x, y) = \delta(x)\,\delta(y), \tag{4.5-12}$$

and substituting Eq. (4.5-12) into Eq. (4.5-7) yields

$$E(f, f_X, f_Y) = \int_{-\infty}^{\infty} \delta(x) \exp(+j2\pi f_X x)\, dx \int_{-\infty}^{\infty} \delta(y) \exp(+j2\pi f_Y y)\, dy, \tag{4.5-13}$$

which is equal to

$$E(f, f_X, f_Y) = 1. \tag{4.5-14}$$

Therefore, from the product theorem [see Eq. (4.5-6)], the far-field beam pattern of the array is

$$D(f, f_X, f_Y) = S(f_X, f_Y), \tag{4.5-15}$$

where $S(f_X, f_Y)$ is given by Eq. (4.5-9). □

EXAMPLE 4.5-2

Consider a planar array of $M \times N$ complex-weighted rectangular pistons, where both M and N are odd. Therefore [see Eq. (3.5-31)],

$$e(f, x, y) = \text{rect}(L_X)\,\text{rect}(L_Y), \tag{4.5-16}$$

where $L_X \neq L_Y$ in general, and [see Eqs. (3.3-14) and (3.5-32)]

$$E(f, f_X, f_Y) = \frac{\sin(\pi L_X f_X)}{\pi f_X} \frac{\sin(\pi L_Y f_Y)}{\pi f_Y} \tag{4.5-17}$$

and

$$E_N(f, f_X, f_Y) = \frac{\sin(\pi L_X f_X)}{\pi L_X f_X} \frac{\sin(\pi L_Y f_Y)}{\pi L_Y f_Y}. \tag{4.5-18}$$

Therefore, from the product theorem [see Eq. (4.5-6)], the unnormalized far-field beam pattern of the array is

$$D(f, f_X, f_Y) = S(f_X, f_Y) \frac{\sin(\pi L_X f_X)}{\pi f_X} \frac{\sin(\pi L_Y f_Y)}{\pi f_Y}, \tag{4.5-19}$$

where $S(f_X, f_Y)$ is given by Eq. (4.5-9).

Recall from our discussion in Section 4.1.1 [see Eqs. (4.1-26) through (4.1-29)] that

$$\sin(\pi L_X f_X) \approx \pi L_X f_X \tag{4.5-20}$$

and

$$\sin(\pi L_Y f_Y) \approx \pi L_Y f_Y \tag{4.5-21}$$

whenever

$$\pi L_X f_X = \pi L_X u/\lambda \leq 0.0873, \tag{4.5-22}$$

or, setting $u = 1.0$,

$$L_X/\lambda \leq 0.0278; \tag{4.5-23}$$

and

$$\pi L_Y f_Y = \pi L_Y v/\lambda \leq 0.0873, \tag{4.5-24}$$

or, setting $v = 1.0$,

$$L_Y/\lambda \leq 0.0278. \tag{4.5-25}$$

Substituting Eqs. (4.5-20) and (4.5-21) into Eq. (4.5-19) yields

$$D(f, f_X, f_Y) \approx L_X L_Y S(f_X, f_Y). \tag{4.5-26}$$

Equation (4.5-26) indicates that if the parameters L_X and L_Y are small compared to a wavelength, then the far-field beam pattern of the array $D(f, f_X, f_Y)$ is, in effect, dependent on $S(f_X, f_Y)$ alone since $E(f, f_X, f_Y)$ is omnidirectional for all practical purposes. □

EXAMPLE 4.5-3

Consider a planar array of $M \times N$ complex-weighted circular pistons, where both M and N are odd. Therefore [see Eq. (3.5-65)],

$$e(f, x, y) = \text{circ}(a) = \begin{cases} 1, & (x^2 + y^2)^{1/2} \leq a, \\ 0, & (x^2 + y^2)^{1/2} > a, \end{cases} \tag{4.5-27}$$

and [see Eqs. (3.5-68) and (3.5-70)]

$$E(f, f_X, f_Y) = a\frac{J_1[2\pi a(f_X^2 + f_Y^2)^{1/2}]}{(f_X^2 + f_Y^2)^{1/2}} \tag{4.5-28}$$

and

$$E_N(f, f_X, f_Y) = 2\frac{J_1[2\pi a(f_X^2 + f_Y^2)^{1/2}]}{2\pi a(f_X^2 + f_Y^2)^{1/2}}. \tag{4.5-29}$$

Therefore, from the product theorem [see Eq. (4.5-6)], the unnormalized far-field beam pattern of the array is

$$D(f, f_X, f_Y) = S(f_X, f_Y)\frac{aJ_1[2\pi a(f_X^2 + f_Y^2)^{1/2}]}{(f_X^2 + f_Y^2)^{1/2}}, \tag{4.5-30}$$

where $S(f_X, f_Y)$ is given by Eq. (4.5-9).

The far-field directivity function of the circular piston $E(f, f_X, f_Y)$ can also be expressed as [see Eq. (3.5-68)]

$$E(f, \theta, \psi) = a\frac{J_1[(2\pi a/\lambda)\sin\theta]}{(\sin\theta)/\lambda}. \tag{4.5-31}$$

Note that if the argument of the Bessel function in Eq. (4.5-31) is less than 1, that is, if

$$(2\pi a/\lambda)\sin\theta < 1, \tag{4.5-32}$$

or, setting $\sin\theta = 1.0$, if

$$2\pi a/\lambda < 1, \tag{4.5-33}$$

then

$$J_1[(2\pi a/\lambda)\sin\theta] \approx (\tfrac{1}{2})(2\pi a/\lambda)\sin\theta \tag{4.5-34}$$

or

$$J_1[(2\pi a/\lambda)\sin\theta] \approx (\pi a/\lambda)\sin\theta. \tag{4.5-35}$$

Substituting Eq. (4.5-35) into Eq. (4.5-31) yields

$$E(f, \theta, \psi) \approx \pi a^2, \tag{4.5-36}$$

and substituting Eq. (4.5-36) into Eq. (4.5-30) yields

$$D(f, f_X, f_Y) \approx \pi a^2 S(f_X, f_Y). \tag{4.5-37}$$

Equation (4.5-37) indicates that if the circumference of the circular piston $2\pi a$ is small compared to a wavelength, then the far-field beam pattern of the array $D(f, f_X, f_Y)$ is, in effect, dependent on $S(f_X, f_Y)$ alone since $E(f, f_X, f_Y)$ is omnidirectional for all practical purposes. □

The product theorem for planar arrays as given by Eq. (4.5-6) is a general expression. A somewhat simplified version of Eq. (4.5-6) can be obtained if the complex weights c_{mn} are *separable*, that is, if

$$c_{mn} = c_m d_n, \tag{4.5-38}$$

where

$$c_m = a_m \exp(+j\theta_m), \tag{4.5-39}$$

$$d_n = b_n \exp(+j\phi_n), \tag{4.5-40}$$

a_m and b_n are real amplitude-weighting terms, and θ_m and ϕ_n are real phase-weighting terms. Substituting Eq. (4.5-38) into Eq. (4.5-5) yields

$$s(x, y) = s_X(x)s_Y(y), \tag{4.5-41}$$

where

$$s_X(x) = \sum_{m=-(M-1)/2}^{(M-1)/2} c_m \, \delta(x - md_X) \tag{4.5-42}$$

and

$$s_Y(y) = \sum_{n=-(N-1)/2}^{(N-1)/2} d_n \, \delta(y - nd_Y). \tag{4.5-43}$$

Since $s(x, y)$ as given by Eq. (4.5-41) is a separable function in the rectangular coordinates x and y,

$$S(f_X, f_Y) = S_X(f_X) S_Y(f_Y), \tag{4.5-44}$$

where

$$S_X(f_X) = \sum_{m=-(M-1)/2}^{(M-1)/2} c_m \exp(+j2\pi f_X m d_X) \tag{4.5-45}$$

and

$$S_Y(f_Y) = \sum_{n=-(N-1)/2}^{(N-1)/2} d_n \exp(+j2\pi f_Y n d_Y). \tag{4.5-46}$$

Therefore, if the complex weights c_{mn} are separable, the product theorem for planar arrays becomes

$$D(f, f_X, f_Y) = S_X(f_X) S_Y(f_Y) E(f, f_X, f_Y). \tag{4.5-47}$$

Note that $S_X(f_X)$ and $S_Y(f_Y)$ as given by Eqs. (4.5-45) and (4.5-46), respectively, are in the form of one-dimensional DSFTs of the complex numbers c_m and d_n, respectively. Thus, $S(f_X, f_Y)$ as given by Eq. (4.5-44) is now equal to the product of two one-dimensional DSFTs instead of being equal to one two-dimensional DSFT as in Eq. (4.5-9).

EXAMPLE 4.5-4

Consider a linear array composed of an odd number M of identical, equally spaced, complex-weighted planar elements, as shown in Fig. 4.5-2. Since the elements are shown lying along the X axis, the far-field directivity function of this array is given by [see Eq. (4.5-47)]

$$D(f, f_X, f_Y) = S_X(f_X) E(f, f_X, f_Y). \tag{4.5-48}$$

□

EXAMPLE 4.5-5

Consider a linear array composed of an odd number N of identical, equally spaced, complex-weighted planar elements, as shown in Fig. 4.5-3. Since the elements are shown lying along the Y axis, the far-field directivity function of

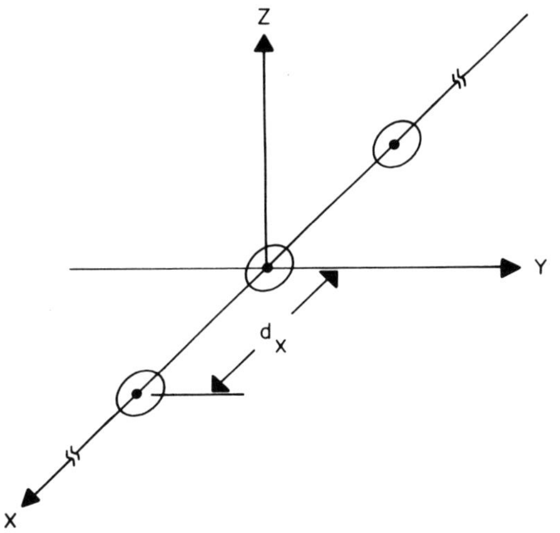

FIG. 4.5-2 Linear array of planar elements lying along the X axis.

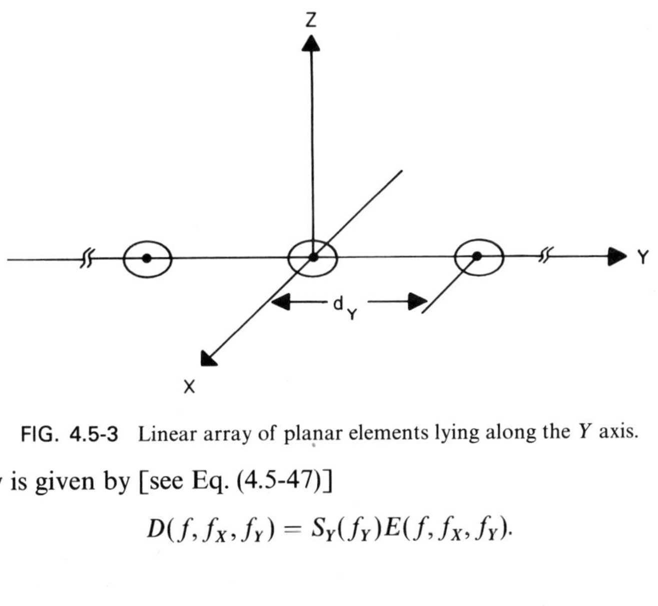

FIG. 4.5-3 Linear array of planar elements lying along the Y axis.

this array is given by [see Eq. (4.5-47)]

$$D(f, f_X, f_Y) = S_Y(f_Y)E(f, f_X, f_Y). \tag{4.5-49}$$

□

EXAMPLE 4.5-6

Consider a linear array composed of three identical, equally spaced, complex-weighted planar elements, as shown in Fig. 4.5-4. Although the far-field directivity function of this array is given by the product theorem

$$D(f, f_X, f_Y) = S(f_X, f_Y)E(f, f_X, f_Y), \tag{4.5-6}$$

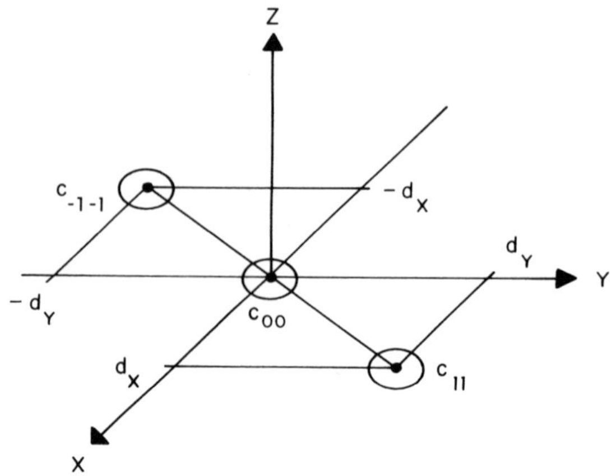

FIG. 4.5-4 Linear array of three planar elements lying in the XY plane with complex weights c_{-1-1}, c_{00}, and c_{11}.

the problem is to find $S(f_X, f_Y)$. By inspecting Fig. 4.5-4, we can write that

$$s(x, y) = c_{-1\,-1}\,\delta(x + d_X)\,\delta(y + d_Y) + c_{00}\,\delta(x)\,\delta(y) + c_{11}\,\delta(x - d_X)\,\delta(y - d_Y). \tag{4.5-50}$$

Therefore [see Eq. (3.3-31)],

$$S(f_X, f_Y) = F_x F_y\{s(x, y)\} = c_{-1\,-1}\exp(-j2\pi f_X d_X)\exp(-j2\pi f_Y d_Y) + c_{00} + c_{11}\exp(+j2\pi f_X d_X)\exp(+j2\pi f_Y d_Y) \tag{4.5-51}$$

or

$$S(f_X, f_Y) = c_{00} + c_{11}\exp[+j2\pi(f_X d_X + f_Y d_Y)] + c_{-1\,-1}\exp[-j2\pi(f_X d_X + f_Y d_Y)]. \tag{4.5-52}$$

□

4.5.2 The Phased Array (Beam Steering)

Consider a planar array of $M \times N$ (odd) identical point sources. From the product theorem, the far-field directivity function of this array is given by

$$D(f, f_X, f_Y) = S(f_X, f_Y), \tag{4.5-15}$$

where, in general, for M and N odd,

$$S(f_X, f_Y) = \sum_{m=-(M-1)/2}^{(M-1)/2} \sum_{n=-(N-1)/2}^{(N-1)/2} c_{mn}\exp[+j2\pi(f_X m d_X + f_Y n d_Y)] \tag{4.5-9}$$

and

$$c_{mn} = a_{mn} \exp(+j\theta_{mn}). \tag{4.5-2}$$

Let $S'(f_X, f_Y)$ represent the far-field beam pattern of the array when it is complex weighted and let $S(f_X, f_Y)$ represent the far-field beam pattern of the array when it is amplitude weighted, that is,

$$S'(f_X, f_Y) = \sum_{m=-(M-1)/2}^{(M-1)/2} \sum_{n=-(N-1)/2}^{(N-1)/2} a_{mn} \exp(+j\theta_{mn}) \times \exp[+j2\pi(f_X m d_X + f_Y n d_Y)] \tag{4.5-53}$$

and

$$S(f_X, f_Y) = \sum_{m=-(M-1)/2}^{(M-1)/2} \sum_{n=-(N-1)/2}^{(N-1)/2} a_{mn} \exp[+j2\pi(f_X m d_X + f_Y n d_Y)]. \tag{4.5-54}$$

If the phase weights θ_{mn} are given by [compare with Eqs. (3.5-16) through (3.5-18)]

$$\theta_{mn} = -2\pi f'_X m d_X - 2\pi f'_Y n d_Y, \qquad m = -(M-1)/2, \ldots, 0, \ldots, (M-1)/2, \quad n = -(N-1)/2, \ldots, 0, \ldots, (N-1)/2, \tag{4.5-55}$$

where

$$f'_X = u'/\lambda = \sin\theta' \cos\psi'/\lambda \tag{4.5-56}$$

and

$$f'_Y = v'/\lambda = \sin\theta' \sin\psi'/\lambda, \tag{4.5-57}$$

then substituting Eq. (4.5-55) into Eq. (4.5-53) yields

$$S'(f_X, f_Y) = \sum_{m=-(M-1)/2}^{(M-1)/2} \sum_{n=-(N-1)/2}^{(N-1)/2} a_{mn} \times \exp\{+j2\pi[(f_X - f'_X) m d_X + (f_Y - f'_Y) n d_Y]\} \tag{4.5-58}$$

or

$$S'(f_X, f_Y) = S(f_X - f'_X, f_Y - f'_Y). \tag{4.5-59}$$

Equation (4.5-59) indicates that the far-field beam pattern $S(f_X, f_Y)$ has been steered in the direction $u = u'$ and $v = v'$ in direction cosine space when the array of point sources is phase weighted according to Eq. (4.5-55). Note that

θ_{mn} as given by Eq. (4.5-55) is a *linear* phase variation (linear in md_X and nd_Y) and that θ_{mn} is an *odd* function of the indices m and n, that is,

$$\theta_{00} = 0$$

and

$$\theta_{-m\,-n} = -\theta_{mn}, \qquad m = 1, 2, \ldots, (M-1)/2,$$
$$n = 1, 2, \ldots, (N-1)/2.$$

Therefore, whether the amplitude weights a_{mn} possess even symmetry or not, that is, whether or not

$$a_{-m\,-n} = a_{mn}, \qquad m = 1, 2, \ldots, (M-1)/2,$$
$$n = 1, 2, \ldots, (N-1)/2,$$

and whether they are separable or not, in order to do beam steering the phase weights θ_{mn} *must* have *odd* symmetry.

Finally, note that a phase shift θ_{mn} is equivalent to a time delay τ_{mn}, that is,

$$\tau_{mn} = \theta_{mn}/(2\pi f) \quad \text{sec.} \tag{4.5-60}$$

Substituting Eqs. (4.5-55) through (4.5-57) into Eq. (4.5-60) yields

$$\tau_{mn} = \tau_m + \tau_n, \tag{4.5-61}$$

where

$$\tau_m = -u' m d_X/c, \qquad m = -(M-1)/2, \ldots, 0, \ldots, (M-1)/2 \tag{4.5-62}$$

and

$$\tau_n = -v' n d_Y/c, \qquad n = -(N-1)/2, \ldots, 0, \ldots, (N-1)/2. \tag{4.5-63}$$

To insure that all the time delays τ_{mn} are *positive* for practical applications, the constant

$$|u'|\frac{(M-1)}{2}\frac{d_X}{c}$$

should be added to all values of τ_m computed from Eq. (4.5-62), and the constant

$$|v'|\frac{(N-1)}{2}\frac{d_Y}{c}$$

should be added to all values of τ_n computed from Eq. (4.5-63).

4.5.3 Concentric Circular Arrays

Recall that the far-field directivity function $D(f, \theta, \psi)$ of a planar aperture function $A(f, r, \phi)$ expressed in terms of the polar coordinates (r, ϕ) is given by [see Eq. (3.5-40)]

$$D(f, \theta, \psi) = \int_0^{2\pi} \int_0^{\infty} A(f, r, \phi) \exp[+j(2\pi r/\lambda) \sin\theta \cos(\psi - \phi)] r \, dr \, d\phi. \tag{4.5-64}$$

Now consider a planar array composed of *concentric circular arrays* of identical, complex-weighted point sources, as shown in Fig. 4.5-5. The aperture function for this array can be written as

$$A(f, r, \phi) = \sum_{m=1}^{M} \sum_{n=1}^{N} c_{mn} \frac{\delta(r - r_m)}{r_m} \delta(\phi - \phi_n), \tag{4.5-65}$$

where M is the total number of concentric circular arrays, N is the total number of point sources per circular array, and

$$c_{mn} = a_{mn} \exp(+j\theta_{mn}) \tag{4.5-2}$$

is the complex weight associated with the point source with polar coordinates (r_m, ϕ_n). Note that M and N can be either even or odd. Substituting Eq. (4.5-65) into Eq. (4.5-64) yields

$$D(f, \theta, \psi) = \sum_{m=1}^{M} \sum_{n=1}^{N} c_{mn} \exp[+j(2\pi r_m/\lambda) \sin\theta \cos(\psi - \phi_n)], \tag{4.5-66}$$

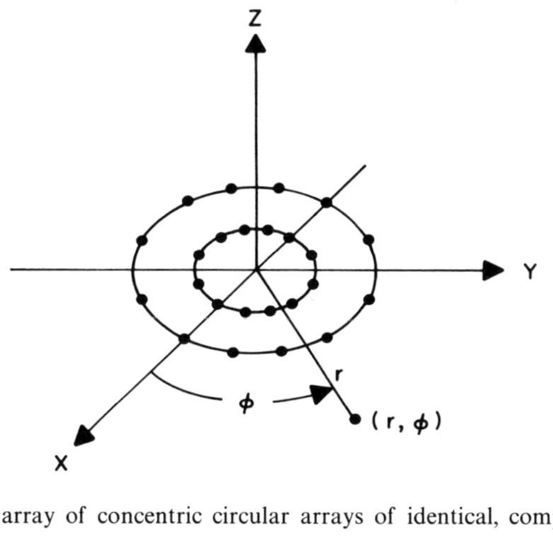

FIG. 4.5-5 Planar array of concentric circular arrays of identical, complex-weighted point sources.

which is the far-field directivity function of a planar array of *concentric circular arrays* of identical, complex-weighted point sources. If the point sources are equally spaced in both the r and ϕ directions, then

$$r_m = m\,\Delta r \tag{4.5-67}$$

and

$$\phi_n = n\,\Delta\phi, \tag{4.5-68}$$

where

$$\Delta r = a/M \tag{4.5-69}$$

is the spacing in meters in the r direction, a is the maximum radial extent of the array in meters, and

$$\Delta\phi = 360^\circ/N \tag{4.5-70}$$

is the spacing in degrees in the ϕ direction.

4.6 PLANAR ARRAYS AND NEAR-FIELD DIRECTIVITY FUNCTIONS

Recall that the near-field directivity function of a complex planar aperture function $A(f, x_0, y_0)$ is given by

$$D(f, r, f_X, f_Y) = \int_{-\infty}^{\infty}\int_{-\infty}^{\infty} A(f, x_0, y_0)\exp[-jk(x_0^2 + y_0^2)/(2r)] \times \exp[+j2\pi(f_X x_0 + f_Y y_0)]\,dx_0\,dy_0. \tag{3.6-7}$$

Now consider a planar array of $M \times N$ (odd) identical, complex-weighted point sources. Since the element function for this array is [see Eq. (4.5-12)]

$$e(f, x_0, y_0) = \delta(x_0)\,\delta(y_0), \tag{4.6-1}$$

the aperture function can be expressed as [see Eqs. (4.5-4) and (4.5-5)]

$$A(f, x_0, y_0) = s(x_0, y_0) ** e(f, x_0, y_0) \tag{4.6-2}$$

$$= s(x_0, y_0) ** \delta(x_0)\,\delta(y_0)$$

$$= s(x_0, y_0) \tag{4.6-3}$$

or

$$A(f, x_0, y_0) = \sum_{m=-(M-1)/2}^{(M-1)/2}\ \sum_{n=-(N-1)/2}^{(N-1)/2} c_{mn}\,\delta(x_0 - md_X)\,\delta(y_0 - nd_Y) \tag{4.6-4}$$

for M and N odd. Substituting Eq. (4.6-4) into Eq. (3.6-7) and making use of the sifting property of the delta function yields

$$D(f, r, f_X, f_Y) = \sum_{m=-(M-1)/2}^{(M-1)/2} \sum_{n=-(N-1)/2}^{(N-1)/2} c_{mn} \exp\{-jk[(md_X)^2 + (nd_Y)^2]/(2r)\} \times \exp[+j2\pi(f_X md_X + f_Y nd_Y)], \tag{4.6-5}$$

where

$$c_{mn} = a_{mn} \exp(+j\theta_{mn}). \tag{4.5-2}$$

Equation (4.6-5) is the near-field directivity function of the planar array under consideration.

4.6.1 Beam Steering and Array Focusing

Consider phase weighting the point sources in the array according to the following rule [compare with Eqs. (3.6-12) through (3.6-14)]:

$$\begin{aligned} \theta_{mn} &= -2\pi f'_X md_X - 2\pi f'_Y nd_Y + k[(md_X)^2 + (nd_Y)^2]/(2r), \\ m &= -(M-1)/2, \ldots, 0, \ldots, (M-1)/2, \\ n &= -(N-1)/2, \ldots, 0, \ldots, (N-1)/2, \end{aligned} \tag{4.6-6}$$

where

$$f'_X = u'/\lambda = \sin\theta' \cos\psi'/\lambda, \tag{4.6-7}$$

$$f'_Y = v'/\lambda = \sin\theta' \sin\psi'/\lambda, \tag{4.6-8}$$

and

$$k = 2\pi f/c. \tag{4.6-9}$$

The first two terms in Eq. (4.6-6) represent linear phase variations (linear in md_X and nd_Y) and are responsible for beam steering, while the third term represents a quadratic phase variation (quadratic in md_X and nd_Y) and is responsible for array focusing. This can be seen by substituting Eqs. (4.5-2) and (4.6-6) into Eq. (4.6-5). Doing so yields

$$D(f, r, f_X, f_Y) = S(f_X - f'_X, f_Y - f'_Y), \tag{4.6-10}$$

where

$$S(f_X, f_Y) = \sum_{m=-(M-1)/2}^{(M-1)/2} \sum_{n=-(N-1)/2}^{(N-1)/2} a_{mn} \exp[+j2\pi(f_X md_X + f_Y nd_Y)] \tag{4.5-9}$$

is the far-field beam pattern of the array when it is only amplitude weighted.

Thus, Eq. (4.6-10) indicates that if we phase weight the array according to Eq. (4.6-6), then we can steer and focus the main lobe of the far-field beam pattern $S(f_X, f_Y)$ given by Eq. (4.5-9) to some location $(r, u = u', v = v')$ in the near-field of the array.

4.7 VOLUME ARRAYS AND FAR-FIELD DIRECTIVITY FUNCTIONS

4.7.1 Cylindrical Arrays

Given a complex aperture function $A(f, x, y, z)$ expressed in the rectangular coordinates (x, y, z), we know that its far-field directivity function is given by the following three-dimensional spatial Fourier transform [see Eqs. (3.2-36) and (3.2-42)]:

$$D(f, f_X, f_Y, f_Z) = \int_{-\infty}^{\infty}\int_{-\infty}^{\infty}\int_{-\infty}^{\infty} A(f, x, y, z) \times \exp[+j2\pi(f_X x + f_Y y + f_Z z)]\, dx\, dy\, dz. \tag{4.7-1}$$

Equation (4.7-1) can be expressed in terms of the cylindrical coordinates (r, ϕ, z), shown in Fig. 4.7-1, by noting that

$$x = r\cos\phi, \tag{4.7-2}$$

$$y = r\sin\phi, \tag{4.7-3}$$

$$z = z, \tag{4.7-4}$$

and

$$dx\, dy\, dz \to r\, dr\, d\phi\, dz. \tag{4.7-5}$$

Therefore,

$$A(f, x, y, z) = A(f, r\cos\phi, r\sin\phi, z) \equiv A(f, r, \phi, z), \tag{4.7-6}$$

and since

$$f_X = u/\lambda = \sin\theta\cos\psi/\lambda, \tag{4.7-7}$$

$$f_Y = v/\lambda = \sin\theta\sin\psi/\lambda, \tag{4.7-8}$$

and

$$f_Z = w/\lambda = \cos\theta/\lambda, \tag{4.7-9}$$

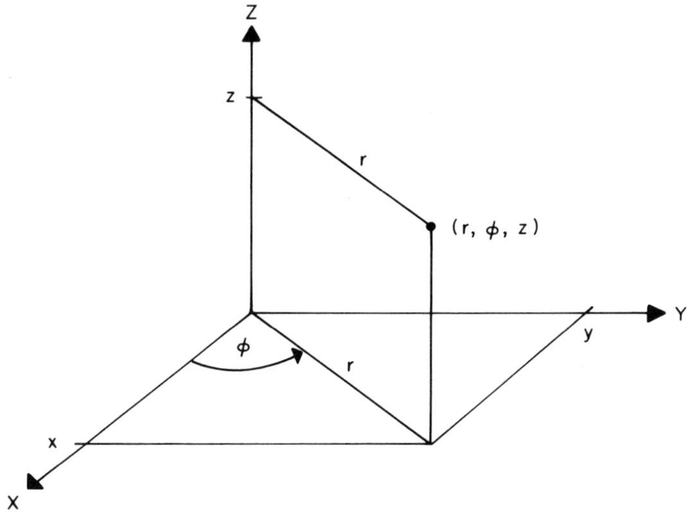

FIG. 4.7-1 The cylindrical coordinates (r, ϕ, z).

substituting Eqs. (4.7-2) through (4.7-9) into Eq. (4.7-1) yields

$$D(f, \theta, \psi) = \int_{-\infty}^{\infty} \int_{0}^{2\pi} \int_{0}^{\infty} A(f, r, \phi, z) \exp\{+j(2\pi/\lambda) \times [r \sin\theta \cos(\psi - \phi) + z \cos\theta]\} r\, dr\, d\phi\, dz. \quad (4.7\text{-}10)$$

Now consider an array of identical, complex-weighted point sources lying on the surface of a cylinder of radius a meters and length L meters, as shown in Fig. 4.7-2. The aperture function for this array can be written as

$$A(f, r, \phi, z) = \frac{\delta(r - a)}{a} \sum_{m=1}^{M} \sum_{n=-(N-1)/2}^{(N-1)/2} c_{mn}\, \delta(\phi - \phi_m)\, \delta(z - z_n), \quad (4.7\text{-}11)$$

where M (either even or odd) is the total number of point sources per circular array, N (odd) is the total number of circular arrays, and

$$c_{mn} = a_{mn} \exp(+j\theta_{mn}) \quad (4.7\text{-}12)$$

is the complex weight associated with the point source with cylindrical coordinates (a, ϕ_m, z_n). Substituting Eq. (4.7-11) into Eq. (4.7-10) yields

$$D(f, \theta, \psi) = \sum_{m=1}^{M} \sum_{n=-(N-1)/2}^{(N-1)/2} c_{mn} \exp\{+j(2\pi/\lambda)[a \sin\theta \cos(\psi - \phi_m) + z_n \cos\theta]\}, \quad (4.7\text{-}13)$$

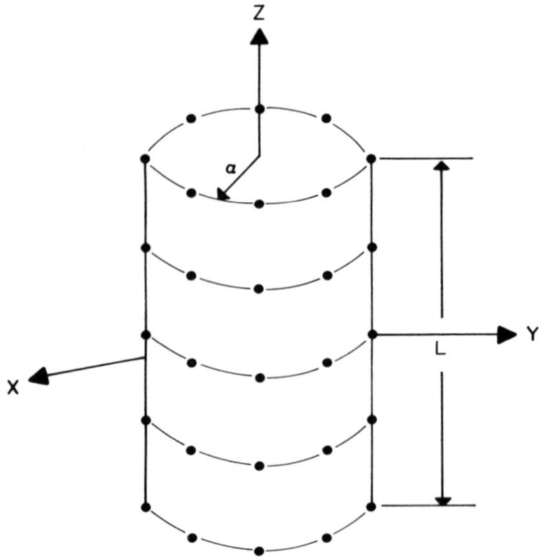

FIG. 4.7-2 Cylindrical array of identical, complex-weighted point sources.

which is the far-field directivity function of a *cylindrical array* of identical, complex-weighted point sources. If the point sources are equally spaced in both the ϕ and Z directions, then

$$\phi_m = m\,\Delta\phi \tag{4.7-14}$$

and

$$z_n = n\,\Delta z, \tag{4.7-15}$$

where

$$\Delta\phi = 360°/M \tag{4.7-16}$$

is the spacing in degrees in the ϕ direction and

$$\Delta z = L/(N-1) \tag{4.7-17}$$

is the spacing in meters in the Z direction. If only certain elements in the cylindrical array are in operation at any given time, the resulting far-field beam pattern can be obtained from Eq. (4.7-13) by simply setting $c_{mn} = 0$ for those remaining elements not in operation.

The near-field directivity function of a cylindrical array of identical, complex-weighted point sources can also be obtained by first expressing the near-field directivity function of a volume aperture given by Eq. (3.4-1) in cylindrical coordinates analogous to the far-field beam pattern expression given by Eq. (4.7-10), substituting the aperture function given by Eq. (4.7-11) into it, and taking advantage of the sifting property of delta functions.

4.7.2 Spherical Arrays

Equation (4.7-1) can also be expressed in terms of the spherical coordinates (r, γ, ϕ), shown in Fig. 4.7-3, by noting that

$$x = r \sin\gamma \cos\phi, \tag{4.7-18}$$

$$y = r \sin\gamma \sin\phi, \tag{4.7-19}$$

$$z = r \cos\gamma, \tag{4.7-20}$$

and

$$dx\, dy\, dz \to r^2 \sin\gamma\, dr\, d\gamma\, d\phi. \tag{4.7-21}$$

Therefore,

$$A(f, x, y, z) = A(f, r\sin\gamma\cos\phi, r\sin\gamma\sin\phi, r\cos\gamma) \equiv A(f, r, \gamma, \phi), \tag{4.7-22}$$

and upon substituting Eqs. (4.7-7) through (4.7-9) and Eqs. (4.7-18) through

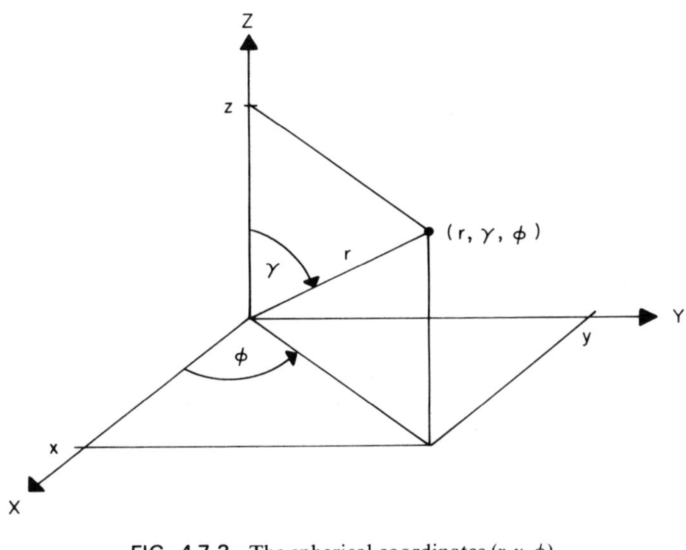

FIG. 4.7-3 The spherical coordinates (r, γ, ϕ).

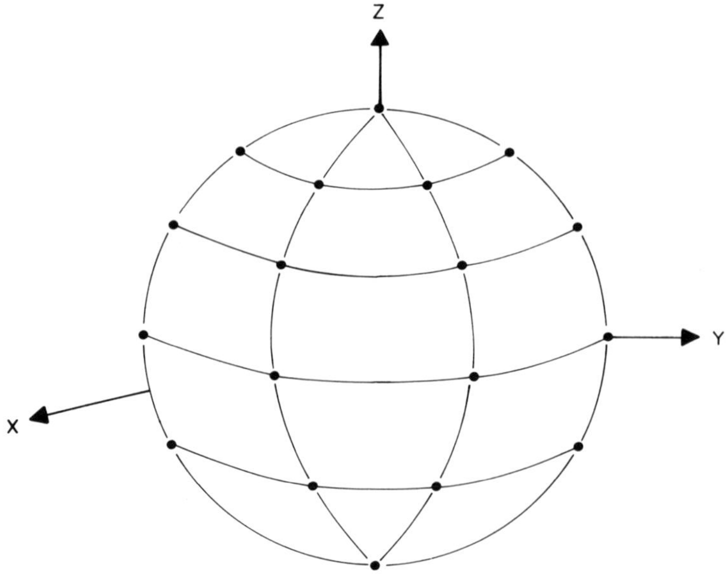

FIG. 4.7-4 Spherical array of identical, complex-weighted point sources.

(4.7-22) into Eq. (4.7-1), we obtain

$$D(f,\theta,\psi) = \int_0^{2\pi}\int_0^{\pi}\int_0^{\infty} A(f,r,\gamma,\phi)\exp\{+j(2\pi r/\lambda) \times [\sin\theta\sin\gamma\cos(\psi-\phi) + \cos\theta\cos\gamma]\} r^2 \sin\gamma \, dr \, d\gamma \, d\phi. \tag{4.7-23}$$

Now consider an array of identical, complex-weighted point sources lying on the surface of a sphere of radius a meters, as shown in Fig. 4.7-4. The aperture function for this array can be written as

$$A(f,r,\gamma,\phi) = \frac{\delta(r-a)}{a^2} \sum_{m=1}^{M}\sum_{n=1}^{N} c_{mn} \frac{\delta(\gamma-\gamma_m)}{\sin\gamma_m} \delta(\phi-\phi_n), \tag{4.7-24}$$

where M and N can be either even or odd and c_{mn}, as given by Eq. (4.7-12), is the complex weight associated with the point source with spherical coordinates (a,γ_m,ϕ_n). Note that in Fig. 4.7-4 a point source is shown to occupy both the north and south pole positions. Substituting Eq. (4.7-24) into Eq. (4.7-23) yields

$$D(f,\theta,\psi) = \sum_{m=1}^{M}\sum_{n=1}^{N} c_{mn} \exp\{+j(2\pi a/\lambda) \times [\sin\theta\sin\gamma_m\cos(\psi-\phi_n) + \cos\theta\cos\gamma_m]\}, \tag{4.7-25}$$

which is the far-field directivity function of a *spherical array* of identical, complex-weighted point sources. If the point sources are equally spaced in both the γ and ϕ directions, then

$$\gamma_m = (m - 1)\,\Delta\gamma \tag{4.7-26}$$

and

$$\phi_n = n\,\Delta\phi, \tag{4.7-27}$$

where

$$\Delta\gamma = 180°/(M - 1) \tag{4.7-28}$$

is the spacing in degrees in the γ direction and

$$\Delta\phi = 360°/N \tag{4.7-29}$$

is the spacing in degrees in the ϕ direction. The factor $m - 1$ is used instead of m in Eq. (4.7-26) to account for the point source at the north pole, where $\gamma = 0°$. If only certain elements in the spherical array are in operation at any given time, the resulting far-field beam pattern can be obtained from Eq. (4.7-25) by simply setting $c_{mn} = 0$ for those remaining elements not in operation.

The near-field directivity function of a spherical array of identical, complex-weighted point sources can also be obtained by first expressing the near-field directivity function of a volume aperture given by Eq. (3.4-1) in spherical coordinates analogous to the far-field beam pattern expression given by Eq. (4.7-23), substituting the aperture function given by Eq. (4.7-24) into it, and taking advantage of the sifting property of delta functions.

PROBLEMS

4-1 Suppose we are given a linear array of *identical*, complex-weighted elements lying along the X axis. The elements are *not* equally spaced. Show that the far-field directivity function of this array is given by the product theorem [see Eq. (4.1-14)]

$$D(f, f_X) = S(f_X)E(f, f_X),$$

where (a) for N even,

$$S(f_X) = \sum_{n=1}^{N/2} \{c_{-n}\exp(-j2\pi f_X d_{-n}) + c_n \exp(+j2\pi f_X d_n)\};$$

and (b) for N odd,

$$S(f_X) = c_0 + \sum_{n=1}^{(N-1)/2} \{c_{-n}\exp(-j2\pi f_X d_{-n}) + c_n \exp(+j2\pi f_X d_n)\}.$$

Note that, in general, $d_{-n} \neq d_n$. However, if the elements are equally spaced, then $S(f_X)$ as given in parts (a) and (b) will reduce to Eqs. (4.1-17) and (4.1-19), respectively, since for equal spacing and N even,

$$d_{-n} = d_n = [(2n - 1)/2]d;$$

and for N odd,

$$d_{-n} = d_n = nd.$$

Hint: For part (a), start with Eq. (4.1-7); for part (b), start with Eq. (4.1-13).

4-2 Compute the *normalized* far-field directivity functions $S_N(f_X)$ of (a) Eq. (4.1-50) and (b) Eq. (4.1-51). Letting $N = 7$ and $d = \lambda/2$, compute the 3-dB beamwidths Δu in direction cosine space of (c) the normalized beam pattern in part (a) and (d) the normalized beam pattern in part (b).

Hint: Use the procedure outlined in Example 3.3-8.

Compute the 3-dB beamwidths $\Delta\theta$ of the vertical beam patterns (e) using your result from part (c) and (f) using your result from part (d).

4-3 Letting $N = 11$ and $d = \lambda/2$, repeat Problem 4-2, parts (c), (d), (e), and (f).

4-4 Use the equations developed in Example 4.1-4 to compute the amplitude weights such that the ratio of the main-lobe level to the sidelobe level of $S(f_X)$ is (a) 20 dB and (b) 40 dB. (c) Using $d = \lambda/2$, what are the beamwidths $\Delta\theta$ of the vertical beam patterns in parts (a) and (b)?

4-5 Suppose we are given a linear array of $N = 6$ identical, equally spaced elements lying along the X axis and it is desired to amplitude weight this array.

(a) Letting the amplitude weights a_n be an even function of the index n, use the Dolph–Chebyshev method to compute the a_n's such that the ratio of the main-lobe level to the sidelobe level of $S(f_X)$ is 30 dB.

(b) Using $d = \lambda/2$, what is the beamwidth $\Delta\theta$ of the vertical beam pattern of $S(f_X)$?

4-6 Consider a linear array of N (even) identical, equally spaced, complex-weighted point sources lying along the X axis. Assume that the amplitude weights a_n do *not* possess symmetry. Show that in order to do beam steering the phase weights θ_n must be given by

$$\theta_n = -2\pi f'_X[(2n - 1)/2]\,d, \qquad n = 1, 2, \ldots, N/2$$

and

$$\theta_{-n} = -\theta_n = 2\pi f'_X[(2n - 1)/2]d, \qquad n = 1, 2, \ldots, N/2.$$

Therefore, whether the amplitude weights a_n possess symmetry or not, the phase weights *must* have *odd* symmetry in order to do beam steering.

4-7 *The Two-Element Interferometer.* Consider a linear array of $N = 2$ complex-weighted point sources, as shown in Fig. P4-7.

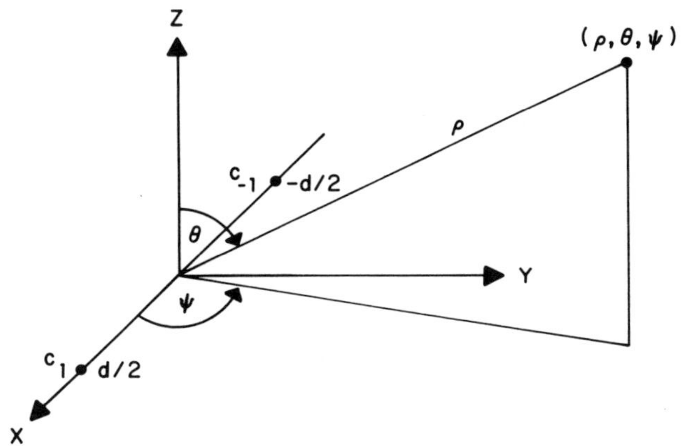

FIG. P4-7

(a) Assume that only amplitude weighting is done, that is, $c_1 = a_1$ and $c_{-1} = a_{-1}$, where $a_{-1} = a_1$. With this even amplitude weighting, the linear array shown in Fig. P4-7 is known as a two-element interferometer. Show that the *unnormalized* far-field beam pattern of this array is given by

$$D(f, \theta, \psi) = 2a_1 \cos[(\pi d/\lambda) \sin\theta \cos\psi].$$

(b) Now we wish to steer this beam pattern to $\theta = \theta'$ and $\psi = \psi'$. What are the equations for the phase weights θ_1 and θ_{-1} that will accomplish this task?

(c) What are the equivalent time delays τ_1 and τ_{-1}?

4-8 *The Dipole.* Consider a linear array of $N = 2$ complex-weighted point sources, as shown in Fig. P4-7.

(a) Assume that only amplitude weighting is done, that is, $c_1 = a_1$ and $c_{-1} = a_{-1}$, where $a_{-1} = -a_1$. With this odd amplitude weighting, the linear array shown in Fig. P4-7 is known as a dipole. Show that the *unnormalized* far-field beam pattern of this array is given by

$$D(f, \theta, \psi) = j2a_1 \sin[(\pi d/\lambda) \sin\theta \cos\psi].$$

(b) Now we wish to steer this beam pattern to $\theta = \theta'$ and $\psi = \psi'$. What are the equations for the phase weights θ_1 and θ_{-1} that will accomplish this task?

(c) What are the equivalent time delays τ_1 and τ_{-1}?

4-9 *The Axial Quadrupole.* Consider a linear array of $N = 3$ complex-weighted point sources, as shown in Fig. P4-9.

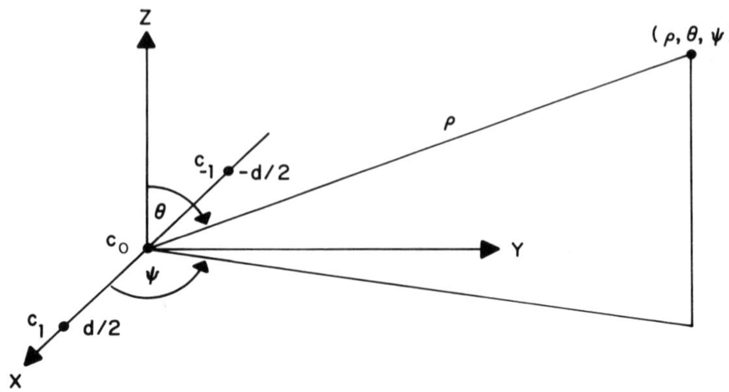

FIG. P4-9

(a) Assume that only amplitude weighting is done, that is,

$$c_1 = a_1 = 1,$$

$$c_0 = a_0 = -2,$$

$$c_{-1} = a_{-1} = 1.$$

With this even amplitude weighting, the linear array shown in Fig. P4-9 is known as an axial quadrupole. Show that the *unnormalized* far-field directivity function of this array is given by

$$D(f, \theta, \psi) = 2\{\cos[(\pi d/\lambda)\sin\theta\cos\psi] - 1\}.$$

(b) Now we wish to steer this beam pattern to $\theta = \theta'$ and $\psi = \psi'$. What are the equations for the phase weights θ_1, θ_0, and θ_{-1} that will accomplish this task?

(c) What are the equivalent time delays τ_1, τ_0, and τ_{-1}?

4-10 Verify Eq. (4.1-115).

4-11 Verify Eq. (4.1-124).

4-12 Equation (4.1-126) can be rewritten as

$$S_N(mf_{X_S}) = \sum_{n=0}^{M-1} a_n \cos\left(\frac{2\pi(m - m')n}{M + Z}\right) \Bigg/ \sum_{n=0}^{M-1} a_n$$

by making the following substitutions. For a linearly phased array [see

Eq. (4.1-99)],

$$\theta_n = -2\pi f_X' nd,$$

and if we let

$$f_X' = m' f_{X_S},$$

where [see Eq. (4.1-123)]

$$f_{X_S} = 1/[(M + Z)d],$$

then

$$\theta_n = -2\pi m' n/(M + Z),$$

where m' is the bin number where the main lobe has been steered. As a result,

$$c_n = a_n \exp[-j2\pi m' n/(M + Z)],$$

and upon substituting this expression into the numerator of Eq. (4.1-126) and taking the real part, we obtain the desired result. Compute and plot $S_N(mf_{X_S})$ as given in this problem for *both* the rectangular and triangular amplitude windows for the following combination of parameters:

(a) $N = 15, Z = 0, m' = 0, m = -7, \ldots, 0, \ldots, 7$
(b) $N = 15, Z = 24, m' = 0, m = -31, \ldots, 0, \ldots, 31$
(c) $N = 15, Z = 24, m' = 15, m = -31, \ldots, 0, \ldots, 31$
(d) Check your results in parts (a) and (b) by evaluating the closed-form expressions for $S_N(f_X)$ obtained in Problems 4-2(a) and 4-2(b) at

$$f_X = mf_{X_S} = \frac{m}{(M + Z)d}$$

for the rectangular and triangular windows, respectively.

Hint: For parts (a) through (c), replace a_0 with $a_0/2$; for parts (a) through (d), $M = [(N - 1)/2] + 1$.

4-13 Repeat Problem 4-12, parts (a), (b), and (c), for the cosine, Hanning, Hamming, and Blackman amplitude windows.

4-14 Compute and plot the magnitude (in decibels) of $S_N(mf_{X_S})$ as given in Problem 4-12 using the amplitude weights obtained from the Dolph–Chebyshev method in Example 4.1-4 and Problems 4-4(a) and 4-4(b). Choose your own combination of values for Z, m', and the range of values for the index m. Compare the actual sidelobe levels with the theoretical design levels.

4-15 The spherical coordinates of a target's location with respect to a linear array of N (odd) identical, equally spaced, complex-weighted point sources lying along the X axis has been determined to be $r_T = 63$ m, $\theta_T = 19°$, and $\psi_T = 51°$.

(a) Is the target in the array's near-field or far-field? Use $L = 3$ m, $c = 1500$ m/sec, and $f = 15$ kHz.

(b) What is the equation for the phase weights θ_n that must be applied across the array if the main lobe of the far-field beam pattern $S(f_X)$ is to exist at coordinates (r_T, θ_T, ψ_T)? Assume that the interelement spacing $d = \lambda/2$. All that is required is *one* equation for θ_n as a function of the index n.

(c) The target changes its coordinates to $r_T = 120$ m, $\theta_T = 77°$, and $\psi_T = 118°$. Repeat parts (a) and (b).

(d) With $d = \lambda/2$, how many elements are in the array?

(e) Repeat parts (a) through (c) for a linear array lying along the Y axis.

4-16 Show that for a linear array of N (even) identical, equally spaced point sources, the criterion for the avoidance of *all* grating lobes under *all* conditions of beam steering is $d < \lambda/2$. Note that this is the same criterion given by Eq. (4.3-15), which was derived for N odd.

4-17 Consider a linear array of $N = 5$ point sources, as shown in Fig. P4-17.

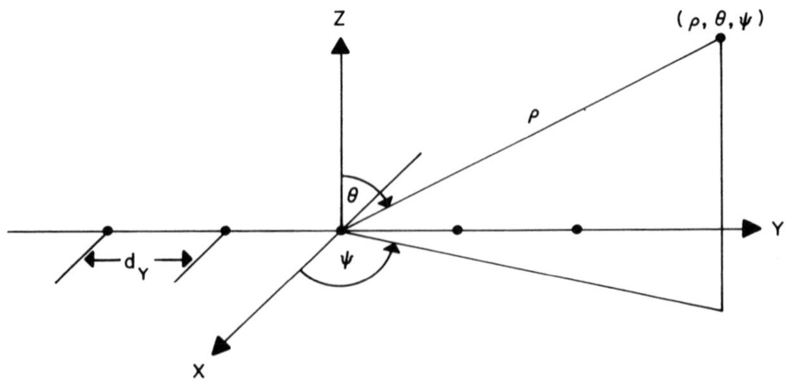

FIG. P4-17

(a) Assume that this array is amplitude weighted by a rectangular window. What is the *normalized* far-field directivity function of this array as a function of θ and ψ?

(b) Assume that this array is amplitude weighted by a triangular window. What is the *normalized* far-field directivity function of this array as a function of θ and ψ?

(c) Now we wish to steer the main lobe of the far-field beam pattern to $\theta = \theta'$ and $\psi = \psi'$. What are the equations of the phase weights θ_n that will accomplish this task? Five equations are required.

(d) What are the equations of the equivalent time delays τ_n?

(e) Assume that this array must be designed to operate at any one of the following three different frequencies: $f = 10$ kHz, 15 kHz, and 25 kHz. What criterion should d_Y satisfy in order to avoid grating lobes (under all conditions of beam steering) regardless of which frequency is used? Use $c = 1500$ m/sec.

4-18 If the complex weights $c_{mn} = a_{mn}\exp(+j\theta_{mn})$ appearing in Eq. (4.5-9) obey the relationship

$$c_{-m-n} = c^*_{mn},$$

what does Eq. (4.5-9) reduce to? Assume that the complex weights are *not* separable.

4-19 Consider a 3 × 5 planar array of amplitude-weighted circular pistons, as shown in Fig. P4-19. Assume that the circular pistons are identical with radius a meters. The amplitude weights are separable with rectangular weighting in the X direction and triangular weighting in the Y direction. What is the *normalized* far-field beam pattern of this array?

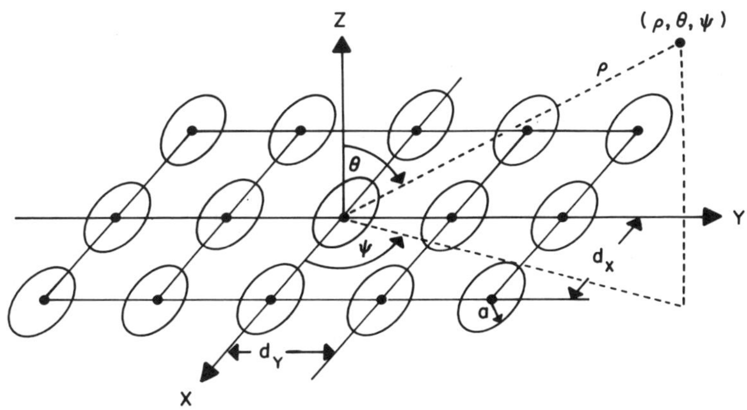

FIG. P4-19

4-20 Consider a linear array of five amplitude-weighted rectangular pistons, as shown in Fig. P4-20. Assume that the rectangular pistons are

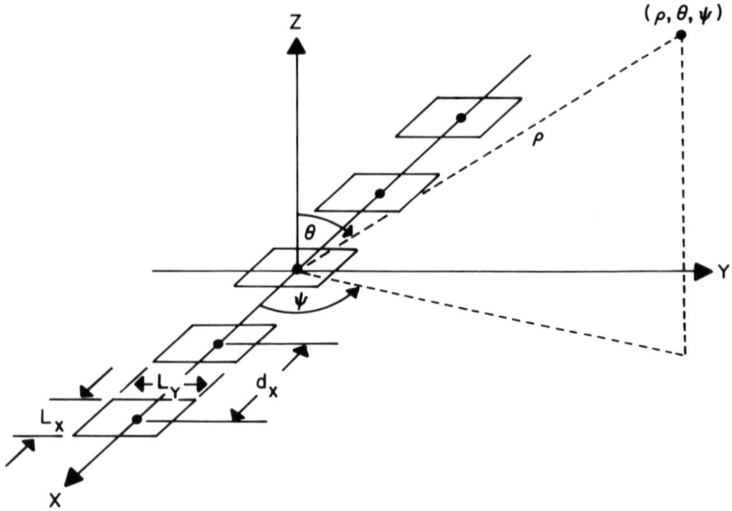

FIG. P4-20

identical with sides L_X and L_Y meters long. What is the *normalized* far-field directivity function of this array if it is amplitude weighted by a triangular window?

4-21 Referring to Example 4.5-6, obtain simplified expressions for $S(f_X, f_Y)$ as given by Eq. (4.5-52) for the following values of the coefficients c_{00}, c_{11}, and $c_{-1\,-1}$:

(a) $c_{00} = 1.0$, $c_{11} = 1.0$, and $c_{-1\,-1} = 1.0$;
(b) $c_{00} = -2.0$, $c_{11} = 1.0$, and $c_{-1\,-1} = 1.0$.

4-22 *The Tesseral Quadrupole.* Consider a planar array of four amplitude-weighted point sources, as shown in Fig. P4-22. With the amplitude weighting as shown, this array is known as a tesseral quadrupole. What is the *unnormalized* far-field directivity function of this array as a function of θ and ψ?

4-23 The spherical coordinates of a target's location with respect to a planar array of $M \times N$ (odd) identical, complex-weighted point sources lying in the XY plane has been determined to be $r_T = 50$ m, $\theta_T = 21°$, and $\psi_T = 49°$.

(a) Is the target in the array's near-field or far-field? Use $L_X = 1.95$ m, $L_Y = 3.0$ m, $c = 1500$ m/sec, and $f = 10$ kHz.

(b) What is the equation for the phase weights θ_{mn} that must be applied across the array if the main lobe of the far-field beam pattern

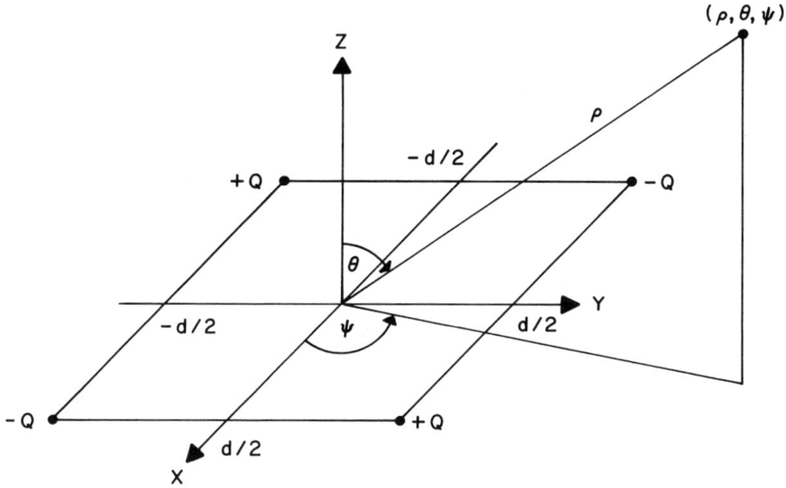

FIG. P4-22

$S(f_X, f_Y)$ is to exist at coordinates (r_T, θ_T, ψ_T)? Assume that the interelement spacings $d_X = \lambda/2$ and $d_Y = \lambda/2$. All that is required is *one* equation for θ_{mn} as a function of the indices m and n.

(c) The target changes its coordinates to $r_T = 80$ m, $\theta_T = 79°$, and $\psi_T = 116°$. Repeat parts (a) and (b).

(d) With $d_X = \lambda/2$ and $d_Y = \lambda/2$, how many elements are in the array in the X direction (M) and Y direction (N)?

BIBLIOGRAPHY

V. M. Albers, *Underwater Acoustics Handbook—II*. The Pennsylvania State University Press, University Park, Pennsylvania, 1965.

E. O. Brigham, *The Fast Fourier Transform*. Prentice-Hall, Englewood Cliffs, New Jersey, 1974.

J. A. Cadzow and H. R. Martens, *Discrete-Time and Computer Control Systems*. Prentice-Hall, Englewood Cliffs, New Jersey, 1970.

J. D. Gaskill, *Linear Systems, Fourier Transforms, and Optics*. Wiley, New York, 1978.

B. Gold and C. M. Rader, *Digital Processing of Signals*. McGraw-Hill, New York, 1969.

F. J. Harris, On the use of windows for harmonic analysis with the discrete Fourier transform. *Proc. IEEE* **66**, 51–83 (1978).

M. Javid and E. Brenner, *Analysis, Transmission, and Filtering of Signals*. Robert E. Krieger Publishing Co., Malabar, Florida, 1981.

L. E. Kinsler, A. R. Frey, A. B. Coppens, and J. V. Sanders, *Fundamentals of Acoustics*, 3rd edition. Wiley, New York, 1982.

M. V. Klein, *Optics*. Wiley, New York, 1970.

W. C. Knight, R. G. Pridham, and S. M. Kay, Digital signal processing for sonar. *Proc. IEEE* **69**, 1451–1506 (1981).

K. Ogata, *Modern Control Engineering.* Prentice-Hall, Englewood Cliffs, New Jersey, 1970.

A. V. Oppenheim and R. W. Schafer, *Digital Signal Processing.* Prentice-Hall, Englewood Cliffs, New Jersey, 1975.

A. Papoulis, *Systems and Transforms with Applications in Optics.* McGraw-Hill, New York, 1968.

A. Papoulis, *Signal Analysis.* McGraw-Hill, New York, 1977.

M. Schwartz and L. Shaw, *Signal Processing: Discrete Spectral Analysis, Detection, and Estimation.* McGraw-Hill, New York, 1975.

H. Stark and F. B. Tuteur, *Modern Electrical Communications.* Prentice-Hall, Englewood Cliffs, New Jersey, 1979.

B. D. Steinberg, *Principles of Aperture and Array System Design.* Wiley, New York, 1976.

S. Temkin, *Elements of Acoustics.* Wiley, New York, 1981.

R. J. Urick, *Principles of Underwater Sound*, 2nd edition. McGraw-Hill, New York, 1975.

5

Signal Processing

5.1 FFT BEAMFORMING FOR PLANAR ARRAYS

Consider the following physical situation. Assume we are given a planar array of $M \times N$ (odd) identical, complex-weighted point sources lying in the XY plane. The array is being used in the *passive mode*, "listening" for sound fields radiated from potential targets of interest. Based on the output electrical signals from the individual elements in the array, it is desired to estimate both the target's direction and the spectral lines (frequency content) of the radiated sound fields. One way to solve this problem is to use the method of FFT beamforming.

Before we begin our discussion of FFT beamforming, let us obtain an expression for the theoretical output frequency and angular spectrum $Y(\eta, \gamma)$ from the planar array. Recall that $Y(\eta, \gamma)$ is given by the multidimensional convolution integral

$$Y(\eta, \gamma) = \int_{-\infty}^{\infty} Y_{\mathrm{M}}(\eta, \boldsymbol{\beta}) D_{\mathrm{R}}(\eta, \gamma - \boldsymbol{\beta})\, d\boldsymbol{\beta}, \tag{3.2-51}$$

where $Y_{\mathrm{M}}(\eta, \boldsymbol{\beta})$ is the frequency and angular spectrum of the output acoustic field from the medium, which is also the acoustic field incident on the receive array, and $D_{\mathrm{R}}(\eta, \boldsymbol{\beta})$ is the far-field directivity function of the receive aperture.

Since it was assumed that the receive aperture is a planar array of $M \times N$ point sources, from the product theorem [see Eq. (4.5-15)], the far-field receive directivity function is given by

$$D_{\mathrm{R}}(\eta, \boldsymbol{\beta}) = D_{\mathrm{R}}(\eta, \beta_X, \beta_Y, \beta_Z) = S'(\beta_X, \beta_Y) \tag{5.1-1}$$

since the array is lying in the XY plane and, in general, for M and N odd [see

Eq. (4.5-9)]

$$S'(\beta_X, \beta_Y) = \sum_{m=-(M-1)/2}^{(M-1)/2} \sum_{n=-(N-1)/2}^{(N-1)/2} c_{mn} \exp[+j2\pi(\beta_X m d_X + \beta_Y n d_Y)] \quad (5.1\text{-}2)$$

and

$$c_{mn} = a_{mn} \exp(+j\theta_{mn}). \quad (4.5\text{-}2)$$

Therefore,

$$D_{\mathrm{R}}(\eta, \boldsymbol{\gamma} - \boldsymbol{\beta}) = D_{\mathrm{R}}(\eta, \gamma_X - \beta_X, \gamma_Y - \beta_Y, \gamma_Z - \beta_Z), \quad (5.1\text{-}3)$$

or, from Eq. (5.1-1),

$$D_{\mathrm{R}}(\eta, \boldsymbol{\gamma} - \boldsymbol{\beta}) = S'(\gamma_X - \beta_X, \gamma_Y - \beta_Y), \quad (5.1\text{-}4)$$

which is one of the integrand terms appearing in Eq. (3.2-51).

Next, consider the remaining integrand term

$$Y_{\mathrm{M}}(\eta, \boldsymbol{\beta}) = F_t F_{\mathbf{r}}\{y_{\mathrm{M}}(t, \mathbf{r})\} \quad (5.1\text{-}5)$$

appearing in Eq. (3.2-51). Let us assume that the acoustic field $y_{\mathrm{M}}(t, \mathbf{r})$ incident on the array is a *general plane-wave field* propagating in the $\pm\hat{n}_0$ direction, that is,

$$y_{\mathrm{M}}(t, \mathbf{r}) = g\left(t \mp \frac{\mathbf{r} \cdot \hat{n}_0}{c}\right), \quad (5.1\text{-}6)$$

where $g(t)$ is an arbitrary function of time,

$$\mathbf{r} = x\hat{x} + y\hat{y} + z\hat{z}, \quad (5.1\text{-}7)$$

$$\hat{n}_0 = u_0\hat{x} + v_0\hat{y} + w_0\hat{z}, \quad (5.1\text{-}8)$$

$$u_0 = \sin\theta_0 \cos\psi_0, \quad (5.1\text{-}9)$$

$$v_0 = \sin\theta_0 \sin\psi_0, \quad (5.1\text{-}10)$$

$$w_0 = \cos\theta_0, \quad (5.1\text{-}11)$$

and c is the speed of sound (assumed constant) in meters per second (see Fig. 5.1-1). Note that $g(t \mp \mathbf{r} \cdot \hat{n}_0/c)$ is a solution of the three-dimensional homogeneous wave equation in rectangular coordinates, that is,

$$\nabla^2 g\left(t \mp \frac{\mathbf{r} \cdot \hat{n}_0}{c}\right) - \frac{1}{c^2}\frac{\partial^2}{\partial t^2} g\left(t \mp \frac{\mathbf{r} \cdot \hat{n}_0}{c}\right) = 0. \quad (5.1\text{-}12)$$

We shall consider both plus and minus $\hat{n}_0$ together since, depending on one's point of view, we either have a general plane-wave field traveling in the $-\hat{n}_0$ direction away from the target, or the target is located in the $+\hat{n}_0$ direction as measured from the center of the array.

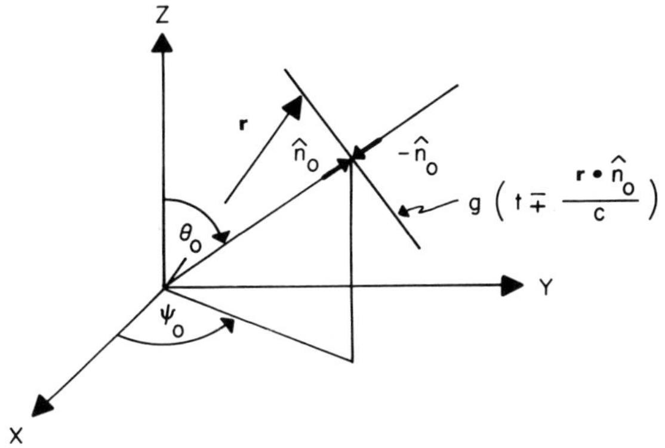

FIG. 5.1-1 General plane-wave field $g(t \mp \mathbf{r} \cdot \hat{n}_0/c)$ propagating in the $\pm\hat{n}_0$ direction.

Therefore, since

$$Y_M(\eta, \mathbf{r}) = F_t\{y_M(t, \mathbf{r})\}, \tag{5.1-13}$$

substituting Eq. (5.1-6) into Eq. (5.1-13) yields

$$Y_M(\eta, \mathbf{r}) = F_t\left\{g\left(t \mp \frac{\mathbf{r} \cdot \hat{n}_0}{c}\right)\right\} \tag{5.1-14}$$

or

$$Y_M(\eta, \mathbf{r}) = G(\eta)\exp\left[\mp j2\pi\eta\left(\frac{\mathbf{r} \cdot \hat{n}_0}{c}\right)\right], \tag{5.1-15}$$

and upon substituting Eqs. (5.1-7) and (5.1-8) into Eq. (5.1-15), we obtain

$$Y_M(\eta, \mathbf{r}) = G(\eta)\exp(\mp j2\pi f_{X_0}x)\exp(\mp j2\pi f_{Y_0}y)\exp(\mp j2\pi f_{Z_0}z), \tag{5.1-16}$$

where $G(\eta)$ is the frequency spectrum of $g(t)$, $c = \eta\lambda$,

$$f_{X_0} = u_0/\lambda, \tag{5.1-17}$$

$$f_{Y_0} = v_0/\lambda, \tag{5.1-18}$$

and

$$f_{Z_0} = w_0/\lambda. \tag{5.1-19}$$

Finally, since

$$Y_M(\eta, \boldsymbol{\beta}) = F_{\mathbf{r}}\{Y_M(\eta, \mathbf{r})\}, \tag{5.1-20}$$

taking the spatial Fourier transform with respect to x, y, and z of Eq. (5.1-16) yields

$$Y_M(\eta, \boldsymbol{\beta}) = G(\eta)\,\delta(\beta_X \mp f_{X_0})\,\delta(\beta_Y \mp f_{Y_0})\,\delta(\beta_Z \mp f_{Z_0}), \tag{5.1-21}$$

where use was made of Eq. (3.3-42). Therefore, substituting Eqs. (5.1-4) and (5.1-21) into Eq. (3.2-51) yields

$$Y(\eta, \gamma_X, \gamma_Y) = \int_{-\infty}^{\infty}\int_{-\infty}^{\infty}\int_{-\infty}^{\infty} G(\eta)S'(\gamma_X - \beta_X, \gamma_Y - \beta_Y)\,\delta(\beta_X \mp f_{X_0}) \times \delta(\beta_Y \mp f_{Y_0})\,\delta(\beta_Z \mp f_{Z_0})\,d\beta_X\,d\beta_Y\,d\beta_Z \quad (5.1\text{-}22)$$

or

$$Y(\eta, \gamma_X, \gamma_Y) = G(\eta)S'(\gamma_X - (\pm f_{X_0}), \gamma_Y - (\pm f_{Y_0})) \quad (5.1\text{-}23)$$

since

$$\int_{-\infty}^{\infty} \delta(\beta_Z \mp f_{Z_0})\,d\beta_Z = 1 \quad (5.1\text{-}24)$$

and where $S'(\cdot,\cdot)$ is given by Eq. (5.1-2),

$$\gamma_X = u/\lambda, \quad (5.1\text{-}25)$$

and

$$\gamma_Y = v/\lambda. \quad (5.1\text{-}26)$$

Equation (5.1-23) represents the *theoretical* output frequency and angular spectrum from a planar array of $M \times N$ (odd) identical, complex-weighted point sources when the acoustic field incident upon the array is a general plane wave. If the complex weights c_{mn} appearing in Eq. (5.1-2) are separable and if linear phase weighting is done for beam-steering purposes, then

$$c_{mn} = a_m \exp(+j\theta_m) b_n \exp(+j\phi_n), \quad (5.1\text{-}27)$$

where

$$\theta_m = -2\pi f'_X m d_X, \qquad m = -(M-1)/2, \ldots, 0, \ldots, (M-1)/2, \quad (5.1\text{-}28)$$

$$\phi_n = -2\pi f'_Y n d_Y, \qquad n = -(N-1)/2, \ldots, 0, \ldots, (N-1)/2, \quad (5.1\text{-}29)$$

$$f'_X = u'/\lambda, \quad (5.1\text{-}30)$$

and

$$f'_Y = v'/\lambda. \quad (5.1\text{-}31)$$

As a result, Eq. (5.1-23) can be rewritten as

$$Y(\eta, \gamma_X, \gamma_Y) = G(\eta)S_X(\gamma_X - [f'_X + (\pm f_{X_0})])S_Y(\gamma_Y - [f'_Y + (\pm f_{Y_0})]), \quad (5.1\text{-}32)$$

where

$$S_X(\gamma_X) = \sum_{m=-(M-1)/2}^{(M-1)/2} a_m \exp(+j2\pi\gamma_X m d_X) \quad (5.1\text{-}33)$$

and

$$S_Y(\gamma_Y) = \sum_{n=-(N-1)/2}^{(N-1)/2} b_n \exp(+j2\pi\gamma_Y n d_Y). \tag{5.1-34}$$

By inspecting Eq. (5.1-32) it can be seen that the angular-spectrum portion of $|Y(\eta, \gamma_X, \gamma_Y)|$ has its maximum value at

$$\gamma_X = f'_X \pm f_{X_0} \tag{5.1-35}$$

and

$$\gamma_Y = f'_Y \pm f_{Y_0}. \tag{5.1-36}$$

If the beam steering is done properly, that is, if

$$f'_X = \mp f_{X_0} \tag{5.1-37}$$

and

$$f'_Y = \mp f_{Y_0}, \tag{5.1-38}$$

then the maximum value of the angular spectrum occurs at $\gamma_X = 0$ and $\gamma_Y = 0$. If *no* beam steering is being done, then $f'_X = 0$ and $f'_Y = 0$, so that Eq. (5.1-32) reduces to

$$Y(\eta, \gamma_X, \gamma_Y) = G(\eta) S_X(\gamma_X - (\pm f_{X_0})) S_Y(\gamma_Y - (\pm f_{Y_0})). \tag{5.1-39}$$

In this case the angular-spectrum portion of $|Y(\eta, \gamma_X, \gamma_Y)|$ has its maximum value at

$$\gamma_X = \pm f_{X_0} \tag{5.1-40}$$

and

$$\gamma_Y = \pm f_{Y_0}. \tag{5.1-41}$$

We shall now discuss obtaining an estimate of the target's direction. An estimate of the azimuthal angle ψ_0 to the target measured in the counter-clockwise direction with respect to the positive X axis of the array can be obtained by forming the ratio

$$(\pm f_{Y_0})/(\pm f_{X_0}) = v_0/u_0 = \tan \psi_0 \tag{5.1-42}$$

or

$$\psi_0 = \tan^{-1}(f_{Y_0}/f_{X_0}). \tag{5.1-43}$$

Similarly, by noting that

$$\sin \theta_0 = \lambda (f^2_{X_0} + f^2_{Y_0})^{1/2}, \tag{5.1-44}$$

an estimate of the angle θ_0 to the target measured with respect to the positive

Z axis of the array is given by

$$\theta_0 = \sin^{-1}[c(f_{X_0}^2 + f_{Y_0}^2)^{1/2}/\eta] \tag{5.1-45}$$

since $c = \eta\lambda$. Note that θ_0 depends on the frequency η, while ψ_0 does not.

The spectral lines or frequency content of the sound field is represented by $G(\eta)$. For example, if $g(t)$ is equal to the superposition of three different time-harmonic components, that is, if

$$g(t) = \alpha_1 \exp(+j2\pi\eta_1 t) + \alpha_2 \exp(+j2\pi\eta_2 t) + \alpha_3 \exp(+j2\pi\eta_3 t),$$

then

$$G(\eta) = \alpha_1\,\delta(\eta - \eta_1) + \alpha_2\,\delta(\eta - \eta_2) + \alpha_3\,\delta(\eta - \eta_3),$$

where α_1, α_2, and α_3 are the amplitudes associated with the frequency components η_1, η_2, and η_3, respectively.

Equations (5.1-23), (5.1-32), and (5.1-39) are different forms of the *theoretical* output frequency and angular spectrum $Y(\eta, \gamma_X, \gamma_Y)$ computed from the convolution integral given by Eq. (3.2-51). Now, suppose that we know nothing about far-field directivity functions or the convolution integral given by Eq. (3.2-51) and that we are faced with the problem of how to process the output data from a planar array of $M \times N$ (odd) identical point sources in order to estimate both the direction and frequency content of the sound fields incident upon the array, radiated from potential targets of interest. As mentioned previously, one way of solving this problem is to use the method of FFT beamforming.

We begin the discussion of this method by noting that the output electrical signal $y(t, \mathbf{r})$ from element (m, n) of the array can be expressed as (see Fig. 5.1-2)

$$y(t, md_X, nd_Y) \tag{5.1-46}$$

since element (m, n) is located at $(x = md_X, y = nd_Y)$, where $m = -(M-1)/2, \ldots, 0, \ldots, (M-1)/2$ and $n = -(N-1)/2, \ldots, 0, \ldots, (N-1)/2$ for M and N odd. Next, let us sample the output signal from element (m, n) in time so that Eq. (5.1-46) becomes

$$y(lT_S, md_X, nd_Y), \tag{5.1-47}$$

where T_S is the sampling period in seconds and, in general, $l = -\infty, \ldots, 0, \ldots, \infty$. In addition, let us complex weight the output signal so that Eq. (5.1-47) is replaced by

$$c_{mn}\,y(lT_S, md_X, nd_Y), \tag{5.1-48}$$

where

$$c_{mn} = a_{mn} \exp(+j\theta_{mn}) \tag{5.1-49}$$

is the complex weight associated with the point source at $(x = md_X, y = nd_Y)$.

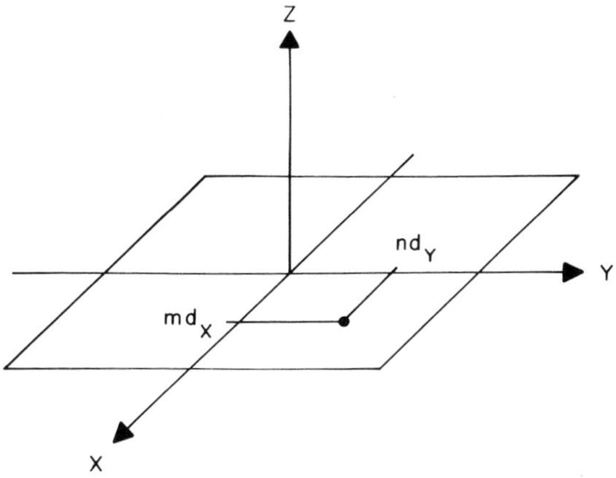

FIG. 5.1-2 Location of point-source element (m, n) of the array expressed in terms of the rectangular coordinates $(x = md_X,\ y = nd_Y)$.

Equation (5.1-48) is a mathematical representation of the complex-weighted output signal from one element in the array. In order to proceed further, we must mathematically represent the *sampled* version of $y(t, \mathbf{r})$ from *all* the elements. Therefore, let the sampled version of $y(t, \mathbf{r})$ be denoted by $y_S(t, \mathbf{r}) = y_S(t, x, y)$, where

$$y_S(t, x, y) = \sum_{m=-(M-1)/2}^{(M-1)/2} \sum_{n=-(N-1)/2}^{(N-1)/2} \sum_{l=-\infty}^{\infty} c_{mn} y(lT_S, md_X, nd_Y) \times \delta(t - lT_S)\, \delta(x - md_X)\, \delta(y - nd_Y). \tag{5.1-50}$$

Equation (5.1-50) is simply a generalization of one-dimensional time-domain impulse sampling. The frequency and angular spectrum of $y_S(t, x, y)$ is given by

$$Y_S(\eta, \gamma_X, \gamma_Y) = F_t F_x F_y \{ y_S(t, x, y) \} \tag{5.1-51}$$

or

$$Y_S(\eta, \gamma_X, \gamma_Y) = \sum_{m=-(M-1)/2}^{(M-1)/2} \sum_{n=-(N-1)/2}^{(N-1)/2} \sum_{l=-\infty}^{\infty} c_{mn} y(lT_S, md_X, nd_Y) \times \exp(-j2\pi\eta l T_S) \exp(+j2\pi\gamma_X md_X) \exp(+j2\pi\gamma_Y nd_Y), \tag{5.1-52}$$

which will be expressed as the *three-dimensional discrete Fourier transform* (*DFT*) of the sequence of numbers or data points $\{c_{mn} y(lT_S, md_X, nd_Y)\}$ later. Note that in the method of FFT beamforming, the complex weights c_{mn} appearing in Eq. (5.1-52) are just complex numbers in a computer algorithm, that is, electronic amplifiers and time-delay circuits do *not* have to be used in order to implement amplitude and phase weighting.

The next step in the analysis is to assume, as before, that the sound field incident upon the array is a general plane wave given by

$$y_M(t, \mathbf{r}) = g\left(t \mp \frac{\mathbf{r} \cdot \hat{n}_0}{c}\right) \tag{5.1-6}$$

As a result, the acoustic field incident upon element (m, n) in the XY plane $(z = 0)$ is given by

$$y_M(t, md_X, nd_Y) = g\left(t \mp \frac{u_0 md_X + v_0 nd_Y}{c}\right), \tag{5.1-53}$$

and if it is further assumed that the output electrical signal from element (m, n) is directly proportional to the input acoustical signal (this is reasonable since we assumed that the transducers are *linear* filters) then, suppressing the constant of proportionality, we can write that

$$y(t, md_X, nd_Y) = y_M(t, md_X, nd_Y), \tag{5.1-54}$$

or, from Eq. (5.1-53),

$$y(lT_S, md_X, nd_Y) = g\left(lT_S \mp \frac{u_0 md_X + v_0 nd_Y}{c}\right). \tag{5.1-55}$$

Thus, the output signals $y(lT_S, md_X, nd_Y)$ from the elements in the array represent *time and spatial samples* of the incident acoustic field.

If we substitute Eq. (5.1-55) into Eq. (5.1-52) and let

$$l'T_S = lT_S \mp \frac{u_0 md_X + v_0 nd_Y}{c}, \tag{5.1-56}$$

then

$$\begin{aligned} Y_S(\eta, \gamma_X, \gamma_Y) = &\sum_{l'=-\infty}^{\infty} g(l'T_S) \exp(-j2\pi\eta l'T_S) \\ &\times \sum_{m=-(M-1)/2}^{(M-1)/2} \sum_{n=-(N-1)/2}^{(N-1)/2} c_{mn} \exp\{+j2\pi[\gamma_X - (\pm f_{X_0})]md_X\} \\ &\qquad \times \exp\{+j2\pi[\gamma_Y - (\pm f_{Y_0})]nd_Y\}, \end{aligned} \tag{5.1-57}$$

or, replacing l' with l and assuming that only L (odd) time samples are available,

$$Y_S(\eta, \gamma_X, \gamma_Y) = \sum_{l=-(L-1)/2}^{(L-1)/2} g(lT_S) \exp(-j2\pi\eta lT_S)\, S'(\gamma_X - (\pm f_{X_0}), \gamma_Y - (\pm f_{Y_0})), \tag{5.1-58}$$

where $S'(\cdot,\cdot)$ is given by Eq. (5.1-2) and is the far-field directivity function of a planar array of $M \times N$ (odd) identical, complex-weighted point sources,

$$f_{X_0} = u_0/\lambda, \tag{5.1-17}$$

and

$$f_{Y_0} = v_0/\lambda \tag{5.1-18}$$

since $c = \eta\lambda$. The only difference between the *sampled* output spectrum $Y_S(\eta, \gamma_X, \gamma_Y)$ given by Eq. (5.1-58) and the *theoretical* output spectrum $Y(\eta, \gamma_X, \gamma_Y)$ given by Eq. (5.1-23) is that the theoretical frequency spectrum $G(\eta)$ is replaced by the discrete time Fourier transform of the time samples $g(lT_S)$ in Eq. (5.1-58). Furthermore, if the complex weights c_{mn} appearing in Eq. (5.1-57) are separable and if linear phase weighting is done for beam-steering purposes [see Eqs. (5.1-27) through (5.1-31)], then Eq. (5.1-58) becomes

$$Y_S(\eta, \gamma_X, \gamma_Y) = \sum_{l=-(L-1)/2}^{(L-1)/2} g(lT_S)\exp(-j2\pi\eta lT_S) \times S_X(\gamma_X - [f_X' + (\pm f_{X_0})])S_Y(\gamma_Y - [f_Y' + (\pm f_{Y_0})]), \tag{5.1-59}$$

where $S_X(\gamma_X)$ and $S_Y(\gamma_Y)$ are given by Eqs. (5.1-33) and (5.1-34), respectively. Compare Eq. (5.1-59) with Eq. (5.1-32).

Therefore, if the acoustic field incident upon the planar array is a *general plane-wave field*, one can estimate both the direction and frequency content of the incident field by processing the output electrical signals from the elements in the array according to Eq. (5.1-52). Equation (5.1-52) can be expressed in terms of a three-dimensional DFT by letting (see Section 4.4.4)

$$\eta = q\eta_S, \tag{5.1-60}$$

$$\gamma_X = r\gamma_{X_S}, \tag{5.1-61}$$

and

$$\gamma_Y = s\gamma_{Y_S}, \tag{5.1-62}$$

where

$$\eta_S = 1/(LT_S), \tag{5.1-63}$$

$$\gamma_{X_S} = 1/(Md_X), \tag{5.1-64}$$

$$\gamma_{Y_S} = 1/(Nd_Y), \tag{5.1-65}$$

and q, r, and s are integers. Substituting Eqs. (5.1-60) through (5.1-65) into

Eq. (5.1-52) and assuming that only L (odd) time samples are available yields

$$Y_S(q,r,s) = \sum_{n=-N'}^{N'} \sum_{m=-M'}^{M'} \sum_{l=-L'}^{L'} c_{mn} y(lT_S, md_X, nd_Y) W_L^{ql} W_M^{rm} W_N^{sn},$$
$$q = -L',\ldots,0,\ldots,L',$$
$$r = -M',\ldots,0,\ldots,M',$$
$$s = -N',\ldots,0,\ldots,N', \tag{5.1-66}$$

where η_S, γ_{X_S}, and γ_{Y_S} were suppressed in the argument of $Y_S(\cdot)$ and where

$$L' = (L - 1)/2, \tag{5.1-67}$$
$$M' = (M - 1)/2, \tag{5.1-68}$$
$$N' = (N - 1)/2, \tag{5.1-69}$$
$$W_L = \exp(-j2\pi/L), \tag{5.1-70}$$
$$W_M = \exp(+j2\pi/M), \tag{5.1-71}$$

and

$$W_N = \exp(+j2\pi/N). \tag{5.1-72}$$

Processing the output data from the array according to Eq. (5.1-66) is therefore known as *FFT beamforming* since one can use a FFT computer algorithm to compute a DFT. The term "beamforming" implies that amplitude and phase weighting are done via the complex weights c_{mn}. And, in order to avoid grating lobes under all conditions of beam steering, the interelement spacings d_X and d_Y must satisfy

$$d_X < \lambda_{\min}/2 \tag{5.1-73}$$

and

$$d_Y < \lambda_{\min}/2, \tag{5.1-74}$$

where $\lambda_{\min}$ is the minimum wavelength associated with the maximum expected frequency.

EXAMPLE 5.1-1

Let us assume, as before, that the acoustic field incident upon the planar array is a general plane-wave field given by Eq. (5.1-6) and, as a result, that the time-sampled output electrical signal from element (m, n) in the array is given by Eq. (5.1-55). Let us further assume, for example purposes, that $g(t)$ can be represented *exactly* by a *finite Fourier series* during the time interval $-T_0/2 \le t \le T_0/2$, that is,

$$g(t) = a_0 + 2 \sum_{q=1}^{K} a_q \cos[2\pi(qf_0)t + \theta_q], \qquad |t| \le T_0/2, \tag{5.1-75}$$

where T_0 is the *fundamental period* in seconds;

$$f_0 = 1/T_0 \tag{5.1-76}$$

is the *fundamental frequency* or *first harmonic* in hertz; a_0 is the *average value* or *dc component* of $g(t)$;

$$c_q = a_q \exp(+j\theta_q) = \frac{1}{T_0}\int_{-T_0/2}^{T_0/2} g(t)\exp(-j2\pi q f_0 t)\,dt \tag{5.1-77}$$

is the *complex Fourier series coefficient* of the *qth harmonic* qf_0 (in hertz), a_q and θ_q being *real* amplitude and phase components, respectively; and K is the total number of harmonics. Equation (5.1-75) is also known as a *trigonometric polynomial.*

Any signal $g(t)$ with *finite energy* in the time interval T_0 can be approximated over the time interval T_0—in the minimum mean-square error sense—by a finite Fourier series representation. The signal $g(t)$ may be real or complex and periodic or nonperiodic. If $g(t)$ is complex, then the complex form of the Fourier series must be used instead of the real trigonometric form given by Eq. (5.1-75). A signal $g(t)$ is said to have finite energy in the time interval T_0 if

$$\int_{T_0} |g(t)|^2\,dt < \infty. \tag{5.1-78}$$

As one increases the total number K of harmonics included in the representation, the mean-square error decreases. If $g(t)$ is periodic with fundamental period T_0, then a finite Fourier series representation is valid over all time.

If $g(t)$ can be represented *exactly* by a finite Fourier series during the time interval T_0, which is the assumption made in this example [see Eq. (5.1-75)], then it can be shown that the complex Fourier series coefficients can be computed from

$$c_q = \frac{1}{L}\sum_{l=-L'}^{L'} g(lT_S)W_L^{ql}, \qquad q = -L', \ldots, 0, \ldots, L', \tag{5.1-79}$$

or

$$c_q = \frac{1}{L}\mathrm{DFT}\{g(lT_S)\}, \qquad q = -L', \ldots, 0, \ldots, L', \tag{5.1-80}$$

where L is the total number of time samples $g(lT_S)$ taken during the time interval T_0 seconds, T_S is the *sampling period* (in seconds per sample), L' is given by Eq. (5.1-67), and W_L is given by Eq. (5.1-70). The relationships between T_0, T_S, K, and L will be derived next.

From the *sampling theorem* we know that in order to avoid *aliasing*, a signal must be sampled at a rate

$$f_S = 1/T_S \geq 2f_{\max}, \tag{5.1-81}$$

where f_S is the *sampling frequency* or *sampling rate* (in samples per second), T_S is the *sampling period* (in seconds per sample), and $f_{\max}$ is the highest-frequency component (in hertz) contained in the frequency spectrum of the signal. The *minimum* sampling rate is called the *Nyquist rate* and is equal to $2f_{\max}$ samples per second. Since our signal $g(t)$ is given by Eq. (5.1-75),

$$f_{\max} = Kf_0 = K/T_0, \tag{5.1-82}$$

so that

$$f_S \geq 2Kf_0 = 2K/T_0 \tag{5.1-83}$$

or

$$T_S \leq T_0/(2K) \tag{5.1-84}$$

Since

$$T_0 = (L-1)T_S, \tag{5.1-85}$$

then

$$L - 1 = T_0/T_S \geq 2K \tag{5.1-86}$$

or

$$L \geq 2K + 1, \tag{5.1-87}$$

where use was made of Eq. (5.1-84). Equation (5.1-87) stipulates the total number of samples that must be taken in order to avoid aliasing. In addition, according to Eq. (5.1-84), to avoid aliasing,

$$\frac{T_S}{T_0} \leq \frac{1}{2K}, \tag{5.1-88}$$

or using Eq. (5.1-85),

$$\frac{T_S}{T_0} = \frac{1}{L-1} \leq \frac{1}{2K}. \tag{5.1-89}$$

Since

$$\frac{1}{L} < \frac{1}{L-1},$$

if we let

$$\frac{T_S}{T_0} = \frac{1}{L}, \tag{5.1-90}$$

then

$$\frac{T_S}{T_0} < \frac{1}{2K},$$

which satisfies Eq. (5.1-88) and thus the sampling theorem. *Therefore, with respect to Eq.* (5.1-79), *L is computed from Eq.* (5.1-87) *and T_S is computed from Eq.* (5.1-90).

In order to evaluate Eq. (5.1-66) numerically, we rewrite it in terms of three one-dimensional DFTs as follows:

$$Y_S(q,r,s) = \sum_{n=-N'}^{N'} Y_S(q,r,n)W_N^{sn}, \qquad s = -N',\ldots,0,\ldots,N', \tag{5.1-91}$$

where

$$Y_S(q,r,n) = \sum_{m=-M'}^{M'} Y_S(q,m,n)W_M^{rm}, \qquad r = -M',\ldots,0,\ldots,M' \tag{5.1-92}$$

and

$$Y_S(q,m,n) = c_m d_n \sum_{l=-L'}^{L'} y(lT_S, md_X, nd_Y)W_L^{ql}, \quad q = -L',\ldots,0,\ldots,L', \tag{5.1-93}$$

where it has been assumed that the complex weights c_{mn} are separable, that is,

$$c_{mn} = c_m d_n = a_m \exp(+j\theta_m) b_n \exp(+j\phi_n), \tag{5.1-94}$$

and $y(lT_S, md_X, nd_Y)$ is given by Eq. (5.1-55). Substituting Eq. (5.1-55) into Eq. (5.1-93) yields

$$Y_S(q,m,n) = c_m d_n \sum_{l=-L'}^{L'} g(lT_S \mp t_0)W_L^{ql}, \qquad q = -L',\ldots,0,\ldots,L', \tag{5.1-95}$$

where

$$t_0 = (u_0 md_X + v_0 nd_Y)/c \quad \text{sec.} \tag{5.1-96}$$

Let us next discuss the physical significance of Eqs. (5.1-91), (5.1-92), and (5.1-95) starting with the innermost DFT calculation given by Eq. (5.1-95).

Since $g(t)$ was assumed to have Fourier series coefficients c_q, then we know from the properties of Fourier series that $g(t \mp t_0)$ will have Fourier series coefficients

$$c_q \exp(\mp j2\pi q f_0 t_0), \tag{5.1-97}$$

where t_0 is given by Eq. (5.1-96). Therefore, by comparing Eq. (5.1-95) with Eq. (5.1-79) and making use of Eq. (5.1-97), Eq. (5.1-95) becomes

$$Y_S(q, m, n) = c_m d_n L c_q \exp(\mp j2\pi q f_0 u_0 m d_X/c) \exp(\mp j2\pi q f_0 v_0 n d_Y/c),$$
$$q = -L', \ldots, 0, \ldots, L', \tag{5.1-98}$$

and by dividing both sides of Eq. (5.1-98) by $a_m b_n L$, which are quantities we know a priori, we obtain the normalized expression

$$Y_{SN}(q, m, n) = Y_S(q, m, n)/(a_m b_n L), \tag{5.1-99}$$

where

$$Y_{SN}(q, m, n) = c_q \exp(+j\theta_m) \exp(\mp j2\pi q f_0 u_0 m d_X/c) \exp(+j\phi_n)$$
$$\times \exp(\mp j2\pi q f_0 v_0 n d_Y/c),$$
$$q = -L', \ldots, 0, \ldots, L'. \tag{5.1-100}$$

Note that besides c_q, both u_0 and v_0 are *unknown* a priori. They are, in fact, those two quantities that need to be determined in order to estimate the target's direction. Equation (5.1-100) is directly proportional to the Fourier series coefficients c_q, and thus to the frequency spectrum of $g(t)$ as measured at element (m, n) in the array. The magnitude of Eq. (5.1-100) is equal to the amplitude a_q of the spectral line at $\eta = q f_0 = q/(L T_S)$ (in hertz) [see Eqs. (5.1-60) and (5.1-63) or Eqs. (5.1-76) and (5.1-90)] since

$$|Y_{SN}(q, m, n)| = |c_q| = a_q, \qquad q = -L', \ldots, 0, \ldots, L'. \tag{5.1-101}$$

The phase of Eq. (5.1-100) is equal to the phase θ_q of the Fourier series coefficient c_q *plus* the phase due to the phase weights θ_m and ϕ_n and the phase due to the two complex exponentials involving u_0 and v_0. Note that if the phase weights θ_m and ϕ_n are chosen properly, Eq. (5.1-100) can be made to equal the complex Fourier series coefficients c_q.

The next DFT calculation to interpret is that given by Eq. (5.1-92). Substituting Eq. (5.1-98) into Eq. (5.1-92) yields

$$Y_S(q, r, n) = L c_q d_n \exp(\mp j2\pi q f_0 v_0 n d_Y/c) \sum_{m=-M'}^{M'} c_m \exp(\mp j2\pi q f_0 u_0 m d_X/c) W_M^{rm},$$
$$r = -M', \ldots, 0, \ldots, M', \tag{5.1-102}$$

and by dividing both sides of Eq. (5.1-102) by $L b_n \sum_{m=-M'}^{M'} a_m$ we obtain the normalized expression

$$Y_{\text{SN}}(q,r,n) = Y_{\text{S}}(q,r,n)\bigg/\left(Lb_n \sum_{m=-M'}^{M'} a_m\right), \tag{5.1-103}$$

where

$$Y_{\text{SN}}(q,r,n) = c_q \exp(+j\phi_n)\exp(\mp j2\pi q f_0 v_0 n d_Y/c)$$
$$\times \sum_{m=-M'}^{M'} c_m \exp(\mp j2\pi q f_0 u_0 m d_X/c) W_M^{rm} \bigg/ \sum_{m=-M'}^{M'} a_m,$$
$$r = -M',\ldots,0,\ldots,M'. \tag{5.1-104}$$

Equation (5.1-104) is directly proportional to the Fourier series coefficients c_q and to the normalized far-field beam pattern or angular spectrum of a linear array of complex-weighted point sources lying along a line parallel to the X axis with constant y coordinate $y = nd_Y$ and operating at a frequency $\eta = qf_0 = q/(LT_S)$ (in hertz). The beam pattern or angular spectrum $Y_{\text{SN}}(q,r,n)$ is a function of the index r, which implies that $Y_{\text{SN}}(q,r,n)$ is a function of the spatial frequency γ_X [see Eqs. (5.1-61) and (5.1-64)] since the c_m values are the complex weights in the X direction. The FFT bin number r can be related to a corresponding value of the direction cosine $u = u_r$ as follows. Substituting Eq. (5.1-64) into Eq. (5.1-61) yields

$$\gamma_X = \frac{r}{Md_X}, \tag{5.1-105}$$

and upon substituting Eq. (5.1-105) into Eq. (5.1-25) and solving for u, we obtain [compare with Eq. (4.1-118)]

$$u = u_r = \frac{r\lambda}{Md_X}, \qquad r = -M',\ldots,0,\ldots,M'. \tag{5.1-106}$$

However, since $c = \eta\lambda$ and $\eta = qf_0 = q/T_0$, solving for λ yields

$$\lambda = cT_0/q, \qquad q \neq 0. \tag{5.1-107}$$

Substituting Eq. (5.1-107) into Eq. (5.1-106) finally yields

$$u_r = \left(\frac{r}{Md_X}\right)\left(\frac{cT_0}{q}\right), \qquad q \neq 0, \quad r = -M',\ldots,0,\ldots,M', \tag{5.1-108}$$

which is the desired result. For example, if *no* beam steering is being done, that is, if $\theta_m = 0$ for all m and $\phi_n = 0$ for all n [see Eq. (5.1-94)], then the maximum value of the *magnitude* of Eq. (5.1-104) will be located at some bin number r that corresponds to a direction cosine value of u_r that is either approximately equal to $\pm u_0$ or exactly equal to $\pm u_0$ [see Eq. (5.1-40)]. This value of u_r then becomes an *estimate* of $\pm u_0$. If $u_r \approx \pm u_0$, then the magnitude of Eq. (5.1-104) will *not* be symmetric about the maximum value, which implies that a better estimate can be obtained by "padding with zeros" for increased resolution. However, if $u_r = \pm u_0$, then the magnitude of Eq. (5.1-104) will be symmetric about the maximum value. If proper beam steering is done, then the angular spectrum will peak at the origin, that is, at bin number $r = 0$ (see p. 157).

The next DFT calculation to interpret is that given by Eq. (5.1-91). Substituting Eq. (5.1-102) into Eq. (5.1-91) yields

$$Y_S(q, r, s) = Lc_q \sum_{m=-M'}^{M'} c_m \exp(\mp j2\pi q f_0 u_0 m d_X/c) W_M^{rm} \times \sum_{n=-N'}^{N'} d_n \exp(\mp j2\pi q f_0 v_0 n d_Y/c) W_N^{sn}, \tag{5.1-109}$$

and by dividing both sides of Eq. (5.1-109) by $L\sum_{m=-M'}^{M'} a_m \sum_{n=-N'}^{N'} b_n$, we obtain the normalized expression

$$Y_{SN}(q, r, s) = Y_S(q, r, s) \Bigg/ \left(L \sum_{m=-M'}^{M'} a_m \sum_{n=-N'}^{N'} b_n \right), \tag{5.1-110}$$

where

$$Y_{SN}(q, r, s) = c_q \left[\sum_{m=-M'}^{M'} c_m \exp(\mp j2\pi q f_0 u_0 m d_X/c) W_M^{rm} \Bigg/ \sum_{m=-M'}^{M'} a_m \right] \times \left[\sum_{n=-N'}^{N'} d_n \exp(\mp j2\pi q f_0 v_0 n d_Y/c) W_N^{sn} \Bigg/ \sum_{n=-N'}^{N'} b_n \right],$$

$$r = -M', \ldots, 0, \ldots, M', \qquad s = -N', \ldots, 0, \ldots, N'. \tag{5.1-111}$$

Equation (5.1-111) is directly proportional to the Fourier series coefficients c_q and to the normalized far-field beam pattern or angular spectrum of a planar array of complex-weighted point sources operating at a frequency $\eta = q f_0 = q/(LT_S)$ (in hertz). The beam pattern or angular spectrum $Y_{SN}(q, r, s)$ is a function of the indices r and s, which implies that $Y_{SN}(q, r, s)$ is a function of the spatial frequencies γ_X and γ_Y [see Eqs. (5.1-61), (5.1-62), (5.1-64), and (5.1-65)] since the c_m and d_n values are the complex weights in the X and Y directions, respectively. The relationship between the FFT bin number r and the value of the direction cosine $u = u_r$ is given by Eq. (5.1-108). The FFT bin number s can be related to a corresponding value of the direction cosine $v = v_s$ as follows. Substituting Eq. (5.1-65) into Eq. (5.1-62) yields

$$\gamma_Y = \frac{s}{N d_Y}, \tag{5.1-112}$$

and upon substituting Eq. (5.1-112) into Eq. (5.1-26) and solving for v, we obtain

$$v = v_s = \frac{s\lambda}{Nd_Y}, \qquad s = -N', \ldots, 0, \ldots, N'. \tag{5.1-113}$$

Substituting Eq. (5.1-107) into Eq. (5.1-113) finally yields

$$v_s = \left(\frac{s}{Nd_Y}\right)\left(\frac{cT_0}{q}\right), \qquad q \neq 0, \quad s = -N', \ldots, 0, \ldots, N', \tag{5.1-114}$$

which is the desired result. For example, if *no* beam steering is being done, then the maximum value of the *magnitude* of Eq. (5.1-111) will be located at some bin numbers r and s that correspond to direction cosine values of u_r and v_s that are either approximately equal to $\pm u_0$ and $\pm v_0$ or exactly equal to $\pm u_0$ and $\pm v_0$ [see Eqs. (5.1-40) and (5.1-41)]. These values of u_r and v_s then become *estimates* of $\pm u_0$ and $\pm v_0$. If $u_r \approx \pm u_0$ and $v_s \approx \pm v_0$, then the magnitude of Eq. (5.1-111) will *not* be symmetric about the maximum value, which implies that better estimates can be obtained by "padding with zeros" for increased resolution. However, if $u_r = \pm u_0$ and $v_s = \pm v_0$, then the magnitude of Eq. (5.1-111) will be symmetric about the maximum value. If proper beam steering is done, then the angular spectrum will peak at the origin, that is, at bin numbers $r = 0$ and $s = 0$ (see pg. 157).

We can also obtain an estimate of $\pm v_0$ by performing the following alternate DFT:

$$Y_S(q, m, s) = \sum_{n=-N'}^{N'} Y_S(q, m, n) W_N^{sn}, \qquad s = -N', \ldots, 0, \ldots, N'. \tag{5.1-115}$$

Substituting Eq. (5.1-98) into Eq. (5.1-115) yields

$$Y_S(q, m, s) = Lc_q c_m \exp(\mp j2\pi q f_0 u_0 m d_X/c) \sum_{n=-N'}^{N'} d_n \exp(\mp j2\pi q f_0 v_0 n d_Y/c) W_N^{sn},$$
$$s = -N', \ldots, 0, \ldots, N', \tag{5.1-116}$$

and by dividing both sides of Eq. (5.1-116) by $La_m \sum_{n=-N'}^{N'} b_n$, we obtain the normalized expression

$$Y_{SN}(q, m, s) = Y_S(q, m, s) \bigg/ \left(La_m \sum_{n=-N'}^{N'} b_n\right), \tag{5.1-117}$$

where

$$Y_{SN}(q, m, s) = c_q \exp(+j\theta_m) \exp(\mp j2\pi q f_0 u_0 m d_X/c)$$
$$\times \sum_{n=-N'}^{N'} d_n \exp(\mp j2\pi q f_0 v_0 n d_Y/c) W_N^{sn} \bigg/ \sum_{n=-N'}^{N'} b_n,$$
$$s = -N', \ldots, 0, \ldots, N'. \tag{5.1-118}$$

TABLE 5.1-1

Estimates of the Frequency Spectrum of $g(t)$ Obtained by Evaluating Eq. (5.1-99) Using Eq. (5.1-95)

			$Y_{SN}(q, m, n)$			
(m, n)	q	η (Hz)	Real	Imaginary	Magnitude	Phase (deg)
	0	0.0	0.5	0.0	0.5	0.0
$(-2, 0)$	1	1000.0	0.917	−0.398	1.0	−23.475
	2	2000.0	1.365	−1.462	2.0	−46.95
	0	0.0	0.5	0.0	0.5	0.0
$(-1, 0)$	1	1000.0	0.993	−0.117	1.0	−6.737
	2	2000.0	1.945	−0.466	2.0	−13.475
	0	0.0	0.5	0.0	0.5	0.0
$(0, 0)$	1	1000.0	0.985	0.174	1.0	10.0
	2	2000.0	1.879	0.684	2.0	20.0
	0	0.0	0.5	0.0	0.5	0.0
$(1, 0)$	1	1000.0	0.893	0.45	1.0	26.737
	2	2000.0	1.19	1.607	2.0	53.475
	0	0.0	0.5	0.0	0.5	0.0
$(2, 0)$	1	1000.0	0.726	0.688	1.0	43.475
	2	2000.0	0.106	1.997	2.0	86.949

Equation (5.1-118) is directly proportional to the Fourier series coefficients c_q and to the normalized far-field beam pattern or angular spectrum of a linear array of complex-weighted point sources lying along a line parallel to the Y axis with constant x coordinate $x = md_X$ and operating at a frequency $\eta = qf_0 = q/(LT_S)$ (in hertz). The FFT bin number s can be related to a corresponding value of the direction cosine $v = v_s$ by using Eq. (5.1-114).

The computer simulation results presented in Tables 5.1-1 through 5.1-5 are based on the following set of parameter values. A planar array of 5×5 point sources was used so that $M = 5$ and $N = 5$. The complex weights c_{mn} were separable (i.e., $c_{mn} = c_m d_n$) and rectangular amplitude weighting was used with *no* phase weighting. Therefore, with $a_m = 1$ and $\theta_m = 0$ for all m and $b_n = 1$ and $\phi_n = 0$ for all n, $c_m = 1$ for all m and $d_n = 1$ for all n [see Eq. (5.1-94)]. Equation (5.1-55) was used for $y(lT_S, md_X, nd_Y)$ with $g(t)$ given by Eq. (5.1-75). The waveform $g(t)$ was assumed to have a fundamental period of $T_0 = 1.0$ msec and $K = 2$ harmonics. As a result, the fundamental frequency $f_0 = 1.0$ kHz [see Eq. (5.1-76)]. The total number of time samples taken per element in the

TABLE 5.1-2

Estimates of the Direction Cosine $u = -u_0$ Obtained by Evaluating Eq. (5.1-103) Using Eqs. (5.1-92) and (5.1-95)

	r	u_r	$Y_{SN}(q, r, n)$ Real	Imaginary	Magnitude	Phase (deg)
$q = 1, n = 0$	−4	−0.8	0.283	0.05	0.288	10.0
	−3	−0.6	0.616	0.109	0.625	10.0
	−2	−0.4	0.877	0.155	0.891	10.0
	−1	−0.2	0.984	0.174	1.0	10.0
	0	0.0	0.903	0.159	0.917	10.0
	1	0.2	0.659	0.116	0.669	10.0
	2	0.4	0.33	0.058	0.335	10.0
$q = 2, n = 0$	−5	−0.5	0.496	0.181	0.528	20.0
	−4	−0.4	1.132	0.412	1.205	20.0
	−3	−0.3	1.647	0.599	1.753	20.0
	−2	−0.2	1.876	0.683	1.996	20.0
	−1	−0.1	1.745	0.635	1.857	20.0
	0	0.0	1.297	0.472	1.381	20.0
	1	0.1	0.676	0.246	0.719	20.0

TABLE 5.1-3

Estimates of the Direction Cosine $u = -u_0$ Obtained by Evaluating Eq. (5.1-110) Using Eqs. (5.1-91), (5.1-92), and (5.1-95)

	r	u_r	$Y_{SN}(q, r, s)$ Real	Imaginary	Magnitude	Phase (deg)
$q = 1, s = 3$	−4	−0.8	0.283	0.050	0.287	10.0
	−3	−0.6	0.614	0.108	0.624	10.0
	−2	−0.4	0.875	0.154	0.889	10.0
	−1	−0.2	0.982	0.173	0.998	10.0
	0	0.0	0.901	0.159	0.915	10.0
	1	0.2	0.658	0.116	0.668	10.0
	2	0.4	0.330	0.058	0.335	10.0
$q = 2, s = 6$	−5	−0.5	0.492	0.179	0.524	20.0
	−4	−0.4	1.124	0.409	1.196	20.0
	−3	−0.3	1.634	0.595	1.739	20.0
	−2	−0.2	1.862	0.678	1.981	20.0
	−1	−0.1	1.732	0.630	1.843	20.0
	0	0.0	1.288	0.469	1.370	20.0
	1	0.1	0.671	0.244	0.714	20.0

TABLE 5.1-4

Estimates of the Direction Cosine $v = -v_0$ Obtained by Evaluating Eq. (5.1-110) Using Eqs. (5.1-91), (5.1-92), and (5.1-95)

			$Y_{SN}(q, r, s)$			
	s	v_s	Real	Imaginary	Magnitude	Phase (deg)
$q = 1, r = -1$	0	0.0	0.353	0.062	0.359	10.0
	1	0.2	0.679	0.120	0.689	10.0
	2	0.4	0.914	0.161	0.928	10.0
	3	0.6	0.982	0.173	0.998	10.0
	4	0.8	0.863	0.152	0.876	10.0
	5	1.0	0.594	0.105	0.603	10.0
	6	1.2	0.261	0.046	0.265	10.0
$q = 2, r = -2$	3	0.3	0.762	0.277	0.811	20.0
	4	0.4	1.370	0.499	1.458	20.0
	5	0.5	1.780	0.648	1.894	20.0
	6	0.6	1.862	0.678	1.981	20.0
	7	0.7	1.588	0.578	1.689	20.0
	8	0.8	1.047	0.381	1.114	20.0
	9	0.9	0.410	0.149	0.437	20.0

TABLE 5.1-5

Estimates of the Direction Cosine $v = -v_0$ Obtained by Evaluating Eq. (5.1-117) Using Eqs. (5.1-115) and (5.1-95)

			$Y_{SN}(q, m, s)$			
	s	v_s	Real	Imaginary	Magnitude	Phase (deg)
$q = 1, m = 0$	0	0.0	0.353	0.062	0.359	10.0
	1	0.2	0.679	0.12	0.69	10.0
	2	0.4	0.914	0.161	0.928	10.0
	3	0.6	0.983	0.173	0.998	10.0
	4	0.8	0.863	0.152	0.877	10.0
	5	1.0	0.594	0.105	0.603	10.0
	6	1.2	0.261	0.046	0.265	10.0
$q = 2, m = 0$	3	0.3	0.764	0.278	0.813	20.0
	4	0.4	1.372	0.5	1.461	20.0
	5	0.5	1.784	0.649	1.898	20.0
	6	0.6	1.865	0.679	1.985	20.0
	7	0.7	1.591	0.579	1.693	20.0
	8	0.8	1.049	0.382	1.116	20.0
	9	0.9	0.411	0.15	0.437	20.0

array was $L = 5$ [see Eq. (5.1-87)], and the sampling period was $T_S = 0.2$ msec [see Eq. (5.1-90)]. Therefore, it was assumed that a data record $T_0 = 1.0$ msec in length was available from each element in the array. The values used for the Fourier series coefficients were $a_0 = 0.5$, ($a_1 = 1.0$, $\theta_1 = 10.0°$), and ($a_2 = 2.0$, $\theta_2 = 20.0°$). The interelement spacings d_X and d_Y were chosen to satisfy [see Eqs. (5.1-73) and (5.1-74)]

$$d_X = d_Y = \lambda_{\min}/2. \tag{5.1-119}$$

Since

$$\lambda_{\min} = c/f_{\max}, \tag{5.1-120}$$

substituting Eq. (5.1-82) into Eq. (5.1-120) yields

$$\lambda_{\min} = cT_0/K, \tag{5.1-121}$$

and upon substituting Eq. (5.1-121) into Eq. (5.1-119) we obtain

$$d_X = d_Y = \frac{cT_0}{2K} = 0.375 \text{ m},$$

where $c = 1500.0$ m/sec was used for the speed of sound. This value for d_X and d_Y was then used in Eq. (5.1-96). The plane wave incident upon the array was assumed to be propagating in the $-\hat{n}_0$ direction with spherical angles $\theta_0 = 143.0°$ and $\psi_0 = 288.0°$ (see Fig. 5.1-3). Substituting $\theta_0 = 143.0°$ and $\psi_0 = 288.0°$ into Eqs. (5.1-9) and (5.1-10) yields $u_0 = 0.186$ and $v_0 = -0.572$. These values of u_0 and v_0 were then used in Eq. (5.1-96). Therefore, since we are not beam steering and since the incident plane wave is travelling in the $-\hat{n}_0$ direction, the magnitudes of the angular spectra should have their maximum values at $u = -u_0 = -0.186$ and $v = -v_0 = +0.572$; which correspond to the direction $+\hat{n}_0$, that is, the direction from the center of the array to the target [see Eqs. (5.1-40) and (5.1-41)].

Finally, if one decides to "pad with zeros" in order to increase the resolution of the angular spectra calculations, then one must replace Eqs. (5.1-64) and (5.1-65) with [see Eq. (4.1-123)]

$$\gamma_{XS} = 1/[(M + ZM)d_X] \tag{5.1-122}$$

and

$$\gamma_{YS} = 1/[(N + ZN)d_Y], \tag{5.1-123}$$

respectively, where ZM and ZN are the number of zeros used. As a result, Eqs. (5.1-71) and (5.1-72) must be replaced with [see Eq. (4.1-122)]

$$W_{M+ZM} = \exp[+j2\pi/(M + ZM)] \tag{5.1-124}$$

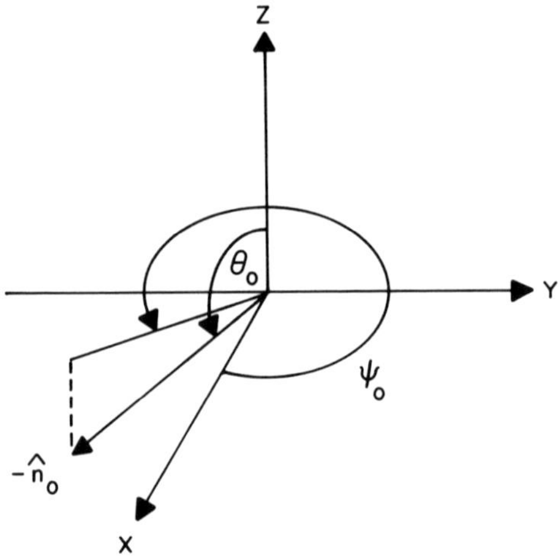

FIG. 5.1-3 Unit vector $-\hat{n}_0$ with associated spherical angles θ_0 and ψ_0. Note that ψ_0 is measured in a counterclockwise direction.

and

$$W_{N+ZN} = \exp[+j2\pi/(N + ZN)], \qquad (5.1\text{-}125)$$

respectively, and Eqs. (5.1-108) and (5.1-114) must be replaced with [see Eq. (4.1-127)]

$$u_r = \left(\frac{r}{(M + ZM)d_X}\right)\left(\frac{cT_0}{q}\right), \qquad q \neq 0,$$
$$r = -(M' + ZM), \ldots, 0, \ldots, (M' + ZM) \qquad (5.1\text{-}126)$$

and

$$v_s = \left(\frac{s}{(N + ZN)d_Y}\right)\left(\frac{cT_0}{q}\right), \qquad q \neq 0,$$
$$s = -(N' + ZN), \ldots, 0, \ldots, (N' + ZN), \qquad (5.1\text{-}127)$$

respectively. Note the extended range of values for the indices r and s. In this example, $ZM = ZN = 15$. The values of the parameters used in this example are summarized as follows:

$$M = 5, \qquad N = 5,$$
$$c_m = 1.0, \qquad m = -2, \ldots, 0, \ldots, 2,$$
$$d_n = 1.0, \qquad n = -2, \ldots, 0, \ldots, 2,$$

$$T_0 = 1.0 \quad \text{msec}, \qquad f_0 = 1.0 \quad \text{kHz},$$
$$K = 2, \qquad L = 5, \qquad T_S = 0.2 \quad \text{msec},$$
$$q = 0: \quad a_q = 0.5,$$
$$q = 1: \quad a_q = 1.0, \qquad \theta_q = 10.0°,$$
$$q = 2: \quad a_q = 2.0, \qquad \theta_q = 20.0°,$$
$$c = 1500.0 \quad \text{m/sec}$$
$$d_X = 0.375 \quad \text{m}, \qquad d_Y = 0.375 \quad \text{m}$$
$$-\hat{n}_0: \quad \theta_0 = 143.0°, \qquad \psi_0 = 288.0°, \qquad u_0 = 0.186, \qquad v_0 = -0.572,$$
$$ZM = 15, \qquad ZN = 15.$$

The data appearing in Table 5.1-1 represent estimates of the frequency spectrum of $g(t)$ as measured at element (m, n) in the array. The data was obtained by evaluating Eq. (5.1-99) using Eq. (5.1-95) and can be interpreted by referring to the discussion concerning Eq. (5.1-100).

The data appearing in Tables 5.1-2 and 5.1-3 represent estimates of the direction cosine $u = -u_0$. The data appearing in Table 5.1-2 was obtained by evaluating Eq. (5.1-103) using Eqs. (5.1-92) and (5.1-95) and can be interpreted by referring to the discussion concerning Eq. (5.1-104). The data appearing in Table 5.1-3 was obtained by evaluating Eq. (5.1-110) using Eqs. (5.1-91), (5.1-92), and (5.1-95) and can be interpreted by referring to the discussion concerning Eq. (5.1-111).

The data appearing in Tables 5.1-4 and 5.1-5 represent estimates of the direction cosine $v = -v_0$. The data appearing in Table 5.1-4 was obtained by evaluating Eq. (5.1-110) using Eqs. (5.1-91), (5.1-92), and (5.1-95) and can be interpreted by referring to the discussion concerning Eq. (5.1-111). The data appearing in Table 5.1-5 was obtained by evaluating Eq. (5.1-117) using Eqs. (5.1-115) and (5.1-95) and can be interpreted by referring to the discussion concerning Eq. (5.1-118).

An estimate of the target's direction $+\hat{n}_0$, as measured from the center of the array to the target and specified by the spherical angles (θ_0, ψ_0), can be obtained as follows (see Fig. 5.1-1). Substituting Eqs. (5.1-17) and (5.1-18) into Eq. (5.1-45) yields

$$\hat{\theta}_0 = \sin^{-1}[(u^2 + v^2)^{1/2}], \tag{5.1-128}$$

and from Eq. (5.1-42)

$$\hat{\psi}_0 = \tan^{-1}(v/u), \tag{5.1-129}$$

where θ_0, ψ_0, u_0, and v_0 have been replaced by $\hat{\theta}_0$, $\hat{\psi}_0$, u, and v, respectively. Substituting the direction cosine estimates $u = -0.2$ from Tables 5.1-2 and 5.1-3 and $v = +0.6$ from Tables 5.1-4 and 5.1-5 into Eq. (5.1-128) yields the estimate

$$\hat{\theta}_0 = 39.232°,$$

which is in error by 2.232° since if $\theta_0 = 143°$ for $-\hat{n}_0$, then θ_0 for $+\hat{n}_0$ should be $180° - 143° = 37°$. Finally, substituting the direction cosine estimates $u = -0.2$ and $v = +0.6$ into Eq. (5.1-129) yields the estimate

$$\hat{\psi}_0 = 108.435°,$$

where use was made of the fact that the target's location should have a negative X coordinate value and a positive Y coordinate value since u is negative and v is positive. This estimate of ψ_0 is in error by 0.435° since if $\psi_0 = 288°$ for $-\hat{n}_0$, then ψ_0 for $+\hat{n}_0$ should be $288° - 180° = 108°$. Keep in mind that the angle ψ_0 is measured in a counterclockwise direction with respect to the positive X axis of the array. Note that more accurate estimates of θ_0 and ψ_0 can be obtained by increasing the number of zeros ZM and ZN used for padding and, hence, increasing the angular resolution. □

5.2 COMPLEX ENVELOPES

In this section we shall introduce the concept of representing a real bandpass signal in terms of its *complex envelope*. As will be shown later, the complex envelope is, in general, a complex signal with a low-pass (baseband) frequency spectrum. One of the main advantages of using the complex envelope is that it provides for a simple representation of amplitude and angle modulation.

Consider an arbitrary, real bandpass signal $x(t)$ whose amplitude spectrum $|X(f)|$ is centered about $f = \pm f_c$ hertz, where

$$X(f) = F_t\{x(t)\} = \int_{-\infty}^{\infty} x(t)\exp(-j2\pi ft)\,dt \tag{5.2-1}$$

and f_c is referred to as the "center" or "carrier" frequency (see Fig. 5.2-1). The *complex envelope* of $x(t)$, denoted by $\tilde{x}(t)$, is defined as

$$\tilde{x}(t) \triangleq [x(t) + j\hat{x}(t)]\exp(-j2\pi f_c t), \tag{5.2-2}$$

where $\hat{x}(t)$ is the *Hilbert transform* of $x(t)$. The expression $x(t) + j\hat{x}(t)$ is known as the *pre-envelope* or the *analytic signal* of the real signal $x(t)$. If we let the pre-

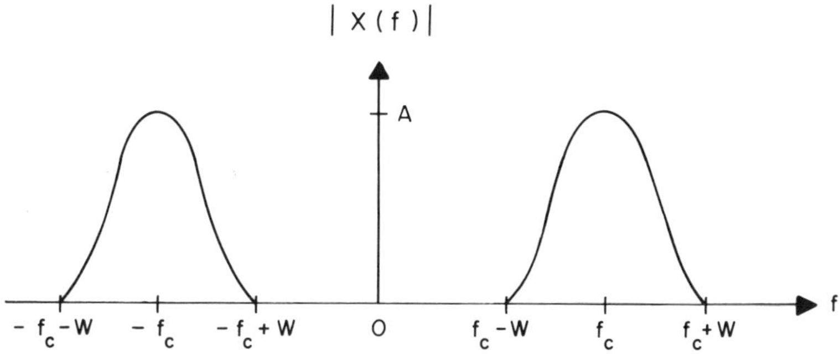

FIG. 5.2-1 Typical bandpass spectrum with bandwidth $2W$ (hertz).

envelope be denoted by $x_P(t)$ where

$$x_P(t) = x(t) + j\hat{x}(t), \tag{5.2-3}$$

then Eq. (5.2-2) can be rewritten as

$$\tilde{x}(t) = x_P(t)\exp(-j2\pi f_c t), \tag{5.2-4}$$

and

$$x_P(t) = \tilde{x}(t)\exp(+j2\pi f_c t). \tag{5.2-5}$$

Note that from Eq. (5.2-3) we have that

$$x(t) = \mathrm{Re}\{x_P(t)\}, \tag{5.2-6}$$

or, upon substituting Eq. (5.2-5) into Eq. (5.2-6),

$$x(t) = \mathrm{Re}\{\tilde{x}(t)\exp(+j2\pi f_c t)\}, \tag{5.2-7}$$

where Re means "take the real part."

The *envelope* of $x(t)$, denoted by $E(t)$, is defined as [see Eq. (5.2-3)]

$$E(t) \triangleq |x_P(t)| = \{x^2(t) + [\hat{x}(t)]^2\}^{1/2}, \tag{5.2-8}$$

or, from Eq. (5.2-5),

$$E(t) \triangleq |x_P(t)| = |\tilde{x}(t)|. \tag{5.2-9}$$

As can be seen from Eq. (5.2-2), in order to compute the complex envelope of $x(t)$, we first need to calculate the Hilbert transform $\hat{x}(t)$. One method of calculating the Hilbert transform is to perform the following inverse Fourier transform:

$$\hat{x}(t) = F_f^{-1}\{-j\,\mathrm{sgn}(f)X(f)\}, \tag{5.2-10}$$

where

$$\operatorname{sgn}(f) = \begin{cases} 1, & f > 0, \\ 0, & f = 0, \\ -1, & f < 0, \end{cases} \tag{5.2-11}$$

is the *signum* or *sign function* and $X(f)$ is given by Eq. (5.2-1). Similarly, from Eq. (5.2-10),

$$\hat{X}(f) = F_t\{\hat{x}(t)\} = -j\operatorname{sgn}(f)X(f). \tag{5.2-12}$$

The Hilbert transform is basically a 90° phase shifter. It is theoretically possible to generate the Hilbert transform of a signal by passing it through a linear circuit that provides a 90° *phase lag* at all positive frequencies and a 90° *phase lead* at all negative frequencies (see Problem 5-3). For example, from Eqs. (5.2-10) and (5.2-11), for $f > 0$,

$$\hat{x}(t) = F_f^{-1}\{-jX(f)\} = F_f^{-1}\{\exp(-j\pi/2)X(f)\},$$

which corresponds to a 90° phase lag, and for $f < 0$,

$$\hat{x}(t) = F_f^{-1}\{+jX(f)\} = F_f^{-1}\{\exp(+j\pi/2)X(f)\},$$

which corresponds to a 90° phase lead. If we designate the Hilbert transform by $H\{\cdot\}$, then it can be shown that

$$H\{\cos(2\pi f_c t + \theta_0)\} = \sin(2\pi f_c t + \theta_0) \tag{5.2-13}$$

and

$$H\{\sin(2\pi f_c t + \theta_0)\} = -\cos(2\pi f_c t + \theta_0), \tag{5.2-14}$$

where θ_0 is a constant phase angle in radians.

Next, let us demonstrate that the complex envelope of a real bandpass signal has a low-pass frequency spectrum. Since

$$\tilde{x}(t) = x_P(t)\exp(-j2\pi f_c t), \tag{5.2-4}$$

$$\tilde{X}(f) = X_P(f + f_c), \tag{5.2-15}$$

where

$$\tilde{X}(f) = F_t\{\tilde{x}(t)\} \quad \text{and} \quad X_P(f) = F_t\{x_P(t)\}.$$

Taking the Fourier transform with respect to t of Eq. (5.2-3) yields

$$X_P(f) = X(f) + j\hat{X}(f), \tag{5.2-16}$$

or, upon substituting Eq. (5.2-12) into Eq. (5.2-16),

$$X_P(f) = [1 + \operatorname{sgn}(f)]X(f). \tag{5.2-17}$$

With the use of Eq. (5.2-11), Eq. (5.2-17) can be expressed as

$$X_P(f) = \begin{cases} 2X(f), & f > 0, \\ X(0) = 0, & f = 0, \\ 0, & f < 0, \end{cases} \tag{5.2-18}$$

where $X(0) = 0$ since $x(t)$ was assumed to be a bandpass signal (see Fig. 5.2-1). The frequency spectrum of the pre-envelope $X_P(f)$, as given by Eq. (5.2-18), is known as a *one-sided spectrum* since it only contains positive frequency components (see Fig. 5.2-2). Also, the maximum value of $|X_P(f)|$ is $2A$ compared with A for $|X(f)|$. Finally, since

$$\tilde{X}(f) = X_P(f + f_c), \tag{5.2-15}$$

we can obtain the curve for $|\tilde{X}(f)|$ by shifting the curve for $|X_P(f)|$ to the left by the amount f_c, as shown in Fig. 5.2-3. Figure 5.2-3 illustrates the fact that

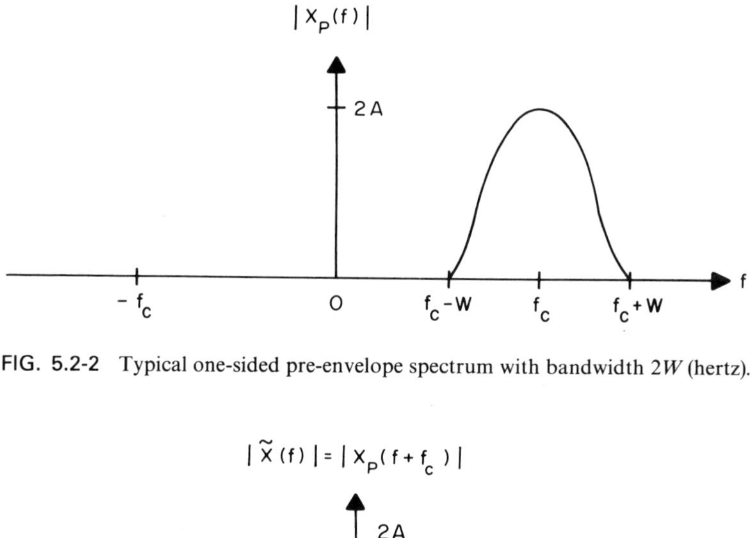

FIG. 5.2-2 Typical one-sided pre-envelope spectrum with bandwidth $2W$ (hertz).

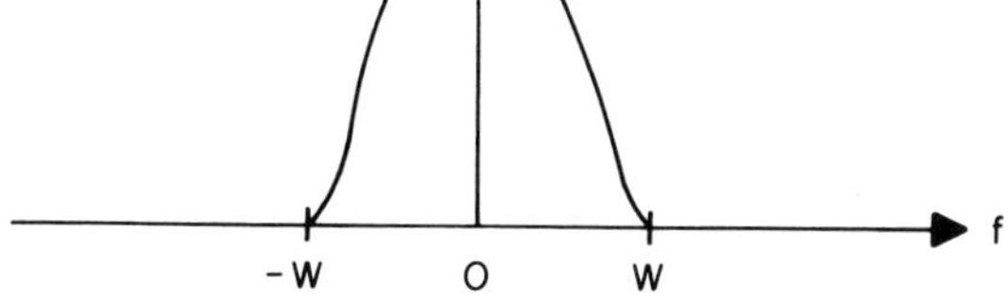

FIG. 5.2-3 Typical low-pass (baseband) complex envelope spectrum with bandwidth W (hertz).

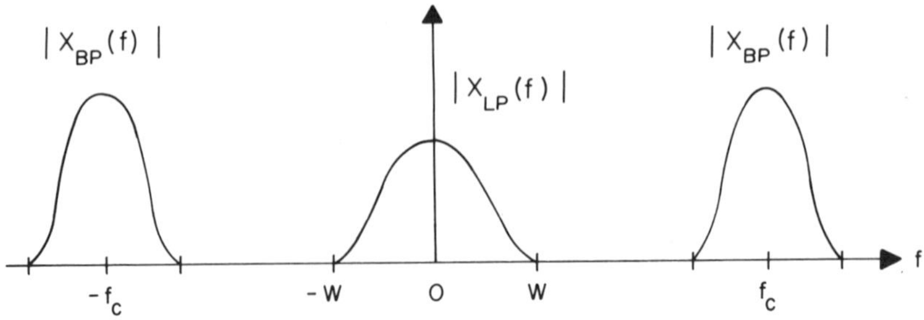

FIG. 5.2-4 The low-pass spectrum $X_{LP}(f)$ and bandpass spectrum $X_{BP}(f)$ do *not* overlap, as shown.

the frequency spectrum of the complex envelope of a real bandpass signal is low pass (baseband), which is another reason for representing a real bandpass signal in terms of its complex envelope. For example, suppose we want to compute the frequency spectrum of the real bandpass signal $x(t)$ by performing a DFT on the time samples $x(lT_S)$, where T_S is the sampling period. If we first transform $x(t)$ into $\tilde{x}(t)$ according to Eq. (5.2-2), then we can sample the low-pass signal $\tilde{x}(t)$ at a much lower rate than the bandpass signal $x(t)$ [see Example 5.2-3]. The resulting DFT of the complex envelope $\tilde{x}(t)$ can be related to the DFT of the bandpass signal $x(t)$ by comparing Figs. 5.2-1 and 5.2-3.

Before we demonstrate two sample complex-envelope calculations, we need the following property of the Hilbert transform. If

$$x(t) = x_{LP}(t)x_{BP}(t), \tag{5.2-19}$$

where $x_{LP}(t)$ is a low-pass signal and $x_{BP}(t)$ is a bandpass signal, and if the amplitude spectra $|X_{LP}(f)|$ and $|X_{BP}(f)|$ do *not* overlap (see Fig. 5.2-4), then

$$\hat{x}(t) = x_{LP}(t)\hat{x}_{BP}(t), \tag{5.2-20}$$

that is, only the bandpass signal is Hilbert transformed.

EXAMPLE 5.2-1

Let us compute the complex envelope of the following double-sideband suppressed-carrier (DSBSC) waveform:

$$x(t) = a(t)\cos(2\pi f_c t), \tag{5.2-21}$$

where $a(t)$ is a low-pass amplitude-modulating signal, $\cos(2\pi f_c t)$ is the carrier waveform, and f_c is the "center" or "carrier" frequency in hertz. The frequency spectrum of $x(t)$ as given by Eq. (5.2-21) is bandpass and is equal to (see Fig. 5.2-5)

$$X(f) = 0.5[A(f - f_c) + A(f + f_c)], \tag{5.2-22}$$

where

$$A(f) = F_t\{a(t)\}. \tag{5.2-23}$$

Provided $f_c > W$,

$$\hat{x}(t) = a(t)\sin(2\pi f_c t), \tag{5.2-24}$$

where use was made of Eqs. (5.2-20) and (5.2-13). Substituting Eqs. (5.2-21) and (5.2-24) into Eq. (5.2-3) yields the pre-envelope

$$x_P(t) = a(t)\exp(+j2\pi f_c t), \tag{5.2-25}$$

and upon substituting Eq. (5.2-25) into Eq. (5.2-4), we finally obtain the

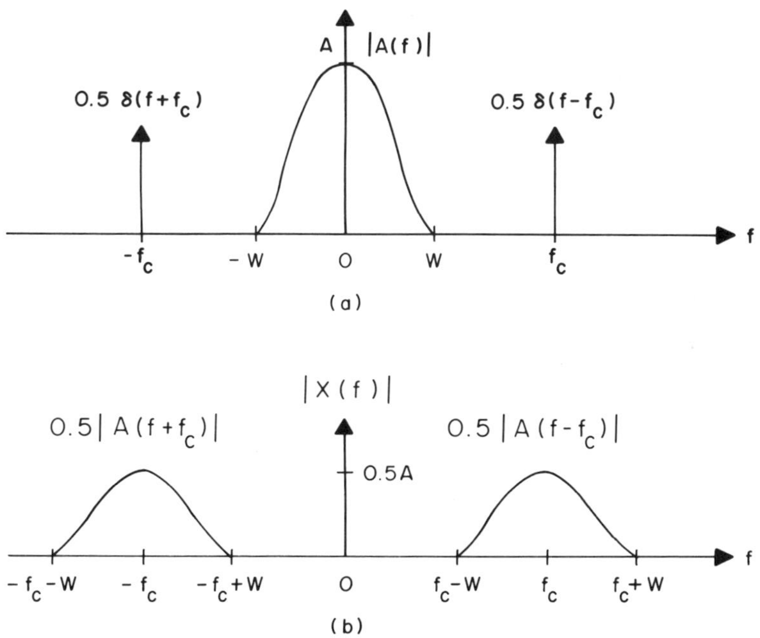

FIG. 5.2-5 (a) Low-pass spectrum $A(f)$ with bandwidth W (hertz) and maximum value A. Also shown is the bandpass frequency spectrum of the carrier waveform $F\{\cos(2\pi f_c t)\} = 0.5\,\delta(f + f_c) + 0.5\,\delta(f - f_c)$. (b) Bandpass spectrum $X(f)$ with bandwidth $2W$ (hertz) and maximum value $0.5A$.

complex envelope

$$\tilde{x}(t) = a(t). \tag{5.2-26}$$

In addition, if we now substitute either Eq. (5.2-25) or Eq. (5.2-26) into Eq. (5.2-9), we obtain

$$E(t) = |a(t)|, \tag{5.2-27}$$

which is the envelope of $x(t)$. □

EXAMPLE 5.2-2

Let us compute the complex envelope of the following amplitude- and angle-modulated carrier:

$$x(t) = a(t)\cos[2\pi f_c t + \theta(t)], \tag{5.2-28}$$

where $a(t)$ and $\theta(t)$ are amplitude- and angle-modulating signals, respectively. The signal $\theta(t)$ is also known as the *phase deviation.* Before we proceed with the calculation of the complex envelope of Eq. (5.2-28), let us briefly review some of the basic terminology associated with angle modulation. This review will be helpful in our discussion of the linear frequency-modulated (LFM) pulse in Section 5.3.2.

The *instantaneous phase* (in radians) of $x(t)$ is defined as

$$\theta_i(t) \triangleq 2\pi f_c t + \theta(t), \tag{5.2-29}$$

and the *instantaneous radian frequency* (in radians per second) of $x(t)$ is defined as

$$\omega_i(t) \triangleq \frac{d}{dt}\theta_i(t) = 2\pi f_c + \frac{d}{dt}\theta(t), \tag{5.2-30}$$

or, equivalently, the *instantaneous frequency* (in hertz)

$$f_i(t) \triangleq \frac{1}{2\pi}\frac{d}{dt}\theta_i(t) = f_c + \frac{1}{2\pi}\frac{d}{dt}\theta(t), \tag{5.2-31}$$

where the time derivative $\theta'(t)$ is known as the *frequency deviation.*

The two basic types of angle modulation are *phase modulation* (PM) and *frequency modulation* (FM). Phase modulation implies that the phase deviation $\theta(t)$ of the carrier is directly proportional to some *message* or *modulating signal* $m(t)$, that is,

$$\theta(t) = D_P m(t), \tag{5.2-32}$$

where D_P is the *phase-deviation constant* in radians per unit of $m(t)$. Frequency modulation implies that the frequency deviation $\theta'(t)$ of the carrier is directly proportional to some message or modulating signal $m(t)$, that is,

$$\theta'(t) = D_F m(t), \tag{5.2-33}$$

where D_F is the *frequency-deviation constant* in radians per second per unit of $m(t)$. As a result, the phase deviation of a frequency-modulated carrier is given by

$$\theta(t) = D_F \int_{t_0}^{t} m(\alpha)\, d\alpha + \theta(t_0), \tag{5.2-34}$$

where $\theta(t_0)$ is the phase deviation at $t = t_0$.

Now, let us return to the problem of computing the complex envelope of Eq. (5.2-28). By making use of the trigonometric identity

$$\cos(\alpha + \beta) = \cos\alpha\cos\beta - \sin\alpha\sin\beta,$$

the waveform $x(t)$ as given by Eq. (5.2-28) can be rewritten as

$$x(t) = x_C(t)\cos(2\pi f_c t) - x_S(t)\sin(2\pi f_c t), \tag{5.2-35}$$

where

$$x_C(t) = a(t)\cos\theta(t) \tag{5.2-36}$$

and

$$x_S(t) = a(t)\sin\theta(t) \tag{5.2-37}$$

are known as the *cosine* and *sine components* of $x(t)$, respectively. The functions $x_C(t)$ and $x_S(t)$ are also known as the *in-phase* and *quadrature-phase components*, respectively. If both $x_C(t)$ and $x_S(t)$ are low-pass (baseband) signals, that is, if

$$|X_C(f)| = 0 \qquad \text{for} \quad |f| > W$$

and

$$|X_S(f)| = 0 \qquad \text{for} \quad |f| > W,$$

then, with the use of Eqs. (5.2-20), (5.2-13), and (5.2-14), the Hilbert transform $\hat{x}(t)$ is given by

$$\hat{x}(t) = x_C(t)\sin(2\pi f_c t) + x_S(t)\cos(2\pi f_c t) \tag{5.2-38}$$

provided $f_c > W$. Substituting Eqs. (5.2-35) and (5.2-38) into Eq. (5.2-3) yields the pre-envelope

$$x_P(t) = [x_C(t) + jx_S(t)]\exp(+j2\pi f_c t), \tag{5.2-39}$$

and upon substituting Eq. (5.2-39) into Eq. (5.2-4), we finally obtain the complex envelope

$$\tilde{x}(t) = x_C(t) + jx_S(t). \tag{5.2-40}$$

Or, in terms of amplitude and phase,

$$\tilde{x}(t) = a(t)\exp[+j\theta(t)], \tag{5.2-41}$$

where it can be shown that

$$a(t) = [x_C^2(t) + x_S^2(t)]^{1/2} \tag{5.2-42}$$

and

$$\theta(t) = \tan^{-1}[x_S(t)/x_C(t)] \tag{5.2-43}$$

by direct substitution of Eqs. (5.2-36) and (5.2-37) into Eqs. (5.2-42) and (5.2-43). In addition, if we now substitute either Eq. (5.2-39) or Eq. (5.2-41) into Eq. (5.2-9), we obtain

$$E(t) = |a(t)|, \tag{5.2-44}$$

which is the envelope of $x(t)$.

The form of the complex envelope given by Eq. (5.2-41) provides for a convenient way to represent real amplitude- and angle-modulated carriers such as the one given by Eq. (5.2-28). Note that the real bandpass signal $x(t)$ specified by Eq. (5.2-28) is also known as a *narrowband waveform* when $a(t)$ and $\theta(t)$ are slowly varying functions of time when compared to $\cos(2\pi f_c t)$, that is, when $a(t)$ and $\theta(t)$ are low-pass signals. Narrowband implies that the ratio of the bandwidth of $|X(f)|$ to f_c is much less than unity, that is, that $(2W/f_c) \ll 1$ (see Fig. 5.2-1). □

The complex-envelope expression given by Eq. (5.2-40) was obtained by evaluating the mathematical definition of Eq. (5.2-2). However, in practical signal-processing applications, Eq. (5.2-40) is obtained by passing the real amplitude- and angle-modulated carrier waveform given by Eq. (5.2-28) through the *quadrature demodulator* system shown in Fig. 5.2-6. As indicated in Fig. 5.2-6, the output signals from this system are the cosine and sine components $x_C(t)$ and $x_S(t)$ given by Eqs. (5.2-36) and (5.2-37), respectively. Therefore, the output from this system is essentially the complex envelope of the input since

$$\tilde{x}(t) = x_C(t) + jx_S(t). \tag{5.2-40}$$

Note that in order for the quadrature demodulator system shown in Fig. 5.2-6 to compute the complex envelope $\tilde{x}(t)$ of the input bandpass signal $x(t)$, the carrier frequency f_c of the input signal must be known. Also, note that although the amplitude- and angle-modulating signals $a(t)$ and $\theta(t)$ of the input signal $x(t)$ may be unknown, they can be determined from $x_C(t)$ and $x_S(t)$ by using Eqs. (5.2-42) and (5.2-43), respectively.

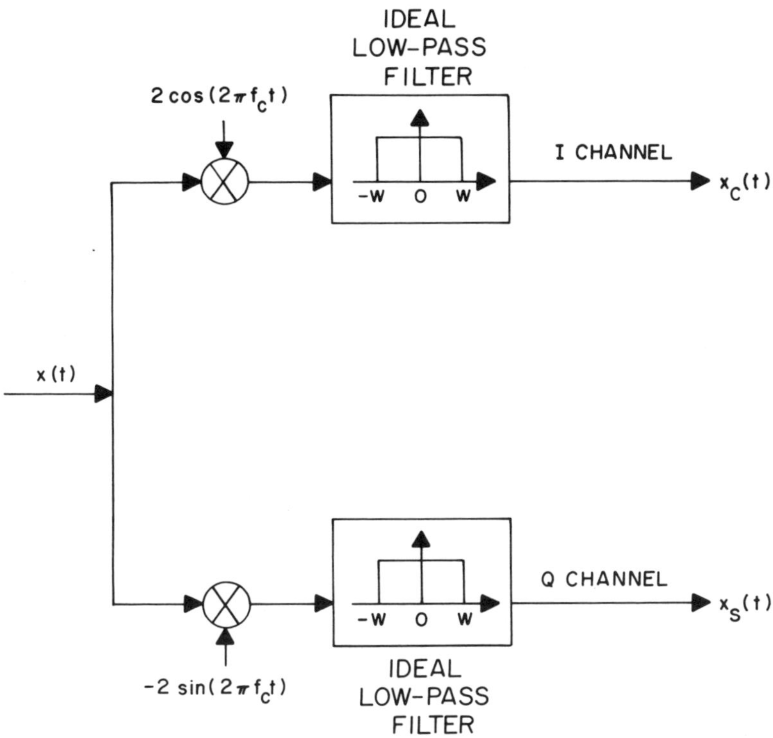

FIG. 5.2-6 Quadrature decomposition of input bandpass signal.

EXAMPLE 5.2-3: BANDPASS SAMPLING THEOREM

Since the complex envelope $\tilde{x}(t)$ is a low-pass (baseband) signal and is bandlimited to W hertz (see Fig. 5.2-3), the *low-pass sampling theorem* states that we can represent $\tilde{x}(t)$ in terms of its sample values $\tilde{x}(nT_S)$ as follows:

$$\tilde{x}(t) = \sum_{n=-\infty}^{\infty} \tilde{x}(nT_S)\,\mathrm{sinc}\left(\frac{t - nT_S}{T_S}\right), \tag{5.2-45}$$

where the *sampling frequency* or *sampling rate* f_S (samples per second) is given by

$$f_S = \frac{1}{T_S} \geq 2W \quad \text{samples/sec}, \tag{5.2-46}$$

T_S is the *sampling period* (seconds per sample), and

$$\mathrm{sinc}(x) \triangleq \sin(\pi x)/(\pi x). \tag{5.2-47}$$

The *minimum* sampling rate is called the *Nyquist rate* and is equal to $2W$ samples per second. Recall that the real bandpass signal $x(t)$ is related to its complex envelope $\tilde{x}(t)$ by Eq. (5.2-7). Substituting Eq. (5.2-45) into Eq. (5.2-7) and using Eq. (5.2-40) yields

$$x(t) = \sum_{n=-\infty}^{\infty} \mathrm{Re}\{[x_C(nT_S) + jx_S(nT_S)]\exp(+j2\pi f_c t)\} \operatorname{sinc}\left(\frac{t - nT_S}{T_S}\right), \quad (5.2\text{-}48)$$

or, upon expanding and taking the real part,

$$x(t) = \sum_{n=-\infty}^{\infty} [x_C(nT_S)\cos(2\pi f_c t) - x_S(nT_S)\sin(2\pi f_c t)] \operatorname{sinc}\left(\frac{t - nT_S}{T_S}\right). \quad (5.2\text{-}49)$$

Equation (5.2-49) is the *bandpass sampling theorem.* It states that we can represent a real bandpass signal $x(t)$ with bandwidth $2W$ hertz (see Fig. 5.2-1) in terms of the sample values $x_C(nT_S)$ and $x_S(nT_S)$ of its cosine and sine components, respectively. Therefore, instead of sampling $x(t)$ at the high rate of $f_S \geq 2(f_c + W)$ samples per second, we need only sample $x_C(t)$ and $x_S(t)$ at the lower rate of $f_S \geq 2W$ samples per second, which is the bandwidth of $x(t)$. However, *two* waveforms must be sampled, namely, $x_C(t)$ and $x_S(t)$. Therefore, during any T second interval, $N \geq 2(2W)T = 4WT$ samples must be taken. □

EXAMPLE 5.2-4

In this example we will relate the *energy* of the real bandpass signal $x(t)$ to the energy of its complex envelope $\tilde{x}(t)$. The energy E_x of the real signal $x(t)$ is, by definition,

$$E_x \triangleq \int_{-\infty}^{\infty} x^2(t)\,dt, \quad (5.2\text{-}50)$$

and similarly, the energy $E_{\tilde{x}}$ of the complex signal $\tilde{x}(t)$ is, by definition,

$$E_{\tilde{x}} \triangleq \int_{-\infty}^{\infty} |\tilde{x}(t)|^2\,dt. \quad (5.2\text{-}51)$$

Using Eq. (5.2-2), we can write that

$$|\tilde{x}(t)|^2 = x^2(t) + [\hat{x}(t)]^2. \quad (5.2\text{-}52)$$

Substituting Eq. (5.2-52) into Eq. (5.2-51) yields

$$E_{\tilde{x}} = E_x + E_{\hat{x}}, \quad (5.2\text{-}53)$$

where

$$E_{\hat{x}} = \int_{-\infty}^{\infty} [\hat{x}(t)]^2\,dt \quad (5.2\text{-}54)$$

is the energy of the Hilbert transform $\hat{x}(t)$. With the use of *Parseval's theorem*, Eq. (5.2-54) can be expressed as

$$E_{\hat{x}} = \int_{-\infty}^{\infty} |\hat{X}(f)|^2 \, df, \tag{5.2-55}$$

and upon substituting Eq. (5.2-12) into Eq. (5.2-55) we obtain

$$E_{\hat{x}} = \int_{-\infty}^{\infty} |X(f)|^2 \, df = E_x, \tag{5.2-56}$$

that is, $x(t)$ and its Hilbert transform $\hat{x}(t)$ have equal energy. Substituting Eq. (5.2-56) into Eq. (5.2-53) finally yields the desired result,

$$E_x = 0.5 E_{\tilde{x}}. \tag{5.2-57}$$

□

EXAMPLE 5.2-5

In this example we shall relate the complex frequency spectrum $X(f)$ of the real bandpass signal $x(t)$ to the complex frequency spectrum $\tilde{X}(f)$ of its complex envelope $\tilde{x}(t)$. We begin by rewriting Eq. (5.2-7) as follows:

$$x(t) = 0.5[\tilde{x}(t)\exp(+j2\pi f_c t) + \tilde{x}^*(t)\exp(-j2\pi f_c t)] \tag{5.2-58}$$

since for any arbitrary complex quantity Z,

$$\text{Re}\{Z\} = 0.5(Z + Z^*).$$

Taking the Fourier transform of both sides of Eq. (5.2-58) with respect to t yields

$$X(f) = 0.5\{\tilde{X}(f - f_c) + \tilde{X}^*[-(f + f_c)]\} \tag{5.2-59}$$

since

$$F_t\{\tilde{x}(t)\exp(\pm j2\pi f_c t)\} = \tilde{X}(f \mp f_c) \tag{5.2-60}$$

and

$$F_t\{\tilde{x}^*(t)\} = \tilde{X}^*(-f). \tag{5.2-61}$$

Equation (5.2-59) is the desired general result. Note that for an amplitude-modulated carrier, $\tilde{x}(t)$ is a real signal since (see Example 5.2-1)

$$\tilde{x}(t) = a(t). \tag{5.2-26}$$

When $\tilde{x}(t)$ is real,

$$\tilde{X}^*(-f) = \tilde{X}(f) = A(f), \tag{5.2-62}$$

and Eq. (5.2-59) reduces to

$$X(f) = 0.5[A(f - f_c) + A(f + f_c)]. \tag{5.2-22}$$

□

The importance of the complex envelope will be demonstrated further in Section 5.3, where we discuss the range and Doppler resolving properties of basic transmit signals via the auto-ambiguity function.

5.3 THE AUTO-AMBIGUITY FUNCTION

5.3.1 Derivation and Interpretation

In order to derive and provide an interpretation of the *auto-ambiguity function*, let us consider the following binary hypothesis testing problem of detecting a slowly fluctuating point target return in the presence of noise:

$$H_1 : \tilde{r}(t) = \tilde{y}_T(t) + \tilde{n}(t), \qquad -\infty < t < \infty, \tag{5.3-1}$$

$$H_0 : \tilde{r}(t) = \tilde{n}(t), \qquad -\infty < t < \infty, \tag{5.3-2}$$

where

$$\tilde{y}_T(t) = c\tilde{x}(t - \tau_A)\exp(+j2\pi\phi_A t) \tag{5.3-3}$$

and

$$c = a\exp(+j\theta). \tag{5.3-4}$$

Hypothesis H_1 states that the complex envelope of the received signal $\tilde{r}(t)$ is equal to the sum of the complex envelopes of the target return $\tilde{y}_T(t)$ and noise $\tilde{n}(t)$. It is assumed that $\tilde{y}_T(t)$ and $\tilde{n}(t)$ are zero-mean, statistically independent, *Gaussian* random processes and that $\tilde{n}(t)$ is wide-sense stationary, low-pass (bandlimited) white noise. Therefore, the real noise signal $n(t)$ must be a zero-mean, Gaussian, wide-sense stationary, bandpass, white noise process. Hypothesis H_0 states that the received signal $\tilde{r}(t)$ is equal to the noise $\tilde{n}(t)$. The target return $\tilde{y}_T(t)$, as given by Eq. (5.3-3), is modeled as a complex-weighted, time- and frequency-shifted replica of the complex envelope of the transmitted signal $\tilde{x}(t)$. The complex weight c given by Eq. (5.3-4) is assumed to be a zero-mean, complex, Gaussian, random variable that accounts for random attenuation and random phase shift. The magnitude a is assumed to be *Rayleigh distributed* and the phase θ is assumed to be *uniform*. Thus, the magnitude and phase are statistically independent random variables. It is also assumed that the parameters τ_A and ϕ_A, which are the *actual* or *true round-trip time delay* and *Doppler shift* of the target, respectively, are *unknown, nonrandom* parameters to be *estimated*.

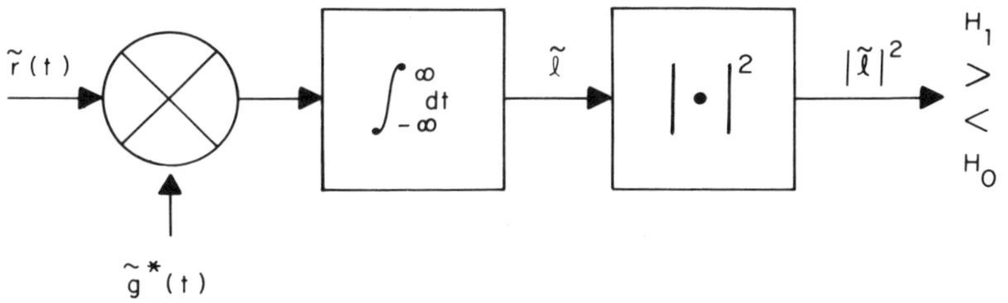

FIG. 5.3-1 Correlator followed by a magnitude-square operation.

Let us process the received signal $\tilde{r}(t)$ with the receiver shown in Fig. 5.3-1. The function $\tilde{g}(t)$ is referred to as the "processing waveform," as of yet unspecified. The receiver performs the following test: choose hypothesis H_1 if

$$|\tilde{l}|^2 > \gamma \tag{5.3-5}$$

and choose H_0 otherwise, where

$$\tilde{l} = \langle \tilde{r}(t), \tilde{g}(t) \rangle = \int_{-\infty}^{\infty} \tilde{r}(t)\tilde{g}^*(t)\, dt \tag{5.3-6}$$

is the *inner product* of $\tilde{r}(t)$ and $\tilde{g}(t)$. The threshold γ is commonly chosen to satisfy a *Neyman–Pearson criterion*, which is to maximize the *probability of detection* for a given *probability of false alarm*.

Next, compute the output signal-to-noise power ratio (SNR) for the receiver shown in Fig. 5.3-1. The SNR is defined as

$$\text{SNR} \triangleq \text{E}\{|\tilde{l}_{\text{T}}|^2\}/\text{E}\{|\tilde{l}_{\text{N}}|^2\}, \tag{5.3-7}$$

where

$$\tilde{l}_{\text{T}} = \langle \tilde{y}_{\text{T}}(t), \tilde{g}(t) \rangle = \int_{-\infty}^{\infty} \tilde{y}_{\text{T}}(t)\tilde{g}^*(t)\, dt \tag{5.3-8}$$

and

$$\tilde{l}_{\text{N}} = \langle \tilde{n}(t), \tilde{g}(t) \rangle = \int_{-\infty}^{\infty} \tilde{n}(t)\tilde{g}^*(t)\, dt. \tag{5.3-9}$$

In order to proceed further, we must specify a functional form for the processing waveform. Therefore, let

$$\tilde{g}(t) = \tilde{x}(t - \hat{\tau})\exp(+j2\pi\hat{\phi}t), \tag{5.3-10}$$

which is identical in form to the target return given by Eq. (5.3-3). The parameters $\hat{\tau}$ and $\hat{\phi}$ are *estimates* of the round-trip time delay and Doppler shift of the target, respectively. Substituting Eqs. (5.3-3) and (5.3-10) into Eq. (5.3-8) and Eq. (5.3-10) into Eq. (5.3-9) yields

$$\tilde{l}_{\mathrm{T}} = c \int_{-\infty}^{\infty} \tilde{x}(t - \tau_{\mathrm{A}})\tilde{x}^*(t - \hat{\tau}) \exp[+j2\pi(\phi_{\mathrm{A}} - \hat{\phi})t]\, dt \qquad (5.3\text{-}11)$$

and

$$\tilde{l}_{\mathrm{N}} = \int_{-\infty}^{\infty} \tilde{n}(t)\tilde{x}^*(t - \hat{\tau}) \exp(-j2\pi\hat{\phi}t)\, dt, \qquad (5.3\text{-}12)$$

respectively. If we now make the change in variable $t' = t - \tau_{\mathrm{A}}$ in Eq. (5.3-11), then

$$\tilde{l}_{\mathrm{T}} = cX(\tau, \phi) \exp(+j2\pi\phi\tau_{\mathrm{A}}), \qquad (5.3\text{-}13)$$

where

$$\begin{aligned} X(\tau, \phi) &= \langle \tilde{x}(t), \tilde{x}(t - \tau) \exp(-j2\pi\phi t) \rangle \\ &= \int_{-\infty}^{\infty} \tilde{x}(t)\tilde{x}^*(t - \tau) \exp(+j2\pi\phi t)\, dt \end{aligned} \qquad (5.3\text{-}14)$$

is the *unnormalized auto-ambiguity function* of $\tilde{x}(t)$,

$$\tau = \hat{\tau} - \tau_{\mathrm{A}}, \qquad (5.3\text{-}15)$$

and

$$\phi = \phi_{\mathrm{A}} - \hat{\phi}. \qquad (5.3\text{-}16)$$

The auto-ambiguity function has its maximum value at $\tau = 0$ and $\phi = 0$, that is,

$$X(0, 0) = \langle \tilde{x}(t), \tilde{x}(t) \rangle = \int_{-\infty}^{\infty} |\tilde{x}(t)|^2\, dt = E_{\tilde{x}}, \qquad (5.3\text{-}17)$$

which is the *energy* of $\tilde{x}(t)$. From Eqs. (5.3-15) and (5.3-16) it can be seen that the values $\tau = 0$ and $\phi = 0$ imply that the estimates $\hat{\tau}$ and $\hat{\phi}$ are equal to the actual or true values τ_{A} and ϕ_{A}. If we take the magnitude square of both sides of Eq. (5.3-14) and use the *Schwarz inequality*

$$\left| \int_{-\infty}^{\infty} u(\alpha)v(\alpha)\, d\alpha \right|^2 \leq \int_{-\infty}^{\infty} |u(\alpha)|^2\, d\alpha \int_{-\infty}^{\infty} |v(\alpha)|^2\, d\alpha, \qquad (5.3\text{-}18)$$

where $u(\alpha)$ and $v(\alpha)$ are any square-integrable complex functions, then

$$|X(\tau, \phi)|^2 \leq \int_{-\infty}^{\infty} |\tilde{x}(t)|^2\, dt \int_{-\infty}^{\infty} |\tilde{x}^*(t - \tau)|^2\, dt, \qquad (5.3\text{-}19)$$

or, from Eq. (5.3-17),

$$|X(\tau, \phi)|^2 \leq |X(0,0)|^2 = E_{\tilde{x}}^2, \tag{5.3-20}$$

which confirms our statement that $X(\tau, \phi)$ has its maximum value at $\tau = 0$ and $\phi = 0$.

The next step in the analysis is to compute the mean-square values of $\tilde{l}_T$ and $\tilde{l}_N$. From Eqs. (5.3-13) and (5.3-4), we obtain

$$|\tilde{l}_T|^2 = a^2 |X(\tau, \phi)|^2, \tag{5.3-21}$$

so that the mean-square value is given by

$$E\{|\tilde{l}_T|^2\} = E\{a^2\} |X(\tau, \phi)|^2. \tag{5.3-22}$$

Similarly, from Eq. (5.3-12),

$$|\tilde{l}_N|^2 = \tilde{l}_N \tilde{l}_N^* = \int_{-\infty}^{\infty} \int_{-\infty}^{\infty} \tilde{n}(t)\tilde{n}^*(t')\tilde{x}^*(t - \hat{\tau})\tilde{x}(t' - \hat{\tau}) \exp[-j2\pi\hat{\phi}(t - t')]\, dt\, dt', \tag{5.3-23}$$

so that

$$E\{|\tilde{l}_N|^2\} = \int_{-\infty}^{\infty} \int_{-\infty}^{\infty} R_{\tilde{n}}(t, t')\tilde{x}^*(t - \hat{\tau})\tilde{x}(t' - \hat{\tau}) \exp[-j2\pi\hat{\phi}(t - t')]\, dt\, dt', \tag{5.3-24}$$

where

$$R_{\tilde{n}}(t, t') = E\{\tilde{n}(t)\tilde{n}^*(t')\} \tag{5.3-25}$$

is the autocorrelation function of $\tilde{n}(t)$. Since it was assumed that $\tilde{n}(t)$ is a wide-sense stationary, bandlimited, white noise process, its power spectral density function $S_{\tilde{n}}(f)$ is given by

$$S_{\tilde{n}}(f) = \begin{cases} N_0, & |f| \leq W_N, \\ 0, & |f| > W_N, \end{cases} \tag{5.3-26}$$

where W_N is the bandwidth in hertz. In addition, since

$$R_{\tilde{n}}(t, t') = R_{\tilde{n}}(\Delta t) = F_f^{-1}\{S_{\tilde{n}}(f)\}, \tag{5.3-27}$$

where $\Delta t = t - t'$, from Eq. (5.3-26),

$$R_{\tilde{n}}(\Delta t) = N_0 \int_{-W_N}^{W_N} \exp(+j2\pi f \Delta t)\, df \tag{5.3-28}$$

or

$$R_{\tilde{n}}(\Delta t) = N_0 \sin(2\pi W_N \Delta t)/(\pi \Delta t). \tag{5.3-29}$$

Now, if W_N is larger than the bandwidth W of $\tilde{x}(t)$, then Eq. (5.3-29) can be approximated by

$$R_{\tilde{n}}(\Delta t) = N_0\, \delta(\Delta t), \tag{5.3-30}$$

so that Eq. (5.3-27) becomes

$$R_{\tilde{n}}(t, t') = N_0\, \delta(t - t'). \tag{5.3-31}$$

Substituting Eq. (5.3-31) into Eq. (5.3-24) yields

$$\mathrm{E}\{|\tilde{l}_N|^2\} = N_0 \int_{-\infty}^{\infty} |\tilde{x}(t - \hat{\tau})|^2\, dt \tag{5.3-32}$$

or

$$\mathrm{E}\{|\tilde{l}_N|^2\} = N_0 E_{\tilde{x}}, \tag{5.3-33}$$

and upon substituting Eqs. (5.3-22) and (5.3-33) into Eq. (5.3-7), we finally obtain the desired result

$$\mathrm{SNR} = \mathrm{E}\{a^2\} E_{\tilde{x}} |X_N(\tau, \phi)|^2 / N_0, \tag{5.3-34}$$

where

$$X_N(\tau, \phi) \triangleq X(\tau, \phi)/E_{\tilde{x}} \tag{5.3-35}$$

is the *normalized* auto-ambiguity function so that [see Eq. (5.3-17)]

$$X_N(0, 0) = 1. \tag{5.3-36}$$

Because of the assumption of Gaussian statistics, the error performance of this receiver for a Neyman–Pearson test is given by

$$P_D = P_{FA}^{1/(1+\mathrm{SNR})}, \tag{5.3-37}$$

where P_D and P_{FA} are the *probabilities of detection and false alarm*, respectively.

Let us now attempt to provide an interpretation of the normalized auto-ambiguity function $X_N(\tau, \phi)$. From Eq. (5.3-34) it can be seen that the SNR is directly proportional to $|X_N(\tau, \phi)|^2$. When $\tau = 0$ and $\phi = 0$, that is, when $\hat{\tau} = \tau_A$ and $\hat{\phi} = \phi_A$, the SNR is a maximum, as would be expected, with value [see Eqs. (5.3-34) and (5.3-36)]

$$\mathrm{SNR}_{max} = \mathrm{E}\{a^2\} E_{\tilde{x}} / N_0. \tag{5.3-38}$$

However, when $\hat{\tau} \neq \tau_A$ and $\hat{\phi} \neq \phi_A$, we would like the SNR to decrease as the errors in the estimates of τ_A and ϕ_A increase. Therefore, in order to insure this kind of performance and, hence, to avoid any ambiguity concerning the estimates of τ_A and ϕ_A, the normalized auto-ambiguity function should have as narrow a main lobe about the origin ($\tau = 0, \phi = 0$) as possible so that as the errors in the estimates of τ_A and ϕ_A increase, the value of $X_N(\tau, \phi)$ decreases rapidly and, hence, the SNR decreases. A small SNR would then indicate that a target with round-trip time delay $\hat{\tau}$ and Doppler shift $\hat{\phi}$ is *not* present.

Now, if it is assumed that τ_A and ϕ_A are *known* constants and if we let

$$\tilde{g}(t) = \tilde{x}(t - \tau_A)\exp(+j2\pi\phi_A t), \tag{5.3-39}$$

then the receiver shown in Fig. 5.3-1 with processing waveform $\tilde{g}(t)$ given by Eq. (5.3-39) is the *optimal, log-likelihood ratio test* for detecting signals with random amplitude and phase, where $|\tilde{l}|^2$ is the *test statistic*. The error performance of this receiver for a Neyman–Pearson test is given by Eq. (5.3-37), with the SNR given by Eq. (5.3-38). For this particular problem, the probability of detection can be increased by simply increasing the transmitted signal's energy $E_{\tilde{x}}$ [see Eq. (5.2-57)].

5.3.2 Common Transmitted Signals

In this section we shall discuss the normalized auto-ambiguity functions of common transmitted signals. Since the auto-ambiguity function $X(\tau, \phi)$ involves the complex envelope of the transmitted signal $\tilde{x}(t)$ [see Eq. (5.3-14)], different signals have different ambiguity functions and, hence, different range and Doppler resolving properties. We begin our discussion with the rectangular-envelope, continuous-wave (cw) pulse.

Rectangular-Envelope cw Pulse. A *cw pulse* is nothing more than a DSBSC waveform [see Eq. (5.2-21)], that is,

$$x(t) = \begin{cases} a(t)\cos(2\pi f_c t), & |t| \le T/2, \\ 0, & |t| > T/2, \end{cases} \tag{5.3-40}$$

where T is the *pulse length* in seconds. By referring to Eqs. (5.2-26) and (5.2-27), it can be seen that the complex envelope and the envelope of Eq. (5.3-40) are given by

$$\tilde{x}(t) = \begin{cases} a(t), & |t| \le T/2, \\ 0, & |t| > T/2, \end{cases} \tag{5.3-41}$$

and

$$E(t) = \begin{cases} |a(t)|, & |t| \le T/2, \\ 0, & |t| > T/2, \end{cases} \tag{5.3-42}$$

respectively. If the amplitude-modulating signal $a(t)$ is equal to A (a constant), then

$$x(t) = \begin{cases} A\cos(2\pi f_c t), & |t| \le T/2, \\ 0, & |t| > T/2, \end{cases} \tag{5.3-43}$$

is known as a *rectangular-envelope, cw pulse* (see Fig. 5.3-2). The complex envelope and envelope of Eq. (5.3-43) can be obtained from Eqs. (5.3-41) and

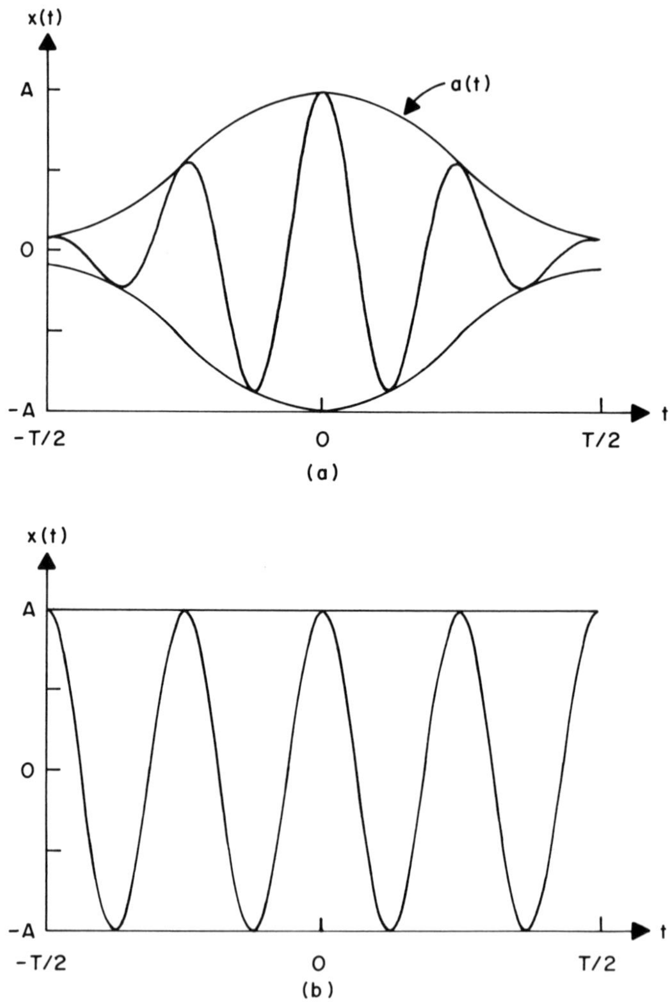

FIG. 5.3-2 The cw pulse with (a) an arbitrary envelope $a(t) \ge 0$ and (b) a rectangular envelope $a(t) = A$ (a constant).

(5.3-42), respectively, by setting $a(t)$ equal to A. The energy of the complex envelope of Eq. (5.3-43) is given by

$$E_{\tilde{x}} = \int_{-\infty}^{\infty} |\tilde{x}(t)|^2 \, dt \tag{5.3-44}$$

$$= \int_{-T/2}^{T/2} A^2 \, dt \tag{5.3-45}$$

$$= A^2 T, \tag{5.3-46}$$

which can be related to the energy of the real signal $x(t)$ by using Eq. (5.2-57).

Finally, it can be shown that the normalized auto-ambiguity function [defined by Eq. (5.3-38)] of the rectangular-envelope, cw pulse is given by

$$X_N(\tau, \phi) = \begin{cases} \left(1 - \dfrac{|\tau|}{T}\right)\left(\dfrac{\sin[\pi\phi T(1 - |\tau|/T)]}{\pi\phi T(1 - |\tau|/T)}\right), & |\tau| \leq T, \\ 0, & |\tau| > T. \end{cases} \tag{5.3-47}$$

By setting $\phi = 0$ in Eq. (5.3-47), we obtain the following expression for the round-trip time-delay (range) profile:

$$X_N(\tau, 0) = \begin{cases} 1 - |\tau|/T, & |\tau| \leq T, \\ 0, & |\tau| > T, \end{cases} \tag{5.3-48}$$

which is the equation for a triangle. Similarly, by setting $\tau = 0$ in Eq. (5.3-47), we obtain the following expression for the Doppler profile:

$$X_N(0, \phi) = \sin(\pi\phi T)/(\pi\phi T). \tag{5.3-49}$$

By inspecting Eqs. (5.3-48) and (5.3-49), it can be seen that the width of the ambiguity function along the τ axis is *directly proportional* to the pulse length T, while the width along the ϕ axis is *inversely proportional* to T. A single parameter, the *pulse length* T, controls both the round-trip time delay (range) and Doppler resolving capabilities of the rectangular-envelope, cw pulse. One can make the ambiguity function arbitrarily narrow in either direction by varying T, but not in both directions simultaneously. As a result, *a long cw pulse has poor range resolution but good Doppler resolution, and a short cw pulse has good range resolution but poor Doppler resolution.* If we are going to improve our range and Doppler estimates *simultaneously*, we must try a more complicated signal that contains several parameters that we can vary. The rectangular-envelope, linear frequency-modulated (LFM) pulse is such a signal and will be discussed next. However, before doing so, it should be mentioned that besides wanting $X_N(\tau, \phi)$ to have as narrow a main lobe about the origin as possible, it is also desirable that $X_N(\tau, \phi)$ have minimum sidelobe levels.

The sidelobe levels of $X_N(\tau, \phi)$ can be reduced by using amplitude functions $a(t)$ other than rectangular, analogous to using different amplitude windows to reduce the sidelobe levels of far-field beam patterns. In fact, one can use the same amplitude windows discussed in Section 3.3.1 for $a(t)$. However, as with the beam patterns, there is a trade-off since, as the sidelobe levels decrease, the width of the main lobe of $X_N(\tau, \phi)$ generally increases.

Rectangular-Envelope LFM Pulse. If we begin with an amplitude- and angle-modulated carrier [see Eq. (5.2-28)]

$$x(t) = \begin{cases} a(t)\cos[2\pi f_c t + \theta(t)], & |t| \le T/2, \\ 0, & |t| > T/2, \end{cases} \tag{5.3-50}$$

where T is the pulse length in seconds, and if we let the angle-modulating signal or phase deviation be given by

$$\theta(t) = bt^2, \tag{5.3-51}$$

so that the frequency deviation

$$\frac{d}{dt}\theta(t) = 2bt \tag{5.3-52}$$

is equal to a *linear* function of time t, then the signal

$$x(t) = \begin{cases} a(t)\cos(2\pi f_c t + bt^2), & |t| \le T/2, \\ 0, & |t| > T/2, \end{cases} \tag{5.3-53}$$

is known as a *LFM pulse*. By referring to Eqs. (5.2-41) and (5.2-44), it can be seen that the complex envelope and the envelope of Eq. (5.3-53) are given by

$$\tilde{x}(t) = \begin{cases} a(t)\exp(+jbt^2), & |t| \le T/2, \\ 0, & |t| > T/2, \end{cases} \tag{5.3-54}$$

and

$$E(t) = \begin{cases} |a(t)|, & |t| \le T/2, \\ 0, & |t| > T/2, \end{cases} \tag{5.3-55}$$

respectively. If the amplitude-modulating signal $a(t)$ is equal to A (a constant), then

$$x(t) = \begin{cases} A\cos(2\pi f_c t + bt^2), & |t| \le T/2, \\ 0, & |t| > T/2, \end{cases} \tag{5.3-56}$$

is known as a *rectangular-envelope, LFM pulse*. The complex envelope and envelope of Eq. (5.3-56) can be obtained from Eqs. (5.3-54) and (5.3-55), respectively, by setting $a(t)$ equal to A. The energy of the complex envelope of Eq. (5.3-56) is given by [see Eqs. (5.3-44) through (5.3-46)]

$$E_{\tilde{x}} = A^2 T. \tag{5.3-46}$$

The instantaneous phase and instantaneous frequency in hertz of Eqs. (5.3-53) and (5.3-56) are given by [see Eqs. (5.2-29) and (5.2-31)]

$$\theta_i(t) = \begin{cases} 2\pi f_c t + bt^2, & |t| \leq T/2, \\ 0, & |t| > T/2, \end{cases} \tag{5.3-57}$$

and

$$f_i(t) = \begin{cases} f_c + (bt/\pi), & |t| \leq T/2, \\ 0, & |t| > T/2, \end{cases} \tag{5.3-58}$$

respectively (see Fig. 5.3-3). From Fig. 5.3-3 it can be seen that the amount of *frequency sweep* is equal to

$$\left(f_c + \frac{bT}{2\pi}\right) - \left(f_c - \frac{bT}{2\pi}\right) = \frac{bT}{\pi} \quad \text{Hz.} \tag{5.3-59}$$

The parameter bT/π is sometimes referred to as the *swept bandwidth*, in hertz. A LFM pulse is also known as a *chirp pulse*. When the phase-deviation constant $b > 0$, a LFM pulse is called an "up chirp," and when $b < 0$, it is called a "down chirp."

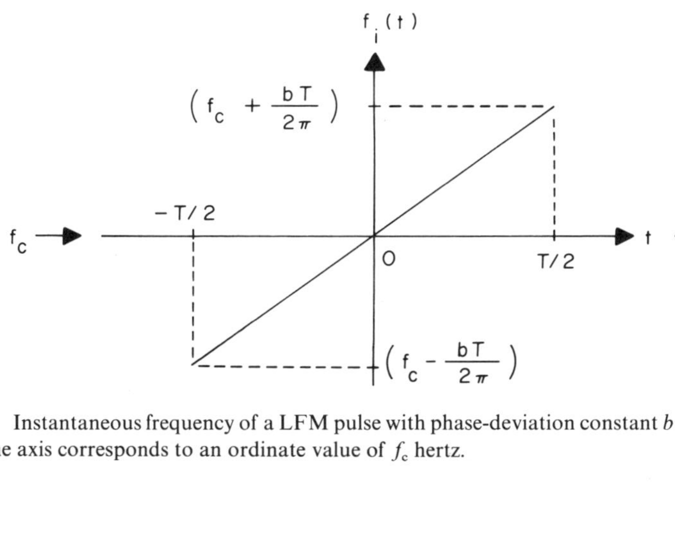

FIG. 5.3-3 Instantaneous frequency of a LFM pulse with phase-deviation constant $b > 0$. Note that the time axis corresponds to an ordinate value of f_c hertz.

Finally, it can be shown that the normalized auto-ambiguity function of the rectangular-envelope, LFM pulse is given by

$$X_N(\tau,\phi) = \begin{cases} \left(1 - \dfrac{|\tau|}{T}\right)\left(\dfrac{\sin[(\pi\phi - b\tau)(T - |\tau|)]}{(\pi\phi - b\tau)(T - |\tau|)}\right), & |\tau| \leq T, \\ 0, & |\tau| > T. \end{cases} \tag{5.3-60}$$

Note that if we let $b = 0$, then Eq. (5.3-60) reduces to Eq. (5.3-47) for the rectangular-envelope, cw pulse. By setting $\phi = 0$ in Eq. (5.3-60), we obtain the following expression for the round-trip time-delay (range) profile:

$$X_N(\tau,0) = \begin{cases} \left(1 - \dfrac{|\tau|}{T}\right)\left(\dfrac{\sin[b\tau(T - |\tau|)]}{b\tau(T - |\tau|)}\right), & |\tau| \leq T, \\ 0, & |\tau| > T. \end{cases} \tag{5.3-61}$$

Similarly, by setting $\tau = 0$ in Eq. (5.3-60), we obtain the following expression for the Doppler profile:

$$X_N(0,\phi) = \sin(\pi\phi T)/(\pi\phi T), \tag{5.3-62}$$

which is identical to the Doppler profile of the rectangular-envelope, cw pulse given by Eq. (5.3-49). The range profile given by Eq. (5.3-61) has its first zero crossings along the τ axis when the argument of the sine function is equal to π radians, that is, when

$$|b|\,|\tau|T - |b|\tau^2 = \pi$$

or

$$\tau^2 - T|\tau| + (\pi/|b|) = 0, \tag{5.3-63}$$

where b and τ were replaced by $|b|$ and $|\tau|$, respectively, since

$$\frac{\sin(-x)}{-x} = \frac{\sin(x)}{x}.$$

Using the quadratic formula to solve Eq. (5.3-63) yields

$$|\tau| = \frac{T}{2} \pm \frac{[T^2 - (4\pi/|b|)]^{1/2}}{2}, \tag{5.3-64}$$

and upon choosing the minus sign, Eq. (5.3-64) can be rewritten as

$$|\tau| = \frac{T}{2}\left(1 - \left[1 - \frac{4\pi}{|b|T^2}\right]^{1/2}\right), \tag{5.3-65}$$

where

$$\frac{4\pi}{|b|T^2} \leq 1 \tag{5.3-66}$$

must be satisfied in order to insure that τ is real and *not* complex. However, if we impose the more stringent condition that

$$\frac{4\pi}{|b|T^2} \leq 0.1, \tag{5.3-67}$$

then we can use a *binomial expansion* of the square root in Eq. (5.3-65). Using the first two terms in the binomial expansion yields [see Eq. (3.2-16)]

$$|\tau| \approx \frac{T}{2}\left(1 - 1 + \frac{2\pi}{|b|T^2}\right)$$

or

$$|\tau| \approx \pi/(|b|T) \quad \text{sec}, \tag{5.3-68}$$

which is the desired result and is equal to the reciprocal of the swept bandwidth [see Eq. (5.3-59) for $b > 0$]. Since the range profile is symmetric about $\tau = 0$,

$$\tau \approx \pm\pi/(|b|T) \quad \text{sec} \tag{5.3-69}$$

are the locations of the first zero crossings along the τ axis when Eq. (5.3-67) is satisfied. The Doppler profile given by Eq. (5.3-62) has its first zero crossings at

$$\phi = \pm 1/T \quad \text{Hz}. \tag{5.3-70}$$

Now we have two parameters, b and T, that we can vary. By inspecting Eqs. (5.3-69) and (5.3-70), it can be seen that the width of the ambiguity function along *both* the τ and ϕ axes is *inversely proportional* to T. We can *simultaneously* control the range and Doppler resolution of a rectangular-envelope, LFM pulse by adjusting the *phase-deviation constant* b and the *pulse length* T, respectively. Note that the Doppler resolving capabilities of a rectangular-envelope, cw pulse and a rectangular-envelope, LFM pulse are identical.

Let us conclude this discussion by noting that the normalized auto-ambiguity function of a rectangular-envelope, LFM pulse given by Eq. (5.3-60) has $\sin(x)/x$ type profiles along every line parallel to and including the ϕ axis, that is, along every line with a constant τ value. For a given value of τ, the maximum value of the corresponding profile is $1 - |\tau|/T$, and this maximum value is located at

$$\phi = b\tau/\pi, \tag{5.3-70}$$

which is the equation of a straight line in the $\tau\phi$ plane that passes through the origin. Therefore, according to Eq. (5.3-70), when $b > 0$ ("up chirp"), $X_N(\tau, \phi)$ will be concentrated mainly in the first and third quadrants of the $\tau\phi$ plane; and when $b < 0$ ("down chirp"), $X_N(\tau, \phi)$ will be concentrated mainly in the second and fourth quadrants. If one wants to cover all four quadrants about the origin, one can first transmit an "up chirp" pulse followed by a "down chirp" pulse or vice versa.

5.4 TIME-COMPRESSION–STRETCH FACTOR, TIME DELAY, AND DOPPLER SHIFT EXPRESSIONS

Let us consider the bistatic scatter geometry shown in Fig. 5.4-1, where T and R represent the transmit and receive apertures (arrays), respectively; S represents the point scatterer or target; $\mathbf{V}_T$, $\mathbf{V}_R$, and $\mathbf{V}_S$ are the *constant* velocity vectors (i.e., *no* accelerations) of the transmit and receive apertures (arrays) and scatterer, respectively; $\hat{n}_T$ and $\hat{n}_R$ are unit vectors in the directions of propagation of the transmitted and scattered waves, respectively; and, finally, R_{0_T} and R_{0_R} are the initial ranges of the scatterer from the transmit and receive apertures (arrays), respectively, when transmission begins at time $t = 0$. Let $\tilde{x}(t)$ represent the complex envelope of the transmitted signal and $\tilde{y}(t)$ the complex envelope of the scattered or return signal. Therefore, with respect to the geometry shown in Fig. 5.4-1, $\tilde{y}(t)$ can be modeled as a time- and frequency-shifted replica of $\tilde{x}(t)$, as follows [compare Eq. (5.3-3) with Eq. (5.4-1)]:

$$\tilde{y}(t) = \tilde{x}[s(t - \tau)] \exp(+j2\pi\phi t), \tag{5.4-1}$$

where s is the dimensionless *time-compression–stretch factor* given by

$$s = \frac{(1 - \beta_{ST})(1 - \beta_{RR})}{(1 - \beta_{SR})(1 - \beta_{TT})}, \tag{5.4-2}$$

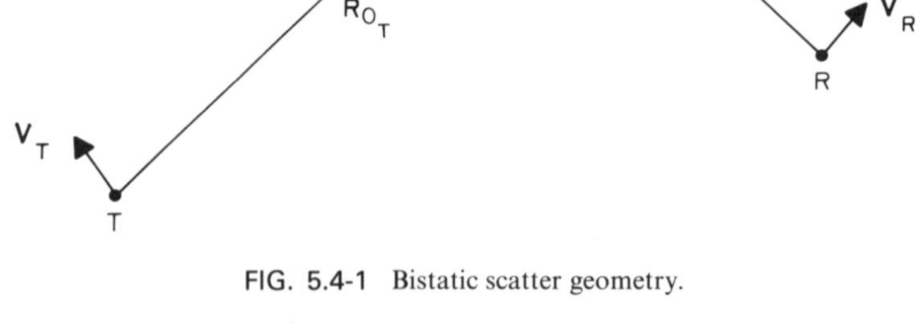

FIG. 5.4-1 Bistatic scatter geometry.

where

$$\beta_{\mathrm{ST}} = \mathbf{V}_{\mathrm{S}} \cdot \hat{n}_{\mathrm{T}}/c, \tag{5.4-3}$$

$$\beta_{\mathrm{RR}} = \mathbf{V}_{\mathrm{R}} \cdot \hat{n}_{\mathrm{R}}/c, \tag{5.4-4}$$

$$\beta_{\mathrm{SR}} = \mathbf{V}_{\mathrm{S}} \cdot \hat{n}_{\mathrm{R}}/c, \tag{5.4-5}$$

and

$$\beta_{\mathrm{TT}} = \mathbf{V}_{\mathrm{T}} \cdot \hat{n}_{\mathrm{T}}/c, \tag{5.4-6}$$

τ is the *time delay* in seconds given by

$$\tau = \frac{[(1-\beta_{\mathrm{SR}})R_{0_{\mathrm{T}}}/(1-\beta_{\mathrm{ST}})] + R_{0_{\mathrm{R}}}}{(1-\beta_{\mathrm{RR}})c}, \tag{5.4-7}$$

ϕ is the *Doppler shift* in hertz given by

$$\phi = -(1-s)f_{\mathrm{c}}, \tag{5.4-8}$$

where f_{c} is the center or carrier frequency of the real transmitted bandpass signal

$$x(t) = \mathrm{Re}\{\tilde{x}(t)\exp(+j2\pi f_{\mathrm{c}} t)\}, \tag{5.2-7}$$

and c is the speed of sound (assumed constant) in meters per second. Note that Eqs. (5.4-3) through (5.4-6) are normalized radial components of velocity—normalized by the speed of sound—and, as a result, they are dimensionless quantities.

Equations (5.4-2) through (5.4-8) are general expressions that pertain to both bistatic and monostatic (backscatter) geometries. For a monostatic geometry, set

$$\mathbf{V}_{\mathrm{R}} = \mathbf{V}_{\mathrm{T}}, \tag{5.4-9}$$

$$\hat{n}_{\mathrm{R}} = -\hat{n}_{\mathrm{T}}, \tag{5.4-10}$$

and

$$R_{0_{\mathrm{R}}} = R_{0_{\mathrm{T}}} = R_0 \tag{5.4-11}$$

in Eqs. (5.4-3) through (5.4-7). Keep in mind that all velocity vectors and unit vectors are three-dimensional in general, that is, they generally have three components.

Note that $s > 1$ indicates *time compression*, that is, signal duration is decreased. As a result, signal bandwidth (in hertz) is increased. Also, when $s > 1$, $\phi > 0$, which indicates a positive Doppler shift [see Eq. (5.4-8)]. Similarly, $s < 1$ indicates *time stretch*, that is, signal duration is increased. As a result, signal bandwidth is decreased. Also, when $s < 1$, $\phi < 0$, which indicates a negative Doppler shift. Finally, $s = 1$ indicates no time compression

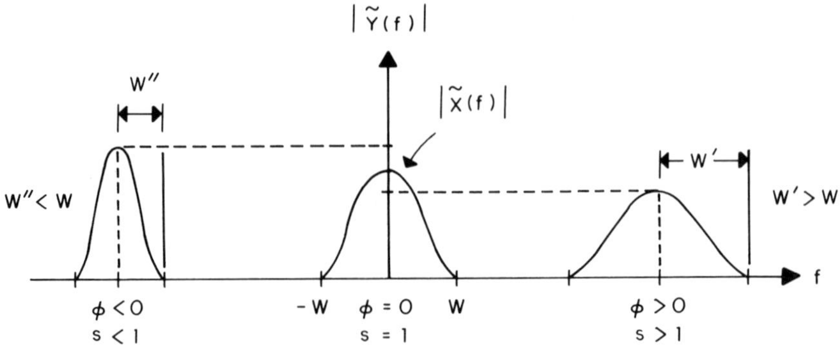

FIG. 5.4-2 Illustration of Eq. (5.4-12) for the three cases $s > 1$, $s < 1$, and $s = 1$.

or time stretch and, as a result, no change in signal bandwidth. Also, when $s = 1$, $\phi = 0$, which indicates no Doppler shift.

The three cases $s > 1$, $s < 1$, and $s = 1$ are illustrated in Fig. 5.4-2, where

$$|\tilde{Y}(f)| = (1/|s|)|\tilde{X}[(f - \phi)/s]| \tag{5.4-12}$$

and ·

$$\tilde{Y}(f) = F_t\{\tilde{y}(t)\},$$

with $\tilde{y}(t)$ given by Eq. (5.4-1). Recall that the complex envelope $\tilde{x}(t)$ is a low-pass (baseband) signal, so that $\tilde{X}(f)$ is centered about $f = 0$ Hz.

EXAMPLE 5.4-1

Referring to the bistatic scatter geometry shown in Fig. 5.4-1, assume that $\mathbf{V}_T = \mathbf{0}$ and $\mathbf{V}_R = \mathbf{0}$. Therefore, the time-compression–stretch factor reduces to

$$s = \frac{1 - \beta_{ST}}{1 - \beta_{SR}} = \frac{1 - (\mathbf{V}_S \cdot \hat{n}_T/c)}{1 - (\mathbf{V}_S \cdot \hat{n}_R/c)} \tag{5.4-13}$$

since $\beta_{TT} = 0$ and $\beta_{RR} = 0$. The time delay is given by

$$\tau = (1/s)\{\tau_0 - [(1 - s)R_{0_R}/c]\}, \tag{5.4-14}$$

where s is given by Eq. (5.4-13) and

$$\tau_0 = (R_{0_T} + R_{0_R})/c. \tag{5.4-15}$$

Finally, the Doppler shift is equal to

$$\phi = -\frac{[(\hat{n}_T - \hat{n}_R) \cdot \mathbf{V}_S]f_c}{c[1 - (\mathbf{V}_S \cdot \hat{n}_R/c)]} \tag{5.4-16}$$

or

$$\phi = -[(\hat{n}_T - \hat{n}_R) \cdot \mathbf{V}_S] f_c/c \tag{5.4-17}$$

since it is reasonable to assume that

$$|\mathbf{V}_S|/c \ll 1. \tag{5.4-18}$$

□

EXAMPLE 5.4-2

Referring to the monostatic (backscatter) geometry shown in Fig. 5.4-3, assume that $\mathbf{V}_T = \mathbf{0}$. Therefore, with the use of Eqs. (5.4-9) through (5.4-11), the time-compression–stretch factor reduces to

$$s = \frac{1 - \beta_{ST}}{1 - \beta_{SR}} = \frac{1 - (\mathbf{V}_S \cdot \hat{n}_T/c)}{1 + (\mathbf{V}_S \cdot \hat{n}_T/c)} \tag{5.4-19}$$

since $\beta_{TT} = 0$ and $\beta_{RR} = 0$. The round-trip time delay is given by

$$\tau = \tau_0[1 + (1/s)]/2, \tag{5.4-20}$$

where s is given by Eq. (5.4-19) and

$$\tau_0 = 2R_0/c. \tag{5.4-21}$$

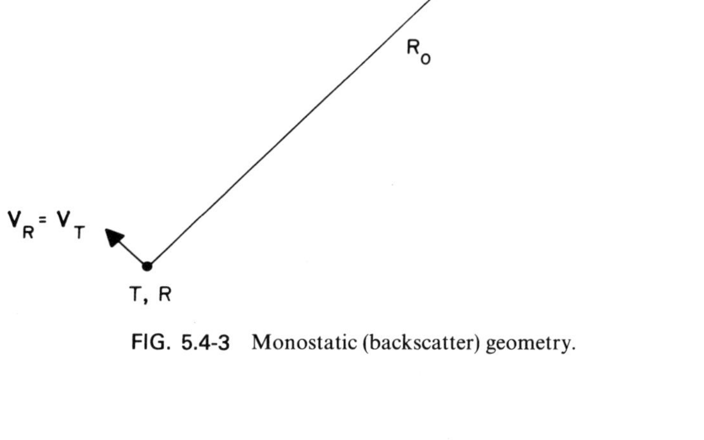

FIG. 5.4-3 Monostatic (backscatter) geometry.

Finally, the Doppler shift is equal to

$$\phi = -\frac{2(\mathbf{V}_S \cdot \hat{n}_T) f_c}{c[1 + (\mathbf{V}_S \cdot \hat{n}_T/c)]} \tag{5.4-22}$$

or

$$\phi = -2(\mathbf{V}_S \cdot \hat{n}_T) f_c/c \tag{5.4-23}$$

since it is reasonable to assume that

$$|\mathbf{V}_S|/c \ll 1. \tag{5.4-18}$$

These results are particularly easy to interpret. For example, when the point scatterer or target is moving *away* from the transmit–receive aperture (array), the radial component of velocity $\mathbf{V}_S \cdot \hat{n}_T > 0$, which means that $s < 1$, $\tau > \tau_0$, and $\phi < 0$, which is physically correct. Similarly, when the point scatterer or target is moving *toward* the transmit–receive aperture (array), the radial component of velocity $\mathbf{V}_S \cdot \hat{n}_T < 0$, which means that $s > 1$, $\tau < \tau_0$, and $\phi > 0$, which is also physically correct. □

PROBLEMS

5-1 Verify Eq. (5.1-12).

5-2 Recall that, in general,

$$Y(\eta, \mathbf{r}) = Y_M(\eta, \mathbf{r}) A_R(\eta, \mathbf{r}). \tag{3.1-11}$$

Assume we are given a planar array of $M \times N$ (odd) identical, complex-weighted point sources lying in the XY plane. In addition, assume that the acoustic field $y_M(t, \mathbf{r})$ incident upon the array is a general plane-wave field [see Eq. (5.1-6)]. Using Eq. (3.1-11) as the starting point, derive the theoretical output frequency and angular spectrum $Y(\eta, \gamma_X, \gamma_Y)$ given by Eq. (5.1-23). Do *not* use the convolution integral given by Eq. (3.2-51).

5-3 If we start with Eq. (5.2-10), we can rewrite the Hilbert transform of $x(t)$ as being equal to the following time-domain convolution expressions: since

$$\hat{x}(t) = F_f^{-1}\{-j\,\mathrm{sgn}(f) X(f)\}, \tag{5.2-10}$$

$$\hat{x}(t) = F_f^{-1}\{-j\,\mathrm{sgn}(f)\} * F_f^{-1}\{X(f)\},$$

or,

$$\hat{x}(t) = h(t) * x(t),$$

where

$$x(t) = F_f^{-1}\{X(f)\}$$

and

$$h(t) = F_f^{-1}\{H(f)\},$$
$$H(f) = -j\,\mathrm{sgn}(f)$$

being the transfer function and $h(t)$ the impulse response of a Hilbert-transform filter. Show that

$$h(t) = \frac{1}{\pi t}.$$

Therefore, we can also write that

$$\hat{x}(t) = \frac{1}{\pi t} * x(t) = \frac{1}{\pi}\int_{-\infty}^{\infty} \frac{x(\tau)}{(t-\tau)}\,d\tau.$$

Hint:

$$H(f) = \lim_{\substack{\alpha \to 0 \\ \alpha > 0}} \begin{cases} -j\exp(-\alpha f), & f > 0, \\ +j\exp(+\alpha f), & f < 0. \end{cases}$$

5-4 Use Eq. (5.2-10) to compute the Hilbert transform of the following signals:

(a) $\cos(2\pi f_c t + \theta_0)$,
(b) $\sin(2\pi f_c t + \theta_0)$.

Compare your answers with Eqs. (5.2-13) and (5.2-14), respectively.

5-5 If the input signal to the system shown in Fig. 5.2-6 is the amplitude- and angle-modulated carrier waveform

$$x(t) = a(t)\cos[2\pi f_c t + \theta(t)] \tag{5.2-28}$$

discussed in Example 5.2-2, show that the output signals from this system are the cosine and sine components $x_C(t)$ and $x_S(t)$ given by Eqs. (5.2-36) and (5.2-37), respectively.

5-6 Let

$$\Delta = \frac{E\{|\tilde{l}|^2 | H_1\} - E\{|\tilde{l}|^2 | H_0\}}{E\{|\tilde{l}|^2 | H_0\}},$$

where, under hypotheses H_1 and H_0, $\tilde{l}$ is given by

$$H_1: \tilde{l} = \tilde{l}_T + \tilde{l}_N,$$
$$H_0: \tilde{l} = \tilde{l}_N,$$

where $\tilde{l}_T$ and $\tilde{l}_N$ are given by Eqs. (5.3-8) and (5.3-9), respectively. If we assume that $\tilde{y}_T(t)$ and $\tilde{n}(t)$ are uncorrelated and that $\tilde{n}(t)$ is zero mean, show that

$$\Delta = \mathrm{E}\{|\tilde{l}_{\mathrm{T}}|^2\}/\mathrm{E}\{|\tilde{l}_{\mathrm{N}}|^2\},$$

which is equal to the SNR definition given by Eq. (5.3-7).

5-7 Design a rectangular-envelope, "down chirp," LFM pulse $x(t)$ with a range resolving capability of ± 1.0 m and a Doppler resolving capability of ± 1.0 Hz as measured by the locations of the first zero crossings of the range and Doppler profiles, respectively, of the corresponding normalized auto-ambiguity function. In addition, $x(t)$ must have unit energy [i.e., $E_x = 1.0$].

(a) What should the pulse length, phase deviation constant, and amplitude of $x(t)$ be in order to satisfy these requirements? Use $\tau = 2r/c$ and $c = 1500$ m/sec. Also, check to make sure Eq. (5.3-67) is satisfied.

(b) What is the resulting swept bandwidth in hertz?

5-8 Verify Eq. (5.4-12).

5-9 Consider a monostatic (backscatter) geometry where $\mathbf{V}_{\mathrm{R}} = \mathbf{V}_{\mathrm{T}} = \mathbf{0}$, $R_0 = 1000$ m, $\hat{n}_{\mathrm{T}} = +(\sqrt{2}/2)\hat{y} + (\sqrt{2}/2)\hat{z}$, $\mathbf{V}_{\mathrm{S}} = -1\hat{x} + 1\hat{y} + 2\hat{z}$, $c = 1500$ m/sec, and $f_{\mathrm{c}} = 10$ kHz.

(a) Compute the time-compression–stretch factor, round-trip time delay, and Doppler shift. Use Eqs. (5.4-22) and (5.4-23) to compute the Doppler shift and compare results. Evaluate Eq. (5.4-18) before using Eq. (5.4-23).

(b) Repeat part (a) using $\mathbf{V}_{\mathrm{S}} = +2\hat{x} - 3\hat{y}$.

5-10 Consider a monostatic (backscatter) geometry where $\mathbf{V}_{\mathrm{R}} = \mathbf{V}_{\mathrm{T}} = -1\hat{y}$, $R_0 = 2000$ m, $\hat{n}_{\mathrm{T}} = -(\sqrt{2}/2)\hat{y} + (\sqrt{2}/2)\hat{z}$, $\mathbf{V}_{\mathrm{S}} = +4\hat{y}$, $c = 1500$ m/sec, and $f_{\mathrm{c}} = 15$ kHz.

(a) Compute the time-compression–stretch factor, round-trip time delay, and Doppler shift.

(b) Repeat part (a) using $\hat{n}_{\mathrm{T}} = +(\sqrt{2}/2)\hat{y} - (\sqrt{2}/2)\hat{z}$.

BIBLIOGRAPHY

A. B. Baggeroer, Sonar signal processing. In *Applications of Digital Signal Processing* (A. V. Oppenheim, ed.), pp. 331–437. Prentice-Hall, Englewood Cliffs, New Jersey, 1978.

E. O. Brigham, *The Fast Fourier Transform*. Prentice-Hall, Englewood Cliffs, New Jersey, 1974.

J. A. Cadzow and H. R. Martens, *Discrete-Time and Computer Control Systems*. Prentice-Hall, Englewood Cliffs, New Jersey, 1970.

C. E. Cook and M. Bernfeld, *Radar Signals*. Academic Press, New York, 1967.

L. W. Couch, II, *Digital and Analog Communication Systems*. Macmillan, New York, 1983.

W. B. Davenport, Jr., *Probability and Random Processes*. McGraw-Hill, New York, 1970.

J. D. Gaskill, *Linear Systems, Fourier Transforms, and Optics*. Wiley, New York, 1978.

B. Gold and C. M. Rader, *Digital Processing of Signals*. McGraw-Hill, New York, 1969.

M. V. Klein, *Optics*. Wiley, New York, 1970.

D. Middleton, A statistical theory of reverberation and similar first-order scattered fields. Part I: waveforms and the general process. *IEEE Trans. Inf. Theory* **IT-13**, 372–392 (1967).

K. Ogata, *Modern Control Engineering*. Prentice-Hall, Englewood Cliffs, New Jersey, 1970.

A. Papoulis, *Signal Analysis*. McGraw-Hill, New York, 1977.

M. Schwartz and L. Shaw, *Signal Processing: Discrete Spectral Analysis, Detection, and Estimation*. McGraw-Hill, New York, 1975.

H. Stark and F. B. Tuteur, *Modern Electrical Communications*. Prentice-Hall, Englewood Cliffs, New Jersey, 1979.

F. G. Stremler, *Introduction to Communication Systems*, 2nd ed. Addison-Wesley, Reading, Massachusetts, 1982.

J. B. Thomas, *An Introduction to Statistical Communication Theory*. Wiley, New York, 1969.

H. L. Van Trees, *Detection, Estimation, and Modulation Theory. Part III*. Wiley, New York, 1971.

A. D. Whalen, *Detection of Signals in Noise*. Academic Press, New York, 1971.

R. E. Ziemer and W. H. Tranter, *Principles of Communications*. Houghton Mifflin, Boston, Massachusetts, 1976.

L. J. Ziomek, *A ScatteringFunction Approach to Underwater Acoustic Detection and Signal Design*. Ph. D. Dissertation, The Pennsylvania State University, University Park, Pennsylvania, 1981.

L. J. Ziomek, Comments on the generalized ambiguity function. *IEEE Trans. Acoust. Speech Signal Process.* **ASSP-30**, 117–119 (1982).

6

Wave Propagation in Inhomogeneous Media

6.1 THE WKB APPROXIMATION

In the first five chapters of this book it has been assumed that the speed of sound in the ocean medium is constant. Now, in this chapter, we shall discuss three different methods of obtaining approximate solutions to the wave equation for variable sound-speed profiles. The first method to be considered is the *geometrical optics approximation*, or *WKB approximation*. The remaining two methods, *ray acoustics* and *the parabolic equation approximation*, are discussed in Sections 6.2 and 6.3, respectively.

Let us begin by considering the solution to the following linear wave equation that describes the propagation of small-amplitude acoustic signals in the ocean:

$$\nabla^2 \varphi(t, \mathbf{r}) - \frac{1}{c^2(\mathbf{r})} \frac{\partial^2}{\partial t^2} \varphi(t, \mathbf{r}) = 0, \tag{6.1-1}$$

where $\varphi(t, \mathbf{r})$ is the *velocity potential* at time t and position $\mathbf{r} = (x, y, z)$ and $c(\mathbf{r})$ is the *variable sound-speed profile*, also shown as a function of position $\mathbf{r}$. The first step in obtaining a solution to Eq. (6.1-1) is to assume that the velocity potential $\varphi(t, \mathbf{r})$ has a time-harmonic dependence, that is,

$$\varphi(t, \mathbf{r}) = \varphi(\mathbf{r}) \exp(+j2\pi ft). \tag{6.1-2}$$

Substituting Eq. (6.1-2) into the time-dependent wave equation given by Eq. (6.1-1) yields the time-independent *Helmholtz wave equation*

$$\nabla^2 \varphi(\mathbf{r}) + k^2(\mathbf{r}) \varphi(\mathbf{r}) = 0, \tag{6.1-3}$$

where

$$k(\mathbf{r}) = 2\pi f/c(\mathbf{r}) \tag{6.1-4}$$

is the *wave number*. This Helmholtz wave equation can also be expressed in terms of the dimensionless *index of refraction*

$$n(\mathbf{r}) = c_0/c(\mathbf{r}), \tag{6.1-5}$$

where c_0 is the *constant, reference speed of sound* at the source location $\mathbf{r}_0 = (x_0, y_0, z_0)$, that is, $c_0 = c(\mathbf{r}_0)$. Note that $n(\mathbf{r}_0) = 1$. If we let

$$k_0 = 2\pi f/c_0 \tag{6.1-6}$$

be the *reference wave number*, then Eq. (6.1-4) can be rewritten as

$$k(\mathbf{r}) = k_0 n(\mathbf{r}). \tag{6.1-7}$$

Therefore, substituting Eq. (6.1-7) into Eq. (6.1-3) yields

$$\nabla^2\varphi(\mathbf{r}) + k_0^2 n^2(\mathbf{r})\varphi(\mathbf{r}) = 0. \tag{6.1-8}$$

The index of refraction $n(\mathbf{r})$ will be treated as a *deterministic function* in this chapter. However, in Chapter 7 $n(\mathbf{r})$ will be decomposed into both *deterministic and random components* so that wave propagation in a *random ocean* can be studied.

The WKB (Wentzel, Kramers, and Brillouin) method addresses the problem of obtaining a solution to the following simplified Helmholtz wave equation:

$$\nabla^2\varphi(x, y, z) + k^2(y)\varphi(x, y, z) = 0, \tag{6.1-9}$$

where

$$k(y) = k_0 n(y) = 2\pi f/c(y), \tag{6.1-10}$$

that is, the speed of sound $c(y)$ is assumed to be only a function of depth y. The rectangular coordinates x, y, and z refer to cross-range, depth, and range, respectively (see Fig. 1.1-2).

We begin the solution to Eq. (6.1-9) by using the method of *separation of variables*, that is, we assume a solution $\varphi(x, y, z)$ of the form

$$\varphi(x, y, z) = X(x)Y(y)Z(z). \tag{6.1-11}$$

Substituting Eq. (6.1-11) into Eq. (6.1-9) yields

$$\frac{1}{Y(y)}\frac{d^2}{dy^2}Y(y) + k^2(y) = -\frac{1}{X(x)}\frac{d^2}{dx^2}X(x) - \frac{1}{Z(z)}\frac{d^2}{dz^2}Z(z). \tag{6.1-12}$$

Since the left-hand side of Eq. (6.1-12) is a function of y and the right-hand side is a function of both x and z, equality is only possible if both sides are equal to a constant, say, K^2. Therefore, the right-hand side of Eq. (6.1-12) is equal to

$$-\frac{1}{X(x)}\frac{d^2}{dx^2}X(x) - \frac{1}{Z(z)}\frac{d^2}{dz^2}Z(z) = K^2 \tag{6.1-13}$$

or

$$-\frac{1}{X(x)}\frac{d^2}{dx^2}X(x) = \frac{1}{Z(z)}\frac{d^2}{dz^2}Z(z) + K^2 = k_X^2 \tag{6.1-14}$$

since equality is possible only if both sides of Eq. (6.1-14) are equal to a constant, say, k_X^2. The constants K^2 and k_X^2 are referred to as *separation constants.* From Eq. (6.1-14) we obtain the differential equation

$$\frac{d^2}{dx^2}X(x) + k_X^2 X(x) = 0, \tag{6.1-15}$$

which has the exact solution

$$X(x) = A_X \exp(-jk_X x) + B_X \exp(+jk_X x), \tag{6.1-16}$$

where A_X and B_X are arbitrary constants. The first term on the right-hand side of Eq. (6.1-16) represents a plane wave traveling in the positive X direction, while the second term represents a plane wave traveling in the negative X direction. Also, from Eq. (6.1-14) we have the additional differential equation

$$\frac{d^2}{dz^2}Z(z) + k_Z^2 Z(z) = 0, \tag{6.1-17}$$

where

$$k_Z^2 = K^2 - k_X^2 \tag{6.1-18}$$

is a constant. Equation (6.1-17) has the exact solution

$$Z(z) = A_Z \exp(-jk_Z z) + B_Z \exp(+jk_Z z), \tag{6.1-19}$$

where A_Z and B_Z are arbitrary constants and the first term on the right-hand side of Eq. (6.1-19) represents a plane wave traveling in the positive Z direction, while the second term represents a plane wave traveling in the negative Z direction. Finally, returning to Eq. (6.1-12), the left-hand side must also be equal to K^2, that is,

$$\frac{1}{Y(y)}\frac{d^2}{dy^2}\,Y(y) + k^2(y) = K^2, \tag{6.1-20}$$

or, with the use of Eq. (6.1-18), we obtain the following differential equation for

$Y(y)$:

$$\frac{d^2}{dy^2} Y(y) + k_Y^2(y) Y(y) = 0, \tag{6.1-21}$$

where

$$k_Y^2(y) = k^2(y) - k_X^2 - k_Z^2 \tag{6.1-22}$$

or [see Eq. (6.1-10)]

$$k_Y^2(y) = k_0^2 n^2(y) - k_X^2 - k_Z^2. \tag{6.1-23}$$

Note that the *propagation vector* $\mathbf{k}(y)$ can be written as

$$\mathbf{k}(y) = k_X \hat{x} + k_Y(y)\hat{y} + k_Z \hat{z}, \tag{6.1-24}$$

where k_X and k_Z are the constant X and Z components, respectively, and $k_Y(y)$ is the Y component and is a function of depth.

In order to obtain a solution to Eq. (6.1-21), we shall follow the WKB method, which begins by assuming that

$$Y(y) = a_Y(y) \exp[+j\theta_Y(y)], \tag{6.1-25}$$

where $a_Y(y)$ and $\theta_Y(y)$ are *real amplitude and phase functions*, respectively. Substituting Eq. (6.1-25) into Eq. (6.1-21) yields

$$a_Y''(y) + [k_Y^2(y) - [\theta_Y'(y)]^2] a_Y(y) + j[a_Y(y)\theta_Y''(y) + 2a_Y'(y)\theta_Y'(y)] = 0, \tag{6.1-26}$$

where the prime and double prime symbols denote first- and second-order derivatives, respectively. Since the left-hand side of Eq. (6.1-26) must equal the right-hand side, which is zero, both the real and imaginary parts of the left-hand side of Eq. (6.1-26) must also be equal to zero, that is,

$$a_Y''(y) + [k_Y^2(y) - [\theta_Y'(y)]^2] a_Y(y) = 0 \tag{6.1-27}$$

and

$$a_Y(y)\theta_Y''(y) + 2a_Y'(y)\theta_Y'(y) = 0. \tag{6.1-28}$$

Rewriting Eq. (6.1-28) as

$$\frac{1}{a_Y(y)} \frac{d}{dy} a_Y(y) = -\frac{1}{2} \frac{1}{\theta_Y'(y)} \frac{d}{dy} \theta_Y'(y) \tag{6.1-29}$$

and multiplying both sides of Eq. (6.1-29) by dy and integrating yields

$$\int \frac{1}{a_Y(y)} da_Y(y) = -\frac{1}{2} \int \frac{1}{\theta_Y'(y)} d\theta_Y'(y) \tag{6.1-30}$$

or

$$\ln[a_Y(y)] = -\tfrac{1}{2}\ln[\theta'_Y(y)] + C, \qquad (6.1\text{-}31)$$

where C is an arbitrary constant of integration. Solving Eq. (6.1-31) for $a_Y(y)$ yields

$$a_Y(y) = A/[\theta'_Y(y)]^{1/2}, \qquad (6.1\text{-}32)$$

where $A = \exp(C)$.

The next step is to rewrite Eq. (6.1-27) as follows:

$$1 + \frac{1}{k_Y^2(y)}\left[\frac{a''_Y(y)}{a_Y(y)} - [\theta'_Y(y)]^2\right] = 0. \qquad (6.1\text{-}33)$$

If it is assumed that

$$|a''_Y(y)/a_Y(y)| \ll [\theta'_Y(y)]^2, \qquad (6.1\text{-}34)$$

then Eq. (6.1-33) reduces to

$$[\theta'_Y(y)]^2 \approx k_Y^2(y) \qquad (6.1\text{-}35)$$

or

$$\frac{d}{dy}\theta_Y(y) \approx \pm k_Y(y). \qquad (6.1\text{-}36)$$

Multiplying both sides of Eq. (6.1-36) by dy and integrating yields the following expression for the real phase function:

$$\theta_Y(y) \approx \pm \int_{y_0}^{y} k_Y(\zeta)\,d\zeta + \theta_Y(y_0), \qquad (6.1\text{-}37)$$

where $\theta_Y(y_0)$ is a constant phase factor. And, finally, since $a_Y(y)$ was also stipulated as being a real function, substituting the right-hand side of Eq. (6.1-36) with the plus sign into Eq. (6.1-32) yields the following expression for the real amplitude function:

$$a_Y(y) \approx A/[k_Y(y)]^{1/2}. \qquad (6.1\text{-}38)$$

Therefore, substituting Eqs. (6.1-37) and (6.1-38) into Eq. (6.1-25) and taking into account both signs appearing in Eq. (6.1-37) yields

$$Y(y) \approx [k_Y(y)]^{-1/2}\left\{A_Y \exp\left[-j\int_{y_0}^{y} k_Y(\zeta)\,d\zeta\right] + B_Y \exp\left[+j\int_{y_0}^{y} k_Y(\zeta)\,d\zeta\right]\right\}, \qquad (6.1\text{-}39)$$

where the constant

$$\exp[+j\theta_Y(y_0)] \qquad (6.1\text{-}40)$$

has been absorbed by the arbitrary complex constants A_Y and B_Y. Equation (6.1-39) is called the *geometrical optics approximation.* It is also known as the approximate solution of Wentzel, Kramers, and Brillouin (WKB) or as the *plane-wave ray solution.*

The first and second terms appearing in Eq. (6.1-39) correspond to waves propagating in the positive and negative Y directions, respectively, since we chose $\exp(+j2\pi ft)$ as our time-harmonic dependence. These waves propagate through the medium independently of one another and, to within the present approximation, there is *no reflection.*

The integral

$$\int_{y_0}^{y} k_Y(\zeta)\, d\zeta \tag{6.1-41}$$

is the *phase change* as the wave travels from an arbitrary point y_0 to the observation point y. The y dependence of the wave amplitude is given by

$$[k_Y(y)]^{-1/2} \tag{6.1-42}$$

This dependence can be explained on the basis of the law of conservation of energy. In fact, since there is no reflected wave, the time-average intensity in the Y direction must be the same for all y (see Example 6.1-2). Note that Eq. (6.1-42) and, as a result, the approximate solution given by Eq. (6.1-39) approach infinity and, hence, are *invalid* as $k_Y(y)$ approaches zero. If $k_Y(y_T) = 0$, then $y = y_T$ is referred to as a *turning point.* More will be said about turning points later.

EXAMPLE 6.1-1: HOMOGENEOUS MEDIUM

Let us consider the case of a homogeneous medium where

$$c(y) = c_0, \tag{6.1-43}$$

and, as a result [see Eqs. (6.1-6) and (6.1-10)],

$$k(y) = 2\pi f/c_0 = k_0, \tag{6.1-44}$$

where c_0 and k_0 are both constants. Substituting Eq. (6.1-44) into Eq. (6.1-22) yields

$$k_Y^2(y) = k_0^2 - k_X^2 - k_Z^2 = k_Y^2, \tag{6.1-45}$$

where k_Y is a constant and, upon substituting Eq. (6.1-45) into Eq. (6.1-21), we obtain

$$\frac{d^2}{dy^2} Y(y) + k_Y^2 Y(y) = 0. \tag{6.1-46}$$

Equation (6.1-46) has the exact solution

$$Y(y) = A_Y \exp(-jk_Y y) + B_Y \exp(+jk_Y y), \tag{6.1-47}$$

where

$$k_Y = (k_0^2 - k_X^2 - k_Z^2)^{1/2}. \tag{6.1-48}$$

Let us now investigate the approximate WKB solution given by Eq. (6.1-39) for the case of a homogeneous medium. Since

$$k_Y(y) = k_Y \tag{6.1-49}$$

for a homogeneous medium where k_Y is given by Eq. (6.1-48), substituting Eq. (6.1-49) into Eq. (6.1-39) yields [compare with Eq. (6.1-47)]

$$Y(y) \approx k_Y^{-1/2}\{A_Y \exp[-jk_Y(y - y_0)] + B_Y \exp[+jk_Y(y - y_0)]\}. \tag{6.1-50}$$

Upon comparing Eq. (6.1-50) with Eq. (6.1-25), we can write that

$$a_Y(y) = A_Y k_Y^{-1/2} \quad \text{or} \quad B_Y k_Y^{-1/2} \tag{6.1-51}$$

and

$$\theta_Y(y) = \pm k_Y(y - y_0). \tag{6.1-52}$$

Therefore,

$$a'_Y(y) = 0, \tag{6.1-53}$$

$$a''_Y(y) = 0, \tag{6.1-54}$$

and

$$\theta'_Y(y) = \pm k_Y. \tag{6.1-55}$$

Substituting Eqs. (6.1-51), (6.1-54), and (6.1-55) into Eq. (6.1-34) yields

$$0 \ll k_Y^2, \tag{6.1-56}$$

which is satisfied so long as k_Y does *not* equal zero. Equation (6.1-34) is a very important relationship since the approximate WKB solution given by Eq. (6.1-39) depends on Eq. (6.1-34) being satisfied. We shall discuss the implications of Eq. (6.1-34) in detail at the end of this section. □

EXAMPLE 6.1-2: INTENSITY CALCULATION

Let us calculate the time-average intensity *vector* associated with the solution to the following wave equation:

$$\nabla^2\varphi(t, x, y, z) - \frac{1}{c^2(y)}\frac{\partial^2}{\partial t^2}\varphi(t, x, y, z) = 0. \tag{6.1-57}$$

The approximate time-harmonic WKB solution to Eq. (6.1-57) is given by

$$\varphi(t, x, y, z) \approx X(x)Y(y)Z(z)\exp(+j2\pi ft), \tag{6.1-58}$$

where Eq. (6.1-11) was substituted into Eq. (6.1-2) and where $X(x)$, $Y(y)$, and $Z(z)$ are given by Eqs. (6.1-16), (6.1-39), and (6.1-19), respectively.

In order to calculate the time-average intensity vector, we must first calculate the acoustic pressure and acoustic fluid (particle) velocity vector. The acoustic pressure $p(t, x, y, z)$ and the acoustic fluid (particle) velocity vector $\mathbf{u}(t, x, y, z)$ can be obtained from the velocity potential $\varphi(t, x, y, z)$ as follows [see Eqs. (1.1-4) and (1.1-5)]:

$$p(t, x, y, z) = -\rho_0 \frac{\partial}{\partial t}\varphi(t, x, y, z) \tag{6.1-59}$$

and

$$\mathbf{u}(t, x, y, z) = \nabla\varphi(t, x, y, z). \tag{6.1-60}$$

Substituting Eq. (6.1-58) into Eqs. (6.1-59) and (6.1-60) yields

$$p(t, x, y, z) \approx -j2\pi f\rho_0 X(x)Y(y)Z(z)\exp(+j2\pi ft) \tag{6.1-61}$$

and

$$\mathbf{u}(t, x, y, z) \approx \nabla[X(x)Y(y)Z(z)]\exp(+j2\pi ft). \tag{6.1-62}$$

Next, assume that the acoustic field is propagating in the positive X, Y, and Z directions, so that

$$X(x) = A_X\exp(-jk_X x), \tag{6.1-63}$$

$$Y(y) \approx A_Y[k_Y(y)]^{-1/2}\exp\left[-j\int_{y_0}^{y} k_Y(\zeta)\,d\zeta\right], \tag{6.1-64}$$

and

$$Z(z) = A_Z\exp(-jk_Z z). \tag{6.1-65}$$

Substituting Eqs. (6.1-63) through (6.1-65) into Eq. (6.1-62) yields

$$\mathbf{u}(t, x, y, z) \approx -jX(x)Y(y)Z(z) \times \left\{k_X\hat{x} + \left[k_Y(y) - j\frac{1}{2k_Y(y)}\frac{d}{dy}k_Y(y)\right]\hat{y} + k_Z\hat{z}\right\}\exp(+j2\pi ft). \tag{6.1-66}$$

Since the pressure and velocity expressions given by Eqs. (6.1-61) and (6.1-66), respectively, have time-harmonic dependence [i.e., $\exp(+j2\pi ft)$ dependence], the time-average intensity vector $\mathbf{I}_{\text{avg}}$ of the field can be computed from the following equation [see Eq. (3.7-13)]:

$$\mathbf{I}_{\text{avg}} = \tfrac{1}{2}\,\text{Re}\{p(x, y, z)\mathbf{u}^*(x, y, z)\}, \tag{6.1-67}$$

where

$$p(t, x, y, z) = p(x, y, z)\exp(+j2\pi f t), \tag{6.1-68}$$

$$\mathbf{u}(t, x, y, z) = \mathbf{u}(x, y, z)\exp(+j2\pi f t), \tag{6.1-69}$$

Re means "take the real part," and the asterisk denotes complex conjugate. By comparing Eqs. (6.1-61) and (6.1-66) with Eqs. (6.1-68) and (6.1-69), respectively, and substituting into Eq. (6.1-67), we obtain the desired result:

$$\mathbf{I}_{\text{avg}} \approx \pi f \rho_0 |A|^2 [k_X k_Y^{-1}(y)\hat{x} + \hat{y} + k_Z k_Y^{-1}(y)\hat{z}]. \tag{6.1-70}$$

In deriving Eq. (6.1-70) it was necessary to evaluate $|X(x)|^2$, $|Y(y)|^2$, and $|Z(z)|^2$, which have the following values:

$$|X(x)|^2 = |A_X|^2, \qquad |Y(y)|^2 \approx |A_Y|^2 k_Y^{-1}(y), \qquad |Z(z)|^2 = |A_Z|^2,$$

so that

$$|X(x)Y(y)Z(z)|^2 \approx |A|^2 k_Y^{-1}(y),$$

where $|A|^2 = |A_X A_Y A_Z|^2$.

By inspecting Eq. (6.1-70), it can be seen that the time-average intensity component in the Y direction is given by

$$\pi f \rho_0 |A|^2 \hat{y} \tag{6.1-71}$$

which is a *constant*, independent of the depth y. This is a plane wave result which agrees with our earlier statements made with respect to the significance of Eq. (6.1-42). □

EXAMPLE 6.1-3: SNELL'S LAW

Let us consider the case of three-dimensional wave propagation, as illustrated in Fig. 6.1-1. In general, the propagation vector $\mathbf{k}(y)$ has three components, that is [see Eq. (6.1-24)],

$$\mathbf{k}(y) = k_X\hat{x} + k_Y(y)\hat{y} + k_Z\hat{z}, \tag{6.1-72}$$

where k_X and k_Z are constants. In order to derive Snell's law, refer to Fig. 6.1-2, where it can be seen that the radial component

$$k_R = (k_X^2 + k_Z^2)^{1/2} \tag{6.1-73}$$

is a constant since k_X and k_Z are constants. It can also be seen that

$$k_R = k(y)\sin\beta(y), \tag{6.1-74}$$

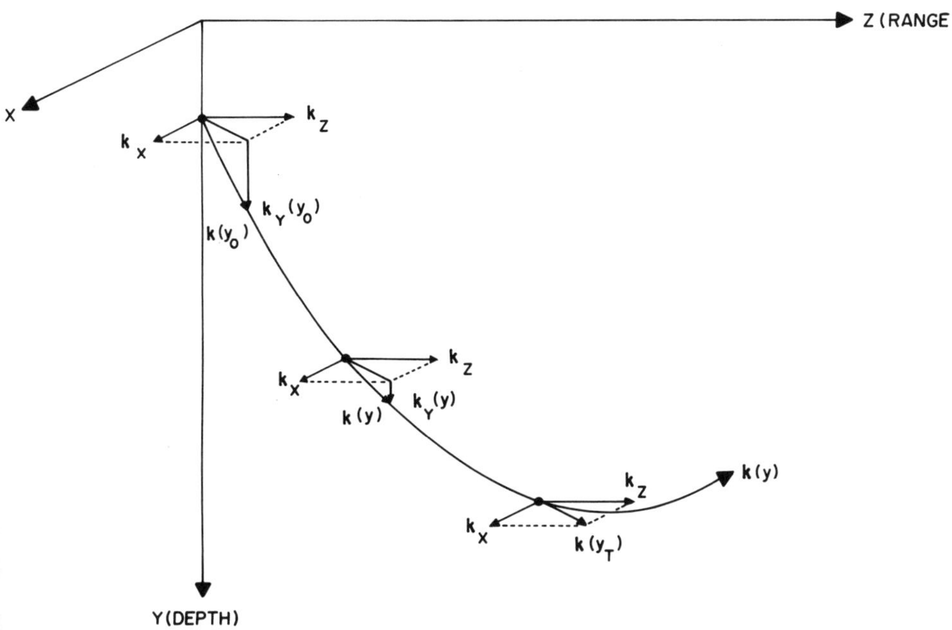

FIG. 6.1-1 Wave propagation in three-dimensional space as depicted by the propagation vector $\mathbf{k}(y)$. Note that $\mathbf{k}(y)$ passes through a turning point at $y = y_T$.

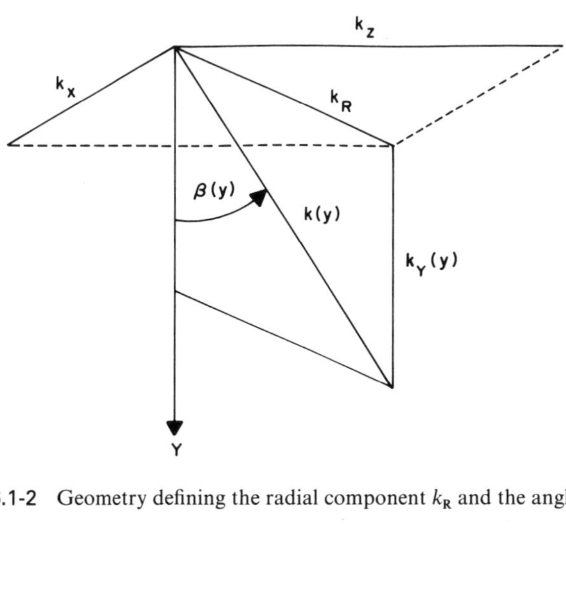

FIG. 6.1-2 Geometry defining the radial component k_R and the angle $\beta(y)$.

where the angle $\beta(y)$ is a function of depth. Since k_R is a constant, we can write that

$$k(y)\sin\beta(y) = k(y_0)\sin\beta(y_0), \tag{6.1-75}$$

and upon substituting Eq. (6.1-10) into Eq. (6.1-75), we obtain

$$\frac{\sin\beta(y)}{c(y)} = \frac{\sin\beta(y_0)}{c(y_0)}, \tag{6.1-76}$$

which is Snell's law. □

As indicated earlier, if the Y component of the propagation vector is equal to zero, that is, if $k_Y(y_T) = 0$, then $y = y_T$ is referred to as a *turning point*. At a turning point the propagation vector is equal to [see Fig. 6.1-1 and Eq. (6.1-72)]

$$\mathbf{k}(y_T) = k_X\hat{x} + k_Z\hat{z}. \tag{6.1-77}$$

In addition, it can be seen from Fig. 6.1-2 that

$$k_Y(y) = k(y)\cos\beta(y), \tag{6.1-78}$$

so that if at $y = y_T$

$$\beta(y_T) = \pi/2, \tag{6.1-79}$$

then

$$k_Y(y_T) = 0. \tag{6.1-80}$$

A wave passing through a turning point is the equivalent of *total reflection*, with a corresponding change in phase of $\pi/2$. As was also mentioned earlier, as $k_Y(y)$ approaches zero, the approximate WKB solution given by Eq. (6.1-39) approaches infinity and, hence, is invalid.

We shall finish our discussion in this section by examining the implications of the assumption given by Eq. (6.1-34), which is responsible for the form of the approximate WKB solution given by Eq. (6.1-39).

In order to evaluate Eq. (6.1-34), we start by noting that the right-hand side is already given by Eq. (6.1-35). And, from Eq. (6.1-38) it can be shown that

$$a_Y''(y) = \frac{A}{2}\left[\frac{3}{2}[k_Y(y)]^{-5/2}[k_Y'(y)]^2 - [k_Y(y)]^{-3/2}k_Y''(y)\right]. \tag{6.1-81}$$

If we rewrite Eq. (6.1-34) as

$$\frac{1}{[\theta_Y'(y)]^2}\left|\frac{a_Y''(y)}{a_Y(y)}\right| \ll 1 \tag{6.1-82}$$

and then substitute Eqs. (6.1-35), (6.1-38), and (6.1-81) into Eq. (6.1-82), we obtain the following inequality:

$$\left| \left[\frac{1}{2k_Y^2(y)} k_Y'(y) \right]^2 - \frac{1}{6} \frac{k_Y''(y)}{k_Y^3(y)} \right| \ll 1, \tag{6.1-83}$$

and, assuming that

$$k_Y''(y) \ll k_Y'(y), \tag{6.1-84}$$

Eq. (6.1-83) reduces to

$$\left| \frac{1}{2k_Y^2(y)} k_Y'(y) \right| \ll 1. \tag{6.1-85}$$

The next step will be to rewrite the left-hand side of Eq. (6.1-85).

Taking the derivative with respect to y of Eq. (6.1-22) yields

$$\frac{d}{dy} k_Y^2(y) = 2k(y)k'(y), \tag{6.1-86}$$

and, since

$$\frac{d}{dy} k_Y^2(y) = 2k_Y(y)k_Y'(y), \tag{6.1-87}$$

equating Eqs. (6.1-86) and (6.1-87) and multiplying and dividing the right-hand side by $k(y)$ yields

$$k_Y'(y) = [k^2(y)k'(y)]/[k_Y(y)k(y)]. \tag{6.1-88}$$

Substituting

$$k(y) = k_0 n(y) = k_0 c_0/c(y) \tag{6.1-89}$$

into Eq. (6.1-88) further yields

$$k_Y'(y) = -[k^2(y)c'(y)]/[k_Y(y)c(y)] \tag{6.1-90}$$

or

$$k_Y'(y)/[2k_Y^2(y)] = -[k^2(y)c'(y)]/[2k_Y^3(y)c(y)]. \tag{6.1-91}$$

Now, rewrite Eq. (6.1-22) as follows:

$$k_Y(y) = k(y)\{1 - [(k_X^2 + k_Z^2)/k^2(y)]\}^{1/2}. \tag{6.1-92}$$

Upon substituting Eqs. (6.1-89) and (6.1-92) into Eq. (6.1-91), Eq. (6.1-85) can finally be rewritten as

$$\left| \frac{c'(y)}{4\pi f\{1 - [(k_X^2 + k_Z^2)/k^2(y)]\}^{3/2}} \right| \ll 1, \tag{6.1-93}$$

where [see Eq. (6.1-22)]

$$k^2(y) = k_X^2 + k_Y^2(y) + k_Z^2. \tag{6.1-94}$$

Equation (6.1-93) indicates that the approximate WKB solution given by Eq. (6.1-39) will be valid (1) for high frequencies f, (2) if the sound-speed profile $c(y)$ is slowly changing, that is, if the *gradient* $|c'(y)| \ll 1$, and (3) if $k_Y(y) \neq 0$, that is, so long as the wave does *not* pass through a turning point.

6.2 RAY ACOUSTICS

6.2.1 Transport and Eikonal Equations

The WKB method discussed in Section 6.1 was used to solve the simplified Helmholtz wave equation specified by Eqs. (6.1-9) and (6.1-10), where the speed of sound $c(y)$ was assumed to be a function of only *one* of the spatial variables, namely, the depth y. As a result, the method of separation of variables could be used to begin the process of obtaining an approximate solution. Now, in this section, we shall investigate the method of *ray acoustics*, which will allow us to obtain an approximate solution to the general Helmholtz wave equation specified by Eqs. (6.1-8) and (6.1-5), where the speed of sound is a function of *all three* spatial variables, that is, $c(\mathbf{r}) = c(x, y, z)$.

Since the speed of sound is now a function of all three spatial variables, we *cannot* use the method of separation of variables to begin the solution process. Instead, we begin by assuming a solution to Eq. (6.1-8) of the form [compare with Eq. (6.1-25)]

$$\varphi(\mathbf{r}) = a(\mathbf{r})\exp[+j\theta(\mathbf{r})] \tag{6.2-1}$$

or

$$\varphi(\mathbf{r}) = a(\mathbf{r})\exp[-jk_0 W(\mathbf{r})], \tag{6.2-2}$$

where $a(\mathbf{r})$ is a *real amplitude function*,

$$\theta(\mathbf{r}) = -k_0 W(\mathbf{r}) \tag{6.2-3}$$

is a *real phase function*, k_0 is the reference wave number given by Eq. (6.1-6), and $\mathbf{r} = (x, y, z)$. Substituting Eq. (6.2-2) into Eq. (6.1-8) yields

$$\nabla^2 a(\mathbf{r}) + k_0^2[n^2(\mathbf{r}) - |\nabla W(\mathbf{r})|^2]a(\mathbf{r}) - jk_0[a(\mathbf{r})\nabla^2 W(\mathbf{r}) + 2\nabla a(\mathbf{r}) \cdot \nabla W(\mathbf{r})] = 0. \tag{6.2-4}$$

Since the left-hand side of Eq. (6.2-4) must equal the right-hand side, which is zero, both the real and imaginary parts on the left-hand side of Eq. (6.2-4) must also be equal to zero, that is,

$$\nabla^2 a(\mathbf{r}) + k_0^2[n^2(\mathbf{r}) - |\nabla W(\mathbf{r})|^2]a(\mathbf{r}) = 0 \tag{6.2-5}$$

and

$$a(\mathbf{r})\nabla^2 W(\mathbf{r}) + 2\nabla a(\mathbf{r}) \cdot \nabla W(\mathbf{r}) = 0. \tag{6.2-6}$$

Recalling that $\theta(\mathbf{r})$ and $W(\mathbf{r})$ are related by Eq. (6.2-3) and referring to Eq. (6.1-7), it can be seen that Eqs. (6.2-4) through (6.2-6) are simply three-dimensional generalizations of the one-dimensional results given by Eqs. (6.1-26) through (6.1-28), respectively, derived using the WKB method. Equation (6.2-6) is known as the *transport equation.* It is used to determine the amplitude term $a(\mathbf{r})$, which we shall do in Section 6.2.4. But first, let us rewrite Eq. (6.2-5) as follows:

$$1 + \frac{1}{k_0^2 n^2(\mathbf{r})}\left[\frac{\nabla^2 a(\mathbf{r})}{a(\mathbf{r})} - k_0^2|\nabla W(\mathbf{r})|^2\right] = 0. \tag{6.2-7}$$

If it is assumed that [compare with Eq. (6.1-34)]

$$\left|\frac{\nabla^2 a(\mathbf{r})}{a(\mathbf{r})}\right| \ll k_0^2|\nabla W(\mathbf{r})|^2, \tag{6.2-8}$$

then Eq. (6.2-7) reduces to

$$|\nabla W(\mathbf{r})|^2 = n^2(\mathbf{r}), \tag{6.2-9}$$

or, equivalently,

$$\left[\frac{\partial}{\partial x}W(\mathbf{r})\right]^2 + \left[\frac{\partial}{\partial y}W(\mathbf{r})\right]^2 + \left[\frac{\partial}{\partial z}W(\mathbf{r})\right]^2 = n^2(\mathbf{r}). \tag{6.2-10}$$

Note that Eqs. (6.2-7) through (6.2-9) are three-dimensional generalizations of the one-dimensional results given by Eqs. (6.1-33) through (6.1-35), respectively, derived using the WKB method. Equation (6.2-9) is known as the *eikonal equation.* It is used to determine $W(\mathbf{r})$ and to develop the *ray equations.* The solution of the eikonal equation and the development of the ray equations will be discussed after Example 6.2-1. The important assumption given by Eq. (6.2-8), which was responsible for the eikonal equation, can be rewritten as follows: substituting Eq. (6.2-9) into Eq. (6.2-8) and making use of Eqs. (6.1-4) and (6.1-7) yields

$$\left|\frac{\nabla^2 a(\mathbf{r})}{a(\mathbf{r})}\right| \ll k^2(\mathbf{r}) = \left[\frac{2\pi f}{c(\mathbf{r})}\right]^2. \tag{6.2-11}$$

Equation (6.2-11) will be satisfied best at high frequencies. Therefore, in order for Eq. (6.2-2) to be a valid solution of the Helmholtz wave equation given by

Eq. (6.1-8), Eq. (6.2-11) must be satisfied, $a(\mathbf{r})$ and $W(\mathbf{r})$ together must satisfy the transport equation given by Eq. (6.2-6), and $W(\mathbf{r})$ must satisfy the eikonal equation given by Eq. (6.2-9).

EXAMPLE 6.2-1: HOMOGENEOUS MEDIUM

Let us consider the case of a homogeneous medium where $c(\mathbf{r}) = c_0$ (a constant). As a result, $n(\mathbf{r}) = 1$ [see Eq. (6.1-5)] and the Helmholtz wave equation given by Eq. (6.1-8) reduces to

$$\nabla^2\varphi(\mathbf{r}) + k_0^2\varphi(\mathbf{r}) = 0, \tag{6.2-12}$$

where k_0 is given by Eq. (6.1-6). A well-known solution of Eq. (6.2-12) is given by

$$\varphi(\mathbf{r}) = A\exp(-jk_0\hat{n}\cdot\mathbf{r}), \tag{6.2-13}$$

which is a *plane wave* (see Fig. 6.2-1) propagating in the $+\hat{n}$ direction [assuming a time-harmonic dependence of $\exp(+j2\pi ft)$], where A is a constant, $\hat{n}$ is a *unit vector* given by

$$\hat{n} = u\hat{x} + v\hat{y} + w\hat{z}, \tag{6.2-14}$$

u, v, and w are *direction cosines* with respect to the positive X, Y, and Z axes, respectively [see Eqs. (3.2-6) through (3.2-8)], and

$$\mathbf{r} = x\hat{x} + y\hat{y} + z\hat{z}. \tag{6.2-15}$$

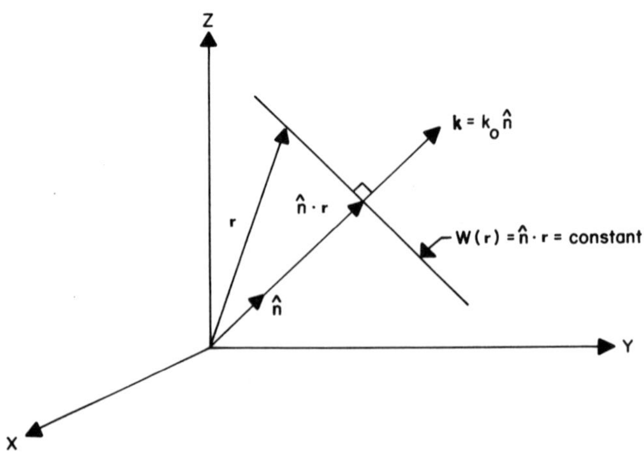

FIG. 6.2-1 Plane wave propagating in the $+\hat{n}$ direction.

By comparing Eqs. (6.2-2) and (6.2-13), it can be seen that the real amplitude function

$$a(\mathbf{r}) = A \tag{6.2-16}$$

is a constant and

$$W(\mathbf{r}) = \hat{n} \cdot \mathbf{r} = ux + vy + wz. \tag{6.2-17}$$

Substituting Eq. (6.2-17) into Eq. (6.2-3) yields the real phase function

$$\theta(\mathbf{r}) = -\mathbf{k} \cdot \mathbf{r}, \tag{6.2-18}$$

where

$$\mathbf{k} = k_0 \hat{n} = k_X \hat{x} + k_Y \hat{y} + k_Z \hat{z} \tag{6.2-19}$$

is the *propagation vector* and

$$k_X = k_0 u, \tag{6.2-20}$$

$$k_Y = k_0 v, \tag{6.2-21}$$

and

$$k_Z = k_0 w \tag{6.2-22}$$

are the components of the propagation vector in the X, Y, and Z directions, respectively. From Eq. (6.2-3) it can be seen that

$$W(\mathbf{r}) = \text{constant} \tag{6.2-23}$$

defines a surface in three-dimensional space of *constant phase*, that is, all field points on this surface will have the same phase. Such surfaces of constant phase are called *wavefronts*. In the case of a homogeneous medium where the sound-speed profile is a constant, the surfaces of constant phase (or wavefronts) are *planar* and are defined by (see Fig. 6.2-1)

$$W(\mathbf{r}) = \hat{n} \cdot \mathbf{r} = \text{constant}. \tag{6.2-24}$$

Also, note from Fig. 6.2-1 that the propagation vector $\mathbf{k}$ is *normal* or *perpendicular* to the wavefront.

Next, let us determine whether or not the plane-wave solution satisfies Eq. (6.2-11) and the transport and eikonal equations. From Eq. (6.2-16) we have that $a(\mathbf{r}) = A$, which is a constant, so that

$$\nabla^2 a(\mathbf{r}) = \nabla^2 A = 0. \tag{6.2-25}$$

From Eq. (6.2-17) we obtain

$$\nabla W(\mathbf{r}) = u\hat{x} + v\hat{y} + w\hat{z}, \tag{6.2-26}$$

or, upon substituting Eq. (6.2-14) into Eq. (6.2-26),

$$\nabla W(\mathbf{r}) = \hat{n}. \tag{6.2-27}$$

Therefore, from Eq. (6.2-26)

$$|\nabla W(\mathbf{r})|^2 = u^2 + v^2 + w^2 = 1 \tag{6.2-28}$$

since u, v, and w are direction cosines or, from Eq. (6.2-27),

$$|\nabla W(\mathbf{r})|^2 = |\hat{n}|^2 = 1 \tag{6.2-29}$$

since $\hat{n}$ is a unit vector. Substituting Eqs. (6.1-6), (6.2-16), and (6.2-25) into Eq. (6.2-11) yields $|0/A| = 0 \ll k_0^2$. Therefore, Eq. (6.2-11) is satisfied. From Eq. (6.2-16) we obtain

$$\nabla a(\mathbf{r}) = \nabla A = 0 \tag{6.2-30}$$

since A is a constant. From Eq. (6.2-26) we obtain

$$\nabla^2 W(\mathbf{r}) = \nabla \cdot \nabla W(\mathbf{r}) = 0 \tag{6.2-31}$$

since u, v, and w are not functions of (x, y, z) in a homogeneous medium. Substituting Eqs. (6.2-16), (6.2-26), (6.2-30), and (6.2-31) into Eq. (6.2-6) yields

$$A(0) + 2(0) \cdot (u\hat{x} + v\hat{y} + w\hat{z}) = 0.$$

Therefore, the transport equation is satisfied. And, finally, substituting Eq. (6.2-28) into Eq. (6.2-9) and making use of the fact that $n(\mathbf{r}) = 1$ for a homogeneous medium, we note that the eikonal equation is also satisfied.

Let us conclude this example by making one additional observation. Using Eq. (6.2-27) the propagation vector $\mathbf{k}$ corresponding to the plane wave shown in Fig. 6.2-1 and given by Eq. (6.2-19) can be rewritten as

$$\mathbf{k} = k_0\hat{n} = k_0\nabla W(\mathbf{r}). \tag{6.2-32}$$

Thus, $\nabla W(\mathbf{r})$ is normal to the planar wavefront defined by Eq. (6.2-24). As a result, $\nabla W(\mathbf{r})$ determines the direction of propagation of the wavefront, and thus the direction of energy propagation. This property of $\nabla W(\mathbf{r})$ for

$$W(\mathbf{r}) = \text{constant} \tag{6.2-23}$$

is true in general and *not* only for plane waves, as we shall discuss next. □

6.2.2 Solution of the Eikonal Equation

The eikonal equation given by Eq. (6.2-9) or, equivalently, by Eq. (6.2-10), is a first-order partial differential equation. Its solution $W(\mathbf{r})$ is referred to as the *eikonal*, which is taken from a Greek word meaning *image*. As was mentioned previously in Example 6.2-1, the equation

$$W(\mathbf{r}) = \text{constant} \tag{6.2-23}$$

defines a surface in three-dimensional space of constant phase known as a wavefront. A wavefront is *not* planar, in general. From the properties of the gradient of a real scalar function, which in this case is the eikonal $W(\mathbf{r})$, $\nabla W(\mathbf{r})$ is *normal* or *perpendicular* to the wavefront defined by Eq. (6.2-23) at *all* points on the wavefront. Each time a different value of the constant in Eq. (6.2-23) is chosen, a different wavefront is defined. *All* field points on a particular wavefront will be *in phase*, although now, in the general case, *not* necessarily of the same amplitude, since the amplitude term $a(\mathbf{r})$ is a function of position $\mathbf{r} = (x, y, z)$. The curves defined by the normals $\nabla W(\mathbf{r})$ to the wavefronts are called *rays* or *ray paths* (see Fig. 6.2-2). Later, in Example 6.2-2, it will be shown that the rays define the direction of energy propagation, as might be expected, since they define the direction of wavefront propagation.

Let us begin the solution of the eikonal equation by expressing the vector $\nabla W(\mathbf{r})$ in rectangular coordinates, that is,

$$\nabla W(\mathbf{r}) = \frac{\partial}{\partial x} W(\mathbf{r})\hat{x} + \frac{\partial}{\partial y} W(\mathbf{r})\hat{y} + \frac{\partial}{\partial z} W(\mathbf{r})\hat{z}. \tag{6.2-33}$$

Since $\nabla W(\mathbf{r})$ is a vector, it can also be expressed in terms of its magnitude and direction as follows:

$$\nabla W(\mathbf{r}) = |\nabla W(\mathbf{r})|\,\hat{n}(\mathbf{r}), \tag{6.2-34}$$

or, from Eq. (6.2-9),

$$\nabla W(\mathbf{r}) = n(\mathbf{r})\hat{n}(\mathbf{r}), \tag{6.2-35}$$

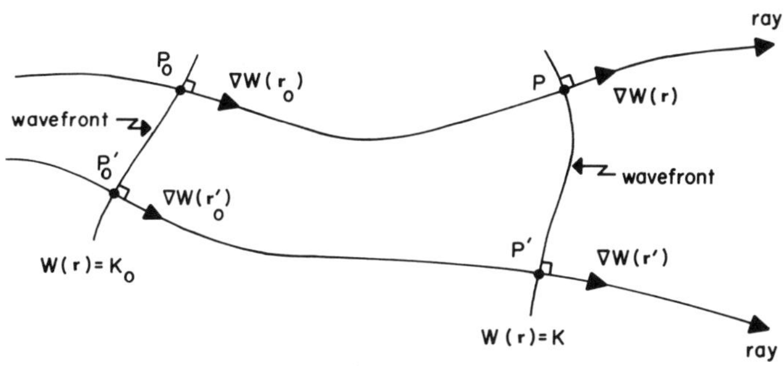

FIG. 6.2-2 Illustration of the concepts of wavefronts and rays. The points P_0, P'_0, P, and P' on the ray paths are specified by the position vectors $\mathbf{r}_0$, $\mathbf{r}'_0$, $\mathbf{r}$, and $\mathbf{r}'$, respectively. The parameters K_0 and K are constants.

where $n(\mathbf{r})$ is the index of refraction defined by Eq. (6.1-5),

$$\hat{n}(\mathbf{r}) = u(\mathbf{r})\hat{x} + v(\mathbf{r})\hat{y} + w(\mathbf{r})\hat{z} \tag{6.2-36}$$

is a *unit vector* in the direction of $\nabla W(\mathbf{r})$, and the *direction cosines* $u(\mathbf{r})$, $v(\mathbf{r})$, and $w(\mathbf{r})$ are now functions of position along the rays and satisfy the relationship

$$u^2(\mathbf{r}) + v^2(\mathbf{r}) + w^2(\mathbf{r}) = 1. \tag{6.2-37}$$

Note that if we multiply both sides of Eq. (6.2-35) by k_0 and make use of Eq. (6.1-7), then the *propagation vector* $\mathbf{k}(\mathbf{r})$ can be written as

$$\mathbf{k}(\mathbf{r}) = k(\mathbf{r})\hat{n}(\mathbf{r}) = k_0 \nabla W(\mathbf{r}). \tag{6.2-38}$$

Compare Eq. (6.2-38) with the "plane wave in a homogeneous medium" result given by Eq. (6.2-32). Next, substitute Eq. (6.2-36) into Eq. (6.2-35) and equate the result with Eq. (6.2-33). Doing so yields the following set of equations:

$$n(\mathbf{r})u(\mathbf{r}) = \frac{\partial}{\partial x} W(\mathbf{r}), \tag{6.2-39}$$

$$n(\mathbf{r})v(\mathbf{r}) = \frac{\partial}{\partial y} W(\mathbf{r}), \tag{6.2-40}$$

and

$$n(\mathbf{r})w(\mathbf{r}) = \frac{\partial}{\partial z} W(\mathbf{r}). \tag{6.2-41}$$

Equations (6.2-39) through (6.2-41) give the x, y, and z components, respectively, of the direction of ray propagation in terms of the index of refraction and the appropriate direction cosines.

Let us next consider an *infinitesimal element of arc length ds* at an arbitrary point P with position vector $\mathbf{r}$ along a ray path, as shown in Fig. 6.2-3. The direction cosines with respect to the positive X, Y, and Z axes are given by the following equations:

$$u(\mathbf{r}) = \cos\alpha(\mathbf{r}) = \sin\theta(\mathbf{r})\cos\psi(\mathbf{r}) = dx/ds, \tag{6.2-42}$$

$$v(\mathbf{r}) = \cos\beta(\mathbf{r}) = \sin\theta(\mathbf{r})\sin\psi(\mathbf{r}) = dy/ds, \tag{6.2-43}$$

and

$$w(\mathbf{r}) = \cos\gamma(\mathbf{r}) = \cos\theta(\mathbf{r}) = dz/ds, \tag{6.2-44}$$

respectively, where

$$ds = [(dx)^2 + (dy)^2 + (dz)^2]^{1/2}. \tag{6.2-45}$$

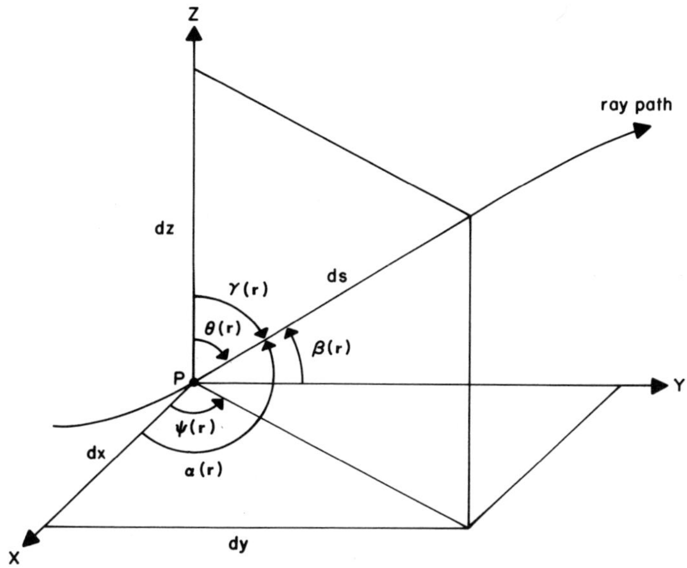

FIG. 6.2-3 An infinitesimal element of arc length ds at an arbitrary point P with position vector $\mathbf{r}$ along a ray path and associated angles.

Therefore, the unit vector given by Eq. (6.2-36) can be rewritten as

$$\hat{n}(\mathbf{r}) = \frac{dx}{ds}\hat{x} + \frac{dy}{ds}\hat{y} + \frac{dz}{ds}\hat{z}, \tag{6.2-46}$$

where

$$\left(\frac{dx}{ds}\right)^2 + \left(\frac{dy}{ds}\right)^2 + \left(\frac{dz}{ds}\right)^2 = 1. \tag{6.2-47}$$

In addition, Eqs. (6.2-39) through (6.2-41) can also be rewritten as follows:

$$n(\mathbf{r})\frac{dx}{ds} = \frac{\partial}{\partial x}W(\mathbf{r}), \tag{6.2-48}$$

$$n(\mathbf{r})\frac{dy}{ds} = \frac{\partial}{\partial y}W(\mathbf{r}), \tag{6.2-49}$$

and

$$n(\mathbf{r})\frac{dz}{ds} = \frac{\partial}{\partial z}W(\mathbf{r}), \tag{6.2-50}$$

respectively. Equations (6.2-48) through (6.2-50) will play an important role in the derivation of the ray equations, as will be discussed later in Section 6.2.3. Next, compute the *directional derivative*

$$\frac{d}{ds}W(\mathbf{r}), \tag{6.2-51}$$

which represents the rate of change of the eikonal $W(\mathbf{r})$ with respect to the *arc length* s along a ray path. Expanding Eq. (6.2-51) using the chain rule yields

$$\frac{d}{ds}W(\mathbf{r}) = \frac{\partial}{\partial x}W(\mathbf{r})\frac{dx}{ds} + \frac{\partial}{\partial y}W(\mathbf{r})\frac{dy}{ds} + \frac{\partial}{\partial z}W(\mathbf{r})\frac{dz}{ds}. \tag{6.2-52}$$

By inspecting Eqs. (6.2-33) and (6.2-46), it can be seen that Eq. (6.2-52) can be written as the *dot product*

$$\frac{d}{ds}W(\mathbf{r}) = \nabla W(\mathbf{r}) \cdot \hat{n}(\mathbf{r}), \tag{6.2-53}$$

and upon substituting Eq. (6.2-35) into Eq. (6.2-53), we obtain

$$\frac{d}{ds}W(\mathbf{r}) = n(\mathbf{r})\hat{n}(\mathbf{r}) \cdot \hat{n}(\mathbf{r})$$

or

$$\frac{d}{ds}W(\mathbf{r}) = n(\mathbf{r}). \tag{6.2-54}$$

The solution of Eq. (6.2-54) can be expressed as

$$W(\mathbf{r}) = \int_{\mathbf{r}_0}^{\mathbf{r}} n(x, y, z)\, ds, \tag{6.2-55}$$

where it is understood that $x = x(s)$, $y = y(s)$, and $z = z(s)$ in the argument of the index of refraction. Equation (6.2-55) is also the solution of the eikonal equation given by Eq. (6.2-9). Besides being referred to as the eikonal, $W(\mathbf{r})$ is also known in optics as the *optical path function*, which is defined as the *optical path length* between two points on a given ray path, for example, points P_0 and P with position vectors $\mathbf{r}_0$ and $\mathbf{r}$, respectively, in Fig. 6.2-2. Also, note that upon substituting Eq. (6.2-55) into Eq. (6.2-3) and using Eq. (6.1-7), we obtain the real phase function

$$\theta(\mathbf{r}) = -\int_{\mathbf{r}_0}^{\mathbf{r}} k(x, y, z)\, ds. \tag{6.2-56}$$

Compare Eq. (6.2-56) with the real phase function given by Eq. (6.1-37) obtained via the WKB method.

EXAMPLE 6.2-2: INTENSITY CALCULATION

Let us calculate the time-average intensity vector associated with the velocity potential

$$\varphi(t,\mathbf{r}) = a(\mathbf{r})\exp[-jk_0 W(\mathbf{r})]\exp(+j2\pi ft), \tag{6.2-57}$$

which is the time-harmonic ray-acoustics solution to the wave equation given by Eq. (6.1-1). Equation (6.2-57) was obtained by substituting Eq. (6.2-2) into Eq. (6.1-2).

In order to calculate the time-average intensity vector, we must first calculate the acoustic pressure $p(t,\mathbf{r})$ and acoustic fluid (particle) velocity vector $\mathbf{u}(t,\mathbf{r})$. Substituting Eq. (6.2-57) into Eqs. (6.1-59) and (6.1-60) yields

$$p(t,\mathbf{r}) = p(\mathbf{r})\exp(+j2\pi ft) \tag{6.2-58}$$

and

$$\mathbf{u}(t,\mathbf{r}) = \mathbf{u}(\mathbf{r})\exp(+j2\pi ft), \tag{6.2-59}$$

where

$$p(\mathbf{r}) = -j2\pi f\rho_0 a(\mathbf{r})\exp[-jk_0 W(\mathbf{r})] \tag{6.2-60}$$

and

$$\mathbf{u}(\mathbf{r}) = -jk_0 a(\mathbf{r})\exp[-jk_0 W(\mathbf{r})]\,\nabla W(\mathbf{r}) + \exp[-jk_0 W(\mathbf{r})]\,\nabla a(\mathbf{r}). \tag{6.2-61}$$

Since the pressure and velocity expressions given by Eqs. (6.2-58) and (6.2-59), respectively, have time-harmonic dependence, the time-average intensity vector $\mathbf{I}_{\text{avg}}$ of the field can be computed by substituting Eqs. (6.2-60) and (6.2-61) into Eq. (6.1-67). Doing so yields the desired result:

$$\mathbf{I}_{\text{avg}} = \tfrac{1}{2}k_0^2\rho_0 c_0\, a^2(\mathbf{r})\,\nabla W(\mathbf{r}). \tag{6.2-62}$$

Thus, $\nabla W(\mathbf{r})$ determines the direction of the time-average intensity vector, and since $\nabla W(\mathbf{r})$ also determines the direction of ray propagation, *the ray paths define the direction of energy propagation.*

Finally, taking the magnitude of both sides of Eq. (6.2-62) yields

$$I_{\text{avg}} = |\mathbf{I}_{\text{avg}}| = \tfrac{1}{2}k_0^2\rho_0 c_0\, a^2(\mathbf{r})\,|\nabla W(\mathbf{r})|, \tag{6.2-63}$$

and by making use of Eqs. (6.2-9), (6.1-5), and (6.2-60), Eq. (6.2-63) can be rewritten as

$$I_{\text{avg}} = \frac{|p(\mathbf{r})|^2}{2\rho_0 c(\mathbf{r})}. \tag{6.2-64}$$

Note that although Eq. (6.2-64) is in the same form as the intensity expression for a plane wave propagating in a homogeneous medium, the acoustic-pressure amplitude $|p(\mathbf{r})|$ and the speed of sound $c(\mathbf{r})$ are now functions of position in the general case of wave propagation in an inhomogeneous medium. □

6.2.3 Ray Equations

In order to develop the ray equations, which enable one to trace the trajectory of a given ray path, take the gradient of both sides of Eq. (6.2-54), that is,

$$\nabla\left(\frac{d}{ds}W(\mathbf{r})\right) = \nabla n(\mathbf{r})$$

or

$$\frac{d}{ds}\nabla W(\mathbf{r}) = \nabla n(\mathbf{r}). \tag{6.2-65}$$

Since $\nabla W(\mathbf{r})$ describes the direction of ray propagation, the expression $d\nabla W(\mathbf{r})/ds$ describes *the rate of change of the direction of ray propagation along a ray path*. This rate of change, as given by Eq. (6.2-65), is equal to the gradient of the index of refraction, that is, $\nabla n(\mathbf{r})$. Thus, Eq. (6.2-65) is a representation of the ray equations in *vector form* since it enables us to trace the trajectory of a given ray path by describing how the ray direction changes.

The ray equations can also be expressed in *scalar form*. Taking the derivative with respect to arc length s of both sides of Eqs. (6.2-48) through (6.2-50) yields

$$\frac{d}{ds}\left(n(\mathbf{r})\frac{dx}{ds}\right) = \frac{d}{ds}\frac{\partial}{\partial x}W(\mathbf{r}) = \frac{\partial}{\partial x}\frac{d}{ds}W(\mathbf{r}), \tag{6.2-66}$$

$$\frac{d}{ds}\left(n(\mathbf{r})\frac{dy}{ds}\right) = \frac{d}{ds}\frac{\partial}{\partial y}W(\mathbf{r}) = \frac{\partial}{\partial z}\frac{d}{ds}W(\mathbf{r}), \tag{6.2-67}$$

and

$$\frac{d}{ds}\left(n(\mathbf{r})\frac{dz}{ds}\right) = \frac{d}{ds}\frac{\partial}{\partial z}W(\mathbf{r}) = \frac{\partial}{\partial z}\frac{d}{ds}W(\mathbf{r}), \tag{6.2-68}$$

and upon substituting Eq. (6.2-54) into Eqs. (6.2-66) through (6.2-68), we obtain

$$\frac{d}{ds}\left(n(\mathbf{r})\frac{dx}{ds}\right) = \frac{\partial}{\partial x}n(\mathbf{r}), \tag{6.2-69}$$

$$\frac{d}{ds}\left(n(\mathbf{r})\frac{dy}{ds}\right) = \frac{\partial}{\partial y}n(\mathbf{r}), \tag{6.2-70}$$

and

$$\frac{d}{ds}\left(n(\mathbf{r})\frac{dz}{ds}\right) = \frac{\partial}{\partial z}n(\mathbf{r}), \tag{6.2-71}$$

which are the ray equations in scalar form. Equations (6.2-69) through (6.2-71) may be considered as a generalized form of Snell's law.

EXAMPLE 6.2-3: SPEED OF SOUND AS A FUNCTION OF DEPTH

Suppose we are given a sound-speed profile that is a function of depth y only, that is, $c(\mathbf{r}) = c(y)$. Therefore, the index of refraction becomes

$$n(\mathbf{r}) = n(y) = c_0/c(y), \tag{6.2-72}$$

where c_0 is the constant, reference speed of sound at the source depth y_0, that is, $c_0 = c(y_0)$. As a result,

$$\frac{\partial}{\partial x}n(\mathbf{r}) = \frac{\partial}{\partial x}n(y) = 0 \tag{6.2-73}$$

and

$$\frac{\partial}{\partial z}n(\mathbf{r}) = \frac{\partial}{\partial z}n(y) = 0, \tag{6.2-74}$$

and upon substituting Eqs. (6.2-73) and (6.2-74) into Eqs. (6.2-69) and (6.2-71), respectively, and using Eq. (6.2-72), the ray equations reduce to

$$\frac{d}{ds}\left(n(y)\frac{dx}{ds}\right) = 0, \tag{6.2-75}$$

$$\frac{d}{ds}\left(n(y)\frac{dy}{ds}\right) = \frac{\partial}{\partial y}n(y) = \frac{d}{dy}n(y), \tag{6.2-76}$$

and

$$\frac{d}{ds}\left(n(y)\frac{dz}{ds}\right) = 0. \tag{6.2-77}$$

Equations (6.2-75) and (6.2-77) imply that

$$n(y)\frac{dx}{ds} = \text{constant} \tag{6.2-78}$$

and

$$n(y)\frac{dz}{ds} = \text{constant}. \tag{6.2-79}$$

If we form the ratio of Eq. (6.2-78) to Eq. (6.2-79), we obtain

$$\frac{dx/ds}{dz/ds} = \frac{u(\mathbf{r})}{w(\mathbf{r})} = \text{constant}, \tag{6.2-80}$$

where $u(\mathbf{r})$ and $w(\mathbf{r})$ are the direction cosines with respect to the positive X and Z axes, respectively, at some point P with position vector $\mathbf{r}$ along a ray path [see Eqs. (6.2-42) through (6.2-44) and Fig. 6.2-3]. The restriction that the ratio of the direction cosines in the X and Z directions must be equal to a constant implies that the ray path is confined to a plane normal to the XZ plane. This can be understood by inspecting Fig. 6.2-4, where it can be seen that the projection of the ray path in the XZ plane is a straight line. Since the direction cosines in the X and Z directions of a straight line in the XZ plane are constant, the ray path implied by Eq. (6.2-80) is confined to a plane that contains this line and is normal to the XZ plane.

Before proceeding further, let us note that if we let $c(y) = c_0$, then $n(y) = 1$ [see Eq. (6.2-72)] and, as a result, Eq. (6.2-76) becomes

$$\frac{d}{ds}\left(\frac{dy}{ds}\right) = 0, \tag{6.2-81}$$

which implies that

$$\frac{dy}{ds} = \cos\beta(\mathbf{r}) = \cos\beta(y) = \text{constant}, \tag{6.2-82}$$

which in turn implies that the angle $\beta(y)$ is constant. With $\beta(y)$ constant for all values of depth y, the ray path shown in Fig. 6.2-4 will be a straight line. Therefore, in the case of a homogeneous medium where $c(\mathbf{r}) = c_0$, *all ray paths will be straight lines.* This result can also be obtained by substituting $n(\mathbf{r}) = 1$ into the ray equations given by Eqs. (6.2-69) through (6.2-71). Doing

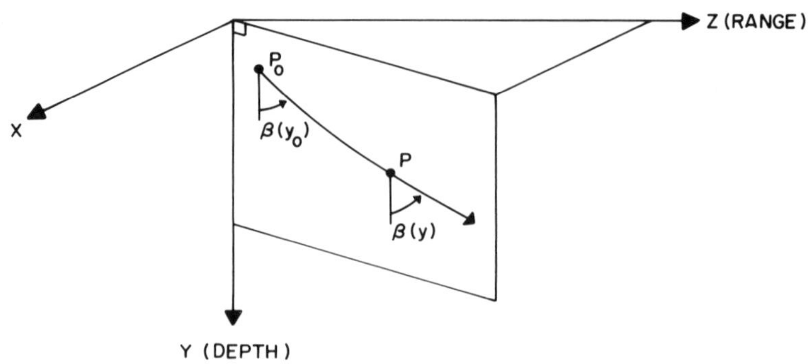

FIG. 6.2-4 Ray path confined to a plane normal to the XZ plane.

so yields the result that all three direction cosines must be constant, which is the description of a straight line in three-dimensional space.

Let us continue with the discussion of the present example by deriving Snell's law. If we multiply both sides of Eqs. (6.2-78) and (6.2-79) by the constant, reference wave number k_0, then

$$k(y)\frac{dx}{ds} = k_X(y) = k_X = \text{constant} \tag{6.2-83}$$

and

$$k(y)\frac{dz}{ds} = k_Z(y) = k_Z = \text{constant} \tag{6.2-84}$$

since [see Eqs. (6.1-10), (6.2-38), and (6.2-46)]

$$k(y) = k_0 n(y) = 2\pi f/c(y) \tag{6.2-85}$$

and

$$\mathbf{k}(y) = k(y)\frac{dx}{ds}\hat{x} + k(y)\frac{dy}{ds}\hat{y} + k(y)\frac{dz}{ds}\hat{z}. \tag{6.2-86}$$

Equations (6.2-83) and (6.2-84) indicate that the components of the propagation vector $\mathbf{k}(y)$ in the X and Z directions are constants. Therefore, Eq. (6.2-86) reduces to Eq. (6.1-72), which appears at the beginning of Example 6.1-3, that is,

$$\mathbf{k}(y) = k_X\hat{x} + k_Y(y)\hat{y} + k_Z\hat{z}, \tag{6.1-72}$$

where

$$k_Y(y) = k(y)\frac{dy}{ds} = k(y)v(y). \tag{6.2-87}$$

Thus, all the analysis performed in Example 6.1-3 is applicable here. Referring back to Example 6.1-3, we find that

$$\frac{\sin\beta(y)}{c(y)} = \frac{\sin\beta(y_0)}{c(y_0)}, \tag{6.1-76}$$

which is Snell's law.

Next, we analyze Eq. (6.2-76) since it is the only ray equation that we have not yet utilized in this example. Since

$$dy/ds = \cos\beta(\mathbf{r}) = \cos\beta(y) \tag{6.2-88}$$

in this example, substituting Eq. (6.2-88) into Eq. (6.2-76) and performing the differentiation with respect to arc length on the left-hand side yields

$$\frac{d}{ds}\beta(y) = -\frac{\sin\beta(y)}{n(y)}\frac{d}{dy}n(y), \tag{6.2-89}$$

where use was made of the following application of the chain rule:

$$\frac{d}{ds}n(y) = \left(\frac{d}{dy}n(y)\right)\frac{dy}{ds} = \left(\frac{d}{dy}n(y)\right)\cos\beta(y). \tag{6.2-90}$$

Since

$$n(y) = c_0/c(y), \tag{6.2-91}$$

$$\frac{d}{dy}n(y) = -\frac{c_0}{c^2(y)}\frac{d}{dy}c(y). \tag{6.2-92}$$

Substituting Eqs. (6.2-91) and (6.2-92) into Eq. (6.2-89) yields

$$\frac{d}{ds}\beta(y) = \frac{\sin\beta(y)}{c(y)}\frac{d}{dy}c(y), \tag{6.2-93}$$

or, using Snell's law [see Eq. (6.1-76)],

$$\frac{d}{ds}\beta(y) = \frac{\sin\beta(y_0)}{c(y_0)}\frac{d}{dy}c(y), \tag{6.2-94}$$

where $\sin\beta(y_0)$ and $c(y_0) = c_0$ are constants. Equation (6.2-94) is a very important result. It states that the *curvature* of a ray, which is defined as $d\beta(y)/ds$, is directly proportional to the sound-speed gradient $dc(y)/dy$. When the sound-speed profile $c(y)$ increases with depth, $dc(y)/dy$ will be positive and, as a result, the curvature will be positive, which means that the ray will curve *upward*. Similarly, when $c(y)$ decreases with depth, $dc(y)/dy$ will be negative and, as a result, the curvature will be negative, which means that the ray will curve *downward*. The rays always curve toward a region of *minimum* sound speed (see Fig. 6.2-5). Note that if $c(y) = c_0$ (a constant), then according to Eq. (6.2-94) the curvature will be *zero*, which means that all ray paths will be straight lines.

Let us finish our discussion of this example by deriving equations for the *travel time* t in seconds and the *range* z in meters along a ray path. Since the ray paths are confined to planes normal to the XZ plane when $c(\mathbf{r}) = c(y)$ (see Fig. 6.2-4), let us restrict ourselves to ray propagation in the YZ plane (see Fig. 6.2-6).

The travel time along a ray path can be expressed as

$$t = \int_0^s \frac{ds}{c(\mathbf{r})} = \int_0^s \frac{ds}{c(y)}, \tag{6.2-95}$$

where s is the arc length and it is understood that $y = y(s)$ in Eq. (6.2-95). From

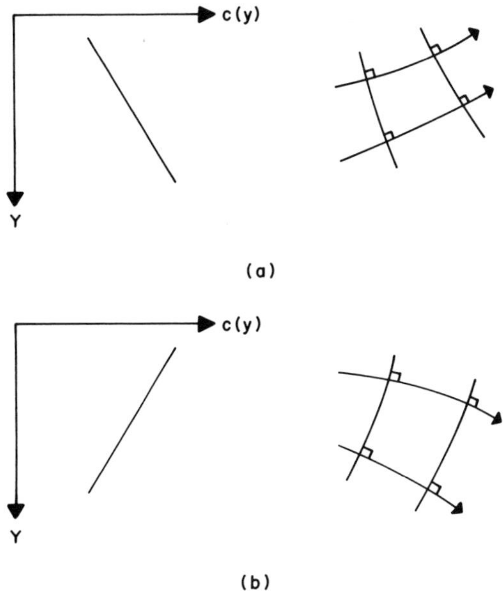

FIG. 6.2-5 (a) Sound-speed profile shown increasing with depth; therefore, rays curve upward. (b) Sound-speed profile shown decreasing with depth; therefore, rays curve downward.

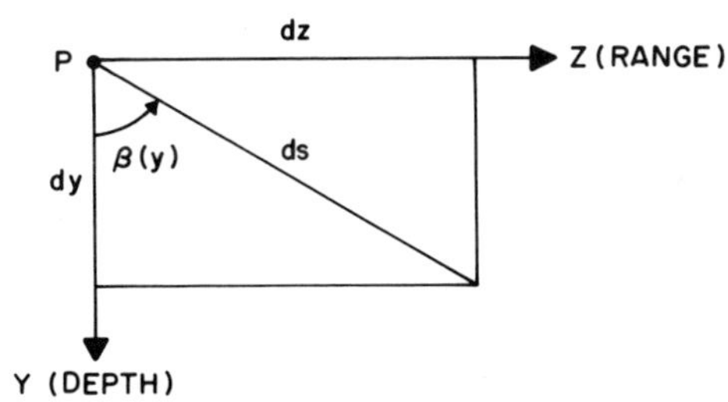

FIG. 6.2-6 An infinitesimal element of arc length *ds* at an arbitrary point *P* along a ray path in the *YZ* plane.

Fig. 6.2-6 it can be seen that

$$\cos \beta(y) = dy/ds \tag{6.2-96}$$

or

$$ds = dy/\cos \beta(y). \tag{6.2-97}$$

Substituting Eq. (6.2-97) into Eq. (6.2-95) yields

$$t = \int_{y_0}^{y} \frac{dy}{c(y) \cos \beta(y)}. \tag{6.2-98}$$

In order to eliminate $\beta(y)$ from Eq. (6.2-98), we use Snell's law as given by Eq. (6.1-76). Since

$$\cos\beta(y) = [1 - \sin^2\beta(y)]^{1/2}, \tag{6.2-99}$$

substituting Eqs. (6.2-99) and (6.1-76) into Eq. (6.2-98) yields

$$t = \int_{y_0}^{y} \frac{dy}{c(y)[1 - b^2c^2(y)]^{1/2}}, \tag{6.2-100}$$

where

$$b = \sin\beta(y_0)/c(y_0). \tag{6.2-101}$$

Equation (6.2-100) is the desired result for the travel time between two points separated by a depth of $(y - y_0)$ meters on a ray path whose initial angle of propagation is $\beta(y_0)$.

Next, let us calculate the range traveled by a ray path. Referring back to Fig. 6.2-6, it can be seen that

$$\tan\beta(y) = dz/dy \tag{6.2-102}$$

or

$$dz = \tan\beta(y)\,dy. \tag{6.2-103}$$

Therefore, the range z can be expressed as

$$z = z_0 + \int_{y_0}^{y} \tan\beta(y)\,dy, \tag{6.2-104}$$

where z_0 is the range coordinate of the source. In order to eliminate $\beta(y)$ from Eq. (6.2-104), we once again use Snell's law. Since

$$\tan\beta(y) = \sin\beta(y)/\cos\beta(y), \tag{6.2-105}$$

substituting Eqs. (6.2-105), (6.1-76), and (6.2-99) into Eq. (6.2-104) yields

$$z = z_0 + b\int_{y_0}^{y} \frac{c(y)}{[1 - b^2c^2(y)]^{1/2}}\,dy, \tag{6.2-106}$$

where b is given by Eq. (6.2-101). Equation (6.2-106) is the desired result for the range between two points separated by a depth of $(y - y_0)$ meters on a ray path whose initial angle of propagation is $\beta(y_0)$. □

EXAMPLE 6.2-4: SPEED OF SOUND AS A FUNCTION OF DEPTH WITH CONSTANT GRADIENT

The simplest type of sound-speed profile is given by

$$c(y) = c_0 + g(y - y_0), \tag{6.2-107}$$

that is, the speed of sound is a *linear function of depth*, where $c_0 = c(y_0)$ and

g is a constant referred to as the *gradient* since $dc(y)/dy = g$. Substituting Eq. (6.2-107) into Eq. (6.2-94) yields

$$\frac{d}{ds}\beta(y) = bg, \tag{6.2-108}$$

where b is given by Eq. (6.2-101). Equation (6.2-108) indicates that the curvature along a ray path is constant. Thus, *the ray path is an arc of a circle.* Since the *radius of curvature* R_c is the reciprocal of the curvature, in our case

$$R_c = \frac{1}{d\beta(y)/ds} = \frac{1}{bg}. \tag{6.2-109}$$

Note that if g is equal to zero, then $c(y) = c_0$ (a constant) and R_c is equal to infinity, which means that all ray paths will be straight lines.

In this example we can obtain equations for both the travel time t and the range z along a ray path by either evaluating the general expressions given by Eqs. (6.2-100) and (6.2-106), respectively, or by taking an alternate approach, as will be discussed next. First, rewrite Snell's law as given by Eq. (6.1-76) as follows:

$$\sin\beta(y) = bc(y), \tag{6.2-110}$$

where b is given by Eq. (6.2-101). Taking the derivative with respect to y of both sides of Eq. (6.2-110) yields

$$\cos\beta(y) = b\frac{c'(y)}{\beta'(y)}, \tag{6.2-111}$$

where $c'(y)$ and $\beta'(y)$ are the first derivatives of $c(y)$ and $\beta(y)$ with respect to y, respectively. Substituting Eq. (6.2-111) into Eq. (6.2-98) and using Eq. (6.2-110) yields

$$t = \int_{\beta(y_0)}^{\beta(y)} \frac{d\beta(y)}{c'(y)\sin\beta(y)}. \tag{6.2-112}$$

Since $c'(y) = g$, Eq. (6.2-112) reduces to

$$t = \frac{1}{g}\ln\left[\frac{\tan[\beta(y)/2]}{\tan[\beta(y_0)/2]}\right], \tag{6.2-113}$$

which is the desired result.

Next, substituting Eqs. (6.2-105) and (6.2-111) into Eq. (6.2-104) yields

$$z = z_0 + \int_{\beta(y_0)}^{\beta(y)} \frac{\sin\beta(y)}{bc'(y)}\,d\beta(y). \tag{6.2-114}$$

Since $c'(y) = g$, and using Eq. (6.2-101), Eq. (6.2-114) reduces to

$$z = z_0 + \left[\frac{c_0}{g \sin \beta(y_0)}\right][\cos \beta(y_0) - \cos \beta(y)], \tag{6.2-115}$$

which is the desired result.

Finally, solving for the depth y in Eq. (6.2-107) yields

$$y = y_0 + \frac{c(y) - c_0}{g}, \tag{6.2-116}$$

and upon substituting Eqs. (6.2-101) and (6.2-110) into Eq. (6.2-116), we obtain

$$y = y_0 + \frac{c_0}{g}\left[\frac{\sin \beta(y)}{\sin \beta(y_0)} - 1\right], \tag{6.2-117}$$

which is the depth in meters of a point on a ray path whose initial angle of propagation is $\beta(y_0)$ and whose angle of propagation at y is $\beta(y)$. □

6.2.4 Amplitude Variation Along Rays

In this section we shall obtain an expression for the real amplitude term $a(\mathbf{r})$. We begin by multiplying the transport equation, as given by Eq. (6.2-6), by $a(\mathbf{r})$. Doing so yields

$$a^2(\mathbf{r}) \nabla^2 W(\mathbf{r}) + 2a(\mathbf{r}) \nabla a(\mathbf{r}) \cdot \nabla W(\mathbf{r}) = \nabla \cdot [a^2(\mathbf{r}) \nabla W(\mathbf{r})] = 0. \tag{6.2-118}$$

Therefore, solving the transport equation is equivalent to solving

$$\nabla \cdot [a^2(\mathbf{r}) \nabla W(\mathbf{r})] = 0, \tag{6.2-119}$$

or, substituting Eq. (6.2-62) into Eq. (6.2-119),

$$\nabla \cdot \mathbf{I}_{\text{avg}} = 0. \tag{6.2-120}$$

The solution of Eq. (6.2-120) can be developed in terms of a *ray tube*, as illustrated in Fig. 6.2-7b. The surfaces S_1 and S_2 shown in Fig. 6.2-7b are assumed to be *very small* portions of the wavefronts that pass through the points $P_1(x_1, y_1, z_1)$ and $P_2(x_2, y_2, z_2)$, respectively, where $\hat{n}_1$ and $\hat{n}_2$ are unit vectors normal to S_1 and S_2, respectively, pointing in the conventional outward direction. The surface S_3 is formed by the outermost ray paths of the family of ray paths shown in Fig. 6.2-7a, where $\hat{n}_3$ is a unit vector normal to S_3 pointing in the conventional outward direction. Recall that the direction of energy propagation, and thus the direction of $\mathbf{I}_{\text{avg}}$, is determined by the direction of the ray paths [see Eq. (6.2-62)]. Since the divergence of the time-average intensity vector $\mathbf{I}_{\text{avg}}$ is equal to zero [see Eq. (6.2-120)], the number of ray paths entering S_1 per unit time is equal to the number of ray paths leaving

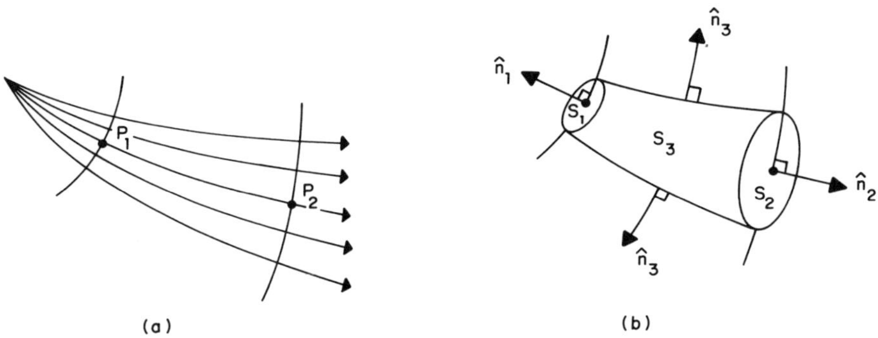

FIG. 6.2-7 (a) Rays and wavefronts. (b) Ray-tube segment formed by the family of ray paths shown in (a).

S_2 per unit time. Thus, Eq. (6.2-120) is a statement of *conservation of energy* since the effects of attenuation due to sound absorption have not been included.

Upon integrating Eq. (6.2-120) over the volume V occupied by the ray-tube segment connecting points P_1 and P_2 and using the *divergence theorem*, we obtain

$$\int_V \nabla \cdot \mathbf{I}_{\text{avg}}\, dV = \oint_S \mathbf{I}_{\text{avg}} \cdot \mathbf{dS} = 0. \tag{6.2-121}$$

Referring to Fig. 6.2-7b, Eq. (6.2-121) can be rewritten as

$$\oint_S \mathbf{I}_{\text{avg}} \cdot \mathbf{dS} = \int_{S_1} \mathbf{I}_{\text{avg}} \cdot \hat{n}_1\, dS_1 + \int_{S_2} \mathbf{I}_{\text{avg}} \cdot \hat{n}_2\, dS_2 + \int_{S_3} \mathbf{I}_{\text{avg}} \cdot \hat{n}_3\, dS_3 = 0. \tag{6.2-122}$$

Since the direction of $\mathbf{I}_{\text{avg}}$ is defined by the direction of the ray paths, $\hat{n}_1$ is in the opposite direction to $\mathbf{I}_{\text{avg}}$, $\hat{n}_2$ is in the same direction as $\mathbf{I}_{\text{avg}}$, and $\hat{n}_3$ is normal to $\mathbf{I}_{\text{avg}}$. Therefore,

$$\mathbf{I}_{\text{avg}} \cdot \hat{n}_1 = -I_{\text{avg}}, \tag{6.2-123}$$

$$\mathbf{I}_{\text{avg}} \cdot \hat{n}_2 = +I_{\text{avg}}, \tag{6.2-124}$$

and

$$\mathbf{I}_{\text{avg}} \cdot \hat{n}_3 = 0, \tag{6.2-125}$$

where the magnitude of the intensity vector I_{avg} is given by Eq. (6.2-63).

Substituting Eqs. (6.2-123) through (6.2-125) into Eq. (6.2-122) yields

$$\int_{S_2} I_{\text{avg}}(\mathbf{r})\,dS_2 = \int_{S_1} I_{\text{avg}}(\mathbf{r})\,dS_1, \tag{6.2-126}$$

where the dependence of the intensity on position is now explicitly indicated. Since we have already assumed that S_1 and S_2 are very small surface areas,

$$I_{\text{avg}}(\mathbf{r}) \approx I_{\text{avg}}(\mathbf{r}_1) \qquad \text{on} \quad S_1 \tag{6.2-127}$$

and

$$I_{\text{avg}}(\mathbf{r}) \approx I_{\text{avg}}(\mathbf{r}_2) \qquad \text{on} \quad S_2, \tag{6.2-128}$$

where $\mathbf{r}_1 = (x_1, y_1, z_1)$ and $\mathbf{r}_2 = (x_2, y_2, z_2)$. Therefore,

$$\int_{S_1} I_{\text{avg}}(\mathbf{r})\,dS_1 \approx I_{\text{avg}}(\mathbf{r}_1) \int_{S_1} dS_1 = I_{\text{avg}}(\mathbf{r}_1)S_1 \tag{6.2-129}$$

and

$$\int_{S_2} I_{\text{avg}}(\mathbf{r})\,dS_2 \approx I_{\text{avg}}(\mathbf{r}_2) \int_{S_2} dS_2 = I_{\text{avg}}(\mathbf{r}_2)S_2, \tag{6.2-130}$$

where S_1 and S_2 are the wavefront surface areas at the ends of the ray-tube segment. Substituting Eqs. (6.2-129) and (6.2-130) into Eq. (6.2-126) yields

$$I_{\text{avg}}(\mathbf{r}_2)S_2 = I_{\text{avg}}(\mathbf{r}_1)S_1, \tag{6.2-131}$$

and upon substituting Eq. (6.2-63) into Eq. (6.2-131) and making use of Eqs. (6.2-9) and (6.1-5), we finally obtain the expression

$$a(\mathbf{r}_2) = a(\mathbf{r}_1)\left[\frac{S_1}{S_2}\frac{c(\mathbf{r}_2)}{c(\mathbf{r}_1)}\right]^{1/2}, \tag{6.2-132}$$

which relates the amplitude $a(\mathbf{r}_2)$ at some point P_2 along a ray path in terms of the amplitude $a(\mathbf{r}_1)$ at some previous point P_1 along the same ray path (see Fig. 6.2-7). Note that if S_2 becomes smaller than S_1, the amplitude, and hence the intensity, will increase since the ray tube is contracting due to the *focusing* of the rays. In the limit as S_2 approaches zero (points at which S_2 approaches zero are known as *focal points*), the amplitude, and hence the intensity, approaches infinity. Thus, the method of ray acoustics is *invalid* at focal points. However, if S_2 becomes larger than S_1, the amplitude, and hence the intensity, will decrease since the ray tube is spreading due to the divergence of the rays. One should compare Eq. (6.2-132) with the expression for the real amplitude term given by Eq. (6.1-38) obtained via the WKB method. Finally, note that a *caustic* is a surface composed entirely of focal points and that as a ray grazes a caustic or, equivalently, passes through a focal point, it undergoes a $\pi/2$ phase shift.

6.3 THE PARABOLIC EQUATION APPROXIMATION

The *parabolic equation approximation* is another method of obtaining an approximate solution to the general Helmholtz wave equation specified by Eqs. (6.1-8) and (6.1-5), where the speed of sound is a function of *all three* spatial variables. We begin by assuming that Eq. (6.1-8) has a solution in the form of a wave propagating in the positive Z direction, that is, assume that

$$\varphi(\mathbf{r}) = g(\mathbf{r}) \exp[-jk_0(z - z_0)], \tag{6.3-1}$$

where $g(\mathbf{r})$ is not a real amplitude term but is, instead, a *complex* quantity, in general, k_0 is the constant, reference wave number given by Eq. (6.1-6), and z_0 is the range coordinate of the source [compare Eq. (6.3-1) with Eqs. (6.1-25) and (6.2-1) or (6.2-2)]. Substituting Eq. (6.3-1) into Eq. (6.1-8) yields

$$\nabla^2 g(\mathbf{r}) - j2k_0 \frac{\partial}{\partial z} g(\mathbf{r}) + k_0^2[n^2(\mathbf{r}) - 1]g(\mathbf{r}) = 0, \tag{6.3-2}$$

where

$$\nabla^2 = \partial^2/\partial x^2 + \partial^2/\partial y^2 + \partial^2/\partial z^2 \tag{6.3-3}$$

is the *Laplacian.* If it is further assumed that

$$\left| \frac{\partial^2}{\partial z^2} g(\mathbf{r}) \right| \ll 2k_0 \left| \frac{\partial}{\partial z} g(\mathbf{r}) \right|, \tag{6.3-4}$$

then Eq. (6.3-2) reduces to

$$\nabla_{\mathrm{T}}^2 g(\mathbf{r}) - j2k_0 \frac{\partial}{\partial z} g(\mathbf{r}) + k_0^2[n^2(\mathbf{r}) - 1]g(\mathbf{r}) = 0, \tag{6.3-5}$$

which is the *parabolic equation approximation* of the Helmholtz wave equation, where

$$\nabla_{\mathrm{T}}^2 = \partial^2/\partial x^2 + \partial^2/\partial y^2 \tag{6.3-6}$$

is referred to as the *transverse Laplacian.* The assumption represented by Eq. (6.3-4) will be satisfied best at high frequencies and is called the *parabolic approximation* since it leads to a *parabolic partial differential equation,* that is, Eq. (6.3-5). Equation (6.3-5) is usually rewritten as

$$\frac{\partial}{\partial z} g(\mathbf{r}) = [-j/(2k_0)]\, \nabla_{\mathrm{T}}^2 g(\mathbf{r}) - j(k_0/2)[n^2(\mathbf{r}) - 1]g(\mathbf{r}) \tag{6.3-7}$$

and must be solved *numerically* for $g(\mathbf{r})$.

EXAMPLE 6.3-1: HOMOGENEOUS MEDIUM

In the case of a homogeneous medium where $n(\mathbf{r}) = 1$, Eq. (6.3-7) reduces to

$$\frac{\partial}{\partial z} g(\mathbf{r}) = [-j/(2k_0)] \nabla_T^2 g(\mathbf{r}). \tag{6.3-8}$$

An exact solution of Eq. (6.3-8) is given by

$$\begin{aligned} g(\mathbf{r}) &= \exp[-jk_X(x - x_0)] \exp[-jk_Y(y - y_0)] \\ &\quad \times \exp\{+j[(k_X^2 + k_Y^2)/(2k_0)](z - z_0)\}, \end{aligned} \tag{6.3-9}$$

where

$$k_X = k_0 u_0 = k_0 \sin\theta_0 \cos\psi_0 \tag{6.3-10}$$

and

$$k_Y = k_0 v_0 = k_0 \sin\theta_0 \sin\psi_0 = k_0 \cos\beta_0 \tag{6.3-11}$$

are the constant components of the propagation vector in the X and Y (depth) directions, respectively, the angles (θ_0, ψ_0) are the constant vertical and azimuthal spherical angles measured with respect to the positive Z (range) and X axes, respectively, the angle β_0 is constant and is measured with respect to the positive Y (depth) axis [see Eqs. (6.2-42) and (6.2-43) and Fig. 6.2-3], and (x_0, y_0, z_0) are the rectangular coordinates of the source. Substituting Eqs. (6.3-9) through (6.3-11) into Eq. (6.3-1) yields

$$\varphi(\mathbf{r}) = \exp[-jk_X(x - x_0)] \exp[-jk_Y(y - y_0)] \exp[-j\hat{k}_Z(z - z_0)], \tag{6.3-12}$$

where

$$\hat{k}_Z = k_0 \hat{w}_0 \tag{6.3-13}$$

and

$$\hat{w}_0 = 1 - [(u_0^2 + v_0^2)/2] \tag{6.3-14}$$

or

$$\hat{w}_0 = 1 - [(\sin^2\theta_0)/2], \tag{6.3-15}$$

which is an approximation or estimate of the direction cosine with respect to the positive Z (range) axis

$$w_0 = \cos\theta_0. \tag{6.3-16}$$

Equations (6.3-14) and (6.3-15) are, in fact, *binomial expansions* of w_0, valid as long as

$$u_0^2 + v_0^2 = \sin^2\theta_0 \ll 1 \tag{6.3-17}$$

or

$$\sin^2 \theta_0 \leq 0.1, \tag{6.3-18}$$

which implies

$$\theta_0 \leq 18.0^\circ. \tag{6.3-19}$$

Figure 6.3-1 is a plot of $\hat{w}_0$ and w_0, as given by Eqs. (6.3-15) and (6.3-16), respectively, versus θ_0. Table 6.3-1 is a tabulation of the approximation error $\Delta = \hat{w}_0 - w_0$ versus θ_0 and the percentage error $\Delta\%$ with respect to w_0.

In the case of a homogeneous medium where $n(\mathbf{r}) = 1$, it is well known that Eq. (6.3-12) will be an exact solution to Eq. (6.1-8) [or, see Eq. (6.2-12) instead] *if* $\hat{k}_Z$ given by Eq. (6.3-13) is replaced by

$$k_Z = k_0 w_0 = k_0 \cos \theta_0, \tag{6.3-20}$$

which is the component of the propagation vector in the Z (range) direction. However, as can be seen from Fig. 6.3-1, $\hat{w}_0$ will be a very good approximation to w_0 so long as $\theta_0 \leq 18.0^\circ$. Therefore, Eq. (6.3-12) will also be a very good approximate solution to the Helmholtz wave equation given by Eq. (6.1-8)

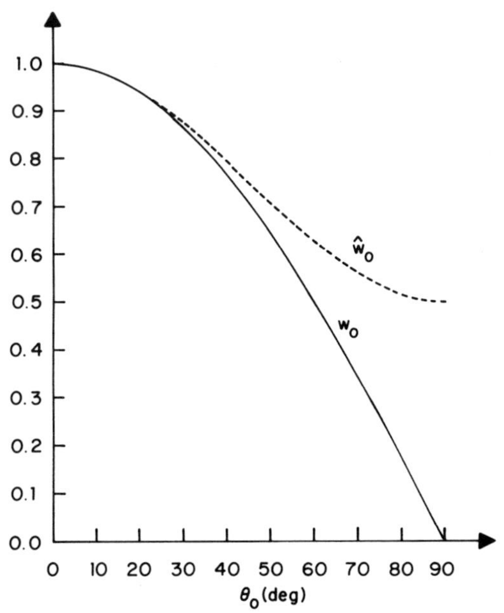

FIG. 6.3-1 Plot of $\hat{w}_0$ (---) and w_0 (——) as given by Eqs. (6.3-15) and (6.3-16), respectively, versus θ_0 (degrees).

Table 6.3-1

Tabulation of the Approximation Error $\Delta = \hat{w}_0 - w_0$ versus θ_0 (degrees) and the Percentage Error $\Delta\%$ with Respect to w_0

θ_0 (degrees)	Δ ($\times 10^{-3}$)	$\Delta\%$
0	0.0	0.0
5	0.007	0.001
10	0.115	0.012
15	0.581	0.06
20	1.818	0.194
25	4.389	0.484
30	8.975	1.036
35	16.353	1.996
40	27.368	3.573
45	42.893	6.066
50	63.8	9.926

when $n(\mathbf{r}) = 1$, so long as Eq. (6.3-19) is satisfied. As a result, the assumption represented by Eq. (6.3-4) is also referred to as the *small-angle approximation.*

Finally, let us conclude this example by listing the following three additional exact solutions to the parabolic partial differential equation given by Eq. (6.3-8):

$$g_1(\mathbf{r}) = j\frac{1}{\lambda_0\,\Delta Z}\exp\left[-j\frac{k_0}{2\,\Delta Z}(x^2 + y^2)\right], \tag{6.3-21}$$

$$g_2(\mathbf{r}) = j\frac{1}{\lambda_0\,\Delta Z}\exp\left\{-j\frac{k_0}{2\,\Delta Z}[(x - x_0)^2 + (y - y_0)^2]\right\}, \tag{6.3-22}$$

and

$$g_3(\mathbf{r}) = j\frac{1}{\lambda_0\,\Delta Z}\int_{-\infty}^{\infty}\int_{-\infty}^{\infty} A(x_0, y_0) \times \exp\left\{-j\frac{k_0}{2\,\Delta Z}[(x - x_0)^2 + (y - y_0)^2]\right\} dx_0\, dy_0, \tag{6.3-23}$$

where $A(x_0, y_0)$ is an arbitrary complex function,

$$k_0 = 2\pi f/c_0 = 2\pi/\lambda_0, \tag{6.3-24}$$

and

$$\Delta Z = z - z_0. \tag{6.3-25}$$

Note that

$$\varphi_3(\mathbf{r}) = g_3(\mathbf{r})\exp[-jk_0(z - z_0)] \qquad (6.3\text{-}26)$$

is an approximate solution to the Helmholtz wave equation given by Eq. (6.1-8) when $n(\mathbf{r}) = 1$, and is, in fact, the *Fresnel diffraction pattern* of the complex function $A(x_0, y_0)$. The derivation of the Fresnel diffraction integral in optics is based on the initial assumption that the angle θ_0 with respect to the Z (range) axis is small. Also, note that Eq. (6.3-26) can be expressed as

$$\varphi_3(x, y, z) = A(x, y) ** \varphi_1(x, y, z), \qquad (6.3\text{-}27)$$

where

$$\varphi_1(\mathbf{r}) = g_1(\mathbf{r})\exp[-jk_0(z - z_0)] \qquad (6.3\text{-}28)$$

and the double asterisk denotes a two-dimensional convolution with respect to x and y. Therefore, $\varphi_1(\mathbf{r})$ given by Eq. (6.3-28) can be interpreted as being the *impulse response* or *Green's function* characterizing Fresnel diffraction in a homogeneous medium. □

BIBLIOGRAPHY

M. Born and E. Wolf, *Principles of Optics*, 6th ed. Pergamon, Oxford, 1980.

L. M. Brekhovskikh, *Waves in Layered Media*, 1st ed. Academic Press, New York, 1960.

L. Brekhovskikh and Y. Lysanov, *Fundamentals of Ocean Acoustics*. Springer-Verlag, Berlin and New York, 1982.

R. H. Clarke, Acoustic and electromagnetic waves propagating in a tenuous random medium. In *Aspects of Signal Processing*, vol. I (edited by G. Tacconi), pp. 61–67. D. Reidel, Dordrecht, Holland, 1977.

J. D. Gaskill, *Linear Systems, Fourier Transforms, and Optics*. Wiley, New York, 1978.

A. W. Goodman, *Analytic Geometry and the Calculus*. Macmillan, New York, 1963.

J. W. Goodman, *Introduction to Fourier Optics*. McGraw-Hill, New York, 1968.

L. E. Kinsler, A. R. Frey, A. B. Coppens, and J. V. Sanders, *Fundamentals of Acoustics*, 3rd ed. Wiley, New York, 1982.

M. V. Klein, *Optics*. Wiley, New York, 1970.

S. T. McDaniel, Application of the parabolic approximation to predict acoustical propagation in the ocean. *Am. J. Phys.* **47**(1), 63–68 (1979).

C. B. Officer, *Introduction to the Theory of Sound Transmission*. McGraw-Hill, New York, 1958.

A. Papoulis, *Systems and Transforms with Applications in Optics*. McGraw-Hill, New York, 1968.

A. D. Pierce, *Acoustics: An Introduction to Its Physical Principles and Applications*. McGraw-Hill, New York, 1981.

F. D. Tappert, The parabolic approximation method. In *Wave Propagation and Underwater Acoustics* (edited by J. B. Keller and J. S. Papadakis), pp. 224–287. Springer-Verlag, Berlin and New York, 1977.

I. Tolstoy and C. S. Clay, *Ocean Acoustics*. McGraw-Hill, New York, 1966.

B. J. Uscinski, *The Elements of Wave Propagation in Random Media*. McGraw-Hill, New York, 1977.

7

Random Ocean Medium Transfer Functions

7.1 COUPLING EQUATIONS AND THE GENERALIZED COHERENCE FUNCTION

We begin this chapter by presenting the equations that couple the transmitted and received electrical signals to the transfer function of the ocean medium via the transmit and receive far-field directivity functions. Referring to Fig. 1.1-3, the frequency and angular spectra of the input acoustic signal to the medium and the output electrical signal from the receive aperture (array) are given, respectively, by the following equations [see Eqs. (3.2-47) and (3.2-51)]:

$$X_{\mathrm{M}}(f, \boldsymbol{\nu}) = \int_{-\infty}^{\infty} X(f, \boldsymbol{\alpha}) D_{\mathrm{T}}(f, \boldsymbol{\nu} - \boldsymbol{\alpha})\, d\boldsymbol{\alpha} \tag{7.1-1}$$

and

$$Y(\eta, \boldsymbol{\gamma}) = \int_{-\infty}^{\infty} Y_{\mathrm{M}}(\eta, \boldsymbol{\beta}) D_{\mathrm{R}}(\eta, \boldsymbol{\gamma} - \boldsymbol{\beta})\, d\boldsymbol{\beta}, \tag{7.1-2}$$

where the frequency and angular spectrum of the output acoustic signal from the medium is given by [see Eq. (2.1-46)]

$$Y_{\mathrm{M}}(\eta, \boldsymbol{\beta}) = \int_{-\infty}^{\infty} \int_{-\infty}^{\infty} X_{\mathrm{M}}(f, \boldsymbol{\nu}) B_{\mathrm{M}}(f, \boldsymbol{\nu}; \eta - f, \boldsymbol{\beta} - \boldsymbol{\nu})\, df\, d\boldsymbol{\nu}, \tag{7.1-3}$$

where [see Eq. (2.1-32)]

$$B_{\mathrm{M}}(f, \boldsymbol{\nu}; \phi, \boldsymbol{\kappa}) = \int_{-\infty}^{\infty} \int_{-\infty}^{\infty} H_{\mathrm{M}}(f, \boldsymbol{\nu}; t, \mathbf{r}) \exp(-j2\pi\phi t) \exp(+j2\pi\boldsymbol{\kappa} \cdot \mathbf{r})\, dt\, d\mathbf{r} \tag{7.1-4}$$

is the *bifrequency function* and $H_M(f, \mathbf{v}; t, \mathbf{r})$ is the *time-variant, space-variant transfer function of the ocean medium.* Equations (7.1-1) through (7.1-4) are the basic *coupling equations.* However, with respect to Eq. (7.1-1), we have the following additional relationships:

$$X_M(f, \mathbf{v}) = F_t F_{\mathbf{r}}\{x_M(t, \mathbf{r})\}$$
$$= \int_{-\infty}^{\infty} \int_V x_M(t, \mathbf{r}) \exp(+j2\pi \mathbf{v} \cdot \mathbf{r})\, dV \exp(-j2\pi f t)\, dt, \quad (7.1\text{-}5)$$

where [see Eq. (3.1-8)]

$$x_M(t, \mathbf{r}) = \int_{-\infty}^{\infty} X(f, \mathbf{r}) A_T(f, \mathbf{r}) \exp(+j2\pi f t)\, df, \quad (7.1\text{-}6)$$

$$X(f, \boldsymbol{\alpha}) = F_t F_{\mathbf{r}}\{x(t, \mathbf{r})\}$$
$$= \int_{-\infty}^{\infty} \int_V x(t, \mathbf{r}) \exp(+j2\pi \boldsymbol{\alpha} \cdot \mathbf{r})\, dV \exp(-j2\pi f t)\, dt \quad (7.1\text{-}7)$$

is the frequency and angular spectrum of the transmitted electrical signal, and [see Eqs. (3.2-41) and (3.2-42)]

$$D_T(f, \boldsymbol{\alpha}) = F_{\mathbf{r}}\{A_T(f, \mathbf{r})\} = \int_V A_T(f, \mathbf{r}) \exp(+j2\pi \boldsymbol{\alpha} \cdot \mathbf{r})\, dV \quad (7.1\text{-}8)$$

is the far-field directivity function (beam pattern) of the complex transmit aperture function $A_T(f, \mathbf{r})$. Also, with respect to Eq. (7.1-6) [see Eq. (3.1-5)],

$$X(f, \mathbf{r}) = F_t\{x(t, \mathbf{r})\} = \int_{-\infty}^{\infty} x(t, \mathbf{r}) \exp(-j2\pi f t)\, dt \quad (7.1\text{-}9)$$

and [see Eqs. (3.2-43) and (3.2-44)]

$$A_T(f, \mathbf{r}) = F_{\boldsymbol{\alpha}}^{-1}\{D_T(f, \boldsymbol{\alpha})\} = \int_{-\infty}^{\infty} D_T(f, \boldsymbol{\alpha}) \exp(-j2\pi \boldsymbol{\alpha} \cdot \mathbf{r})\, d\boldsymbol{\alpha}. \quad (7.1\text{-}10)$$

Similarly, with respect to Eq. (7.1-2), we have the following additional relationships:

$$Y(\eta, \boldsymbol{\gamma}) = F_t F_{\mathbf{r}}\{y(t, \mathbf{r})\}$$
$$= \int_{-\infty}^{\infty} \int_V y(t, \mathbf{r}) \exp(+j2\pi \boldsymbol{\gamma} \cdot \mathbf{r})\, dV \exp(-j2\pi \eta t)\, dt, \quad (7.1\text{-}11)$$

where

$$y(t, \mathbf{r}) = \int_{-\infty}^{\infty} Y_M(\eta, \mathbf{r}) A_R(\eta, \mathbf{r}) \exp(+j2\pi \eta t)\, d\eta, \quad (7.1\text{-}12)$$

the frequency and angular spectrum of the output acoustic signal from the medium is given by

$$Y_{\mathrm{M}}(\eta, \boldsymbol{\beta}) = F_t F_{\mathbf{r}}\{ y_{\mathrm{M}}(t, \mathbf{r})\}$$
$$= \int_{-\infty}^{\infty} \int_V y_{\mathrm{M}}(t, \mathbf{r}) \exp(+j2\pi \boldsymbol{\beta} \cdot \mathbf{r})\, dV \exp(-j2\pi\eta t)\, dt, \tag{7.1-13}$$

where [see Eq. (2.1-25)]

$$y_{\mathrm{M}}(t, \mathbf{r}) = \int_{-\infty}^{\infty} \int_{-\infty}^{\infty} X_{\mathrm{M}}(f, \mathbf{v}) H_{\mathrm{M}}(f, \mathbf{v}; t, \mathbf{r}) \exp(+j2\pi f t) \exp(-j2\pi \mathbf{v} \cdot \mathbf{r})\, df\, d\mathbf{v}, \tag{7.1-14}$$

and

$$D_{\mathrm{R}}(\eta, \boldsymbol{\beta}) = F_{\mathbf{r}}\{A_{\mathrm{R}}(\eta, \mathbf{r})\} = \int_V A_{\mathrm{R}}(\eta, \mathbf{r}) \exp(+j2\pi \boldsymbol{\beta} \cdot \mathbf{r})\, dV \tag{7.1-15}$$

is the far-field directivity function (beam pattern) of the complex receive aperture function $A_{\mathrm{R}}(\eta, \mathbf{r})$. Finally, with respect to Eq. (7.1-12),

$$Y_{\mathrm{M}}(\eta, \mathbf{r}) = F_t\{ y_{\mathrm{M}}(t, \mathbf{r})\} = \int_{-\infty}^{\infty} y_{\mathrm{M}}(t, \mathbf{r}) \exp(-j2\pi\eta t)\, dt \tag{7.1-16}$$

and

$$A_{\mathrm{R}}(\eta, \mathbf{r}) = F_{\boldsymbol{\beta}}^{-1}\{D_{\mathrm{R}}(\eta, \boldsymbol{\beta})\} = \int_{-\infty}^{\infty} D_{\mathrm{R}}(\eta, \boldsymbol{\beta}) \exp(-j2\pi \boldsymbol{\beta} \cdot \mathbf{r})\, d\boldsymbol{\beta}. \tag{7.1-17}$$

Substituting Eq. (7.1-14) into Eq. (7.1-16) yields

$$Y_{\mathrm{M}}(\eta, \mathbf{r}) = \int_{-\infty}^{\infty} \int_{-\infty}^{\infty} X_{\mathrm{M}}(f, \mathbf{v}) B_{\mathrm{M}}(f, \mathbf{v}; \eta - f, \mathbf{r}) \exp(-j2\pi \mathbf{v} \cdot \mathbf{r})\, df\, d\mathbf{v}, \tag{7.1-18}$$

where

$$B_{\mathrm{M}}(f, \mathbf{v}; \phi, \mathbf{r}) = \int_{-\infty}^{\infty} H_{\mathrm{M}}(f, \mathbf{v}; t, \mathbf{r}) \exp(-j2\pi\phi t)\, dt, \tag{7.1-19}$$

which is *not* the same as the bifrequency function [see Eq. (7.1-4)]. The components of the vectors $\boldsymbol{\alpha}$, $\mathbf{v}$, $\boldsymbol{\beta}$, and γ are *spatial frequencies* with units of cycles per meter, and $\mathbf{r} = (x, y, z)$.

A moment's reflection leads one to the conclusion that Eq. (7.1-12) is well suited for space–time signal-processing applications. By inspecting Eqs. (7.1-18), (7.1-19), and (7.1-1), it can be seen that the output signal given by Eq. (7.1-12) can be expressed as a function of time and space in terms of the transmitted electrical signal, the transmit aperture, the transfer function of the ocean medium, and the receive aperture. Equation (7.1-2), however, represents the theoretical output "spectrum," albeit dependent on the same system functions.

It is probably obvious by now that the weak link in the coupling equations is the ocean medium transfer function $H_{\mathrm{M}}(f, \mathbf{v}; t, \mathbf{r})$ and, if the transfer function is random, its corresponding autocorrelation function [see Eq. (2.2-2)]

$$R_{H_{\mathrm{M}}}(f, f', \mathbf{v}, \mathbf{v}'; t, t', \mathbf{r}, \mathbf{r}') = \mathrm{E}\{H_{\mathrm{M}}(f, \mathbf{v}; t, \mathbf{r}) H^{*}_{\mathrm{M}}(f', \mathbf{v}'; t', \mathbf{r}')\}. \quad (7.1\text{-}20)$$

If one is not able to specify a *realistic* functional form for $H_{\mathrm{M}}(f, \mathbf{v}; t, \mathbf{r})$, then the equations presented in this section represent an exercise in the methods of linear systems theory. This leads us to the main purpose of this chapter, which is to demonstrate the derivation of two random ocean medium transfer functions based on the WKB and parabolic equation approximations. The transfer function derivations are discussed in Sections 7.2.1 and 7.3.1, respectively.

Finally, note that the autocorrelation function given by Eq. (7.1-20) can be referred to as a *generalized coherence function* since it is a generalization of the two-frequency correlation function or two-frequency mutual coherence function based on linear, time-variant filter theory. Equation (7.1-20) represents the *correlation* between the two *output fields* $H_{\mathrm{M}}(f, \mathbf{v}; t, \mathbf{r})$ and $H_{\mathrm{M}}(f', \mathbf{v}'; t', \mathbf{r}')$ at two different times (t and t') and at two different spatial locations ($\mathbf{r}$ and $\mathbf{r}'$) when two small-amplitude time-harmonic plane-wave *input fields* are transmitted at two different frequencies (f and f') and in two different *initial* directions ($\mathbf{v}$ and $\mathbf{v}'$).

The effects of a time-variant, space-variant, random medium on small-amplitude wave propagation can be described in terms of *coherence time*, *coherence bandwidth*, *spatial coherence*, and *angular coherence*. This information can be obtained from the generalized coherence function given by Eq. (7.1-20).

The value of the time difference $\Delta t = t - t'$ at which the autocorrelation function $R_{H_{\mathrm{M}}}(f, f, \mathbf{v}, \mathbf{v}; t, t', \mathbf{r}, \mathbf{r})$ decreases to a specified level is called the *coherence time*. It is a measure of the time correlation of a field at a frequency f and at a spatial location $\mathbf{r}$ when the initial direction of propagation is $\mathbf{v}$. The reciprocal of the coherence time is a measure of the amount of frequency spreading (in hertz) to be expected.

The value of the frequency difference $\Delta f = f - f'$ at which the autocorrelation function $R_{H_{\mathrm{M}}}(f, f', \mathbf{v}, \mathbf{v}; t, t, \mathbf{r}, \mathbf{r})$ decreases to a specified level is called the *coherence bandwidth*. It is a measure of the frequency correlation of a field at time t and at a spatial location $\mathbf{r}$ when the initial direction of propagation is $\mathbf{v}$. The reciprocal of the coherence bandwidth is a measure of the amount of spread to be expected in time-delay values.

The value of the spatial difference $\Delta\mathbf{r} = \mathbf{r} - \mathbf{r}' = (\Delta x, \Delta y, \Delta z)$ at which the autocorrelation function $R_{H_M}(f, f, \mathbf{v}, \mathbf{v}; t, t, \mathbf{r}, \mathbf{r}')$ decreases to a specified level is called the *spatial coherence.* It is a measure of the spatial correlation of a field at time t and at a frequency f when the initial direction of propagation is $\mathbf{v}$. The quantities $\Delta x = x - x'$, $\Delta y = y - y'$, and $\Delta z = z - z'$ are referred to as the *scale sizes* of the medium in the X, Y, and Z directions, respectively. The reciprocal of the scale sizes are measures of the amount of spatial-frequency spreading (in cycles per meter) to be expected in the X, Y, and Z directions. Since spatial frequencies are related to direction cosines and, hence, to the directions of wave propagation, spatial-frequency spread corresponds to angular spread or scatter.

And, in conclusion, the value of the angular difference $\Delta\mathbf{v} = \mathbf{v} - \mathbf{v}' = (\Delta f_X, \Delta f_Y, \Delta f_Z)$ at which the autocorrelation function $R_{H_M}(f, f, \mathbf{v}, \mathbf{v}'; t, t, \mathbf{r}, \mathbf{r})$ decreases to a specified level is called the *angular coherence.* It is a measure of the correlation between two fields at time t, frequency f, and spatial location $\mathbf{r}$ when the initial directions of propagation are $\mathbf{v}$ and $\mathbf{v}'$. The reciprocal of the quantities $\Delta f_X = f_X - f_X'$, $\Delta f_Y = f_Y - f_Y'$, and $\Delta f_Z = f_Z - f_Z'$ are measures of the spatial spread or physical dimensions of the source in the X, Y, and Z directions, respectively.

The result of all this spreading is, of course, distortion of the shape of the transmitted pulse.

7.2 THE WKB APPROXIMATION REVISITED

7.2.1 Transfer Function Derivation

We shall now proceed to derive a transfer function that models the bistatic communication channel geometry shown in Fig. 1.1-2. The communication channel is regarded to be the ocean volume between the apertures, so surface and bottom scattering effects are not included. Both apertures are stationary (not in motion), and it is assumed that no discrete point scatterers (bubbles, fish, etc.) are in the volume between the apertures. No motion implies that the resulting transfer function will be time-invariant.

The propagation of *small-amplitude* acoustic signals in the ocean from the transmit aperture to the receive aperture can be described by the following *linear*, inhomogeneous, scalar wave equation [see Eq. (1.1-1)]:

$$\nabla^2\varphi(t, \mathbf{r}) - \frac{1}{c^2(\mathbf{r})}\frac{\partial^2}{\partial t^2}\varphi(t, \mathbf{r}) = x_M(t, \mathbf{r}), \tag{7.2-1}$$

where $\varphi(t, \mathbf{r})$ is the velocity potential at time t and position $\mathbf{r} = (x, y, z)$, $x_M(t, \mathbf{r})$ is the source distribution [see Fig. 1.1-3 and Eq. (7.1-6)], and $c(\mathbf{r})$ is the speed of sound in the ocean. Since the coupling equations discussed in Section 7.1

already allow for an arbitrary $x_M(t, \mathbf{r})$, with corresponding frequency and angular spectrum $X_M(f, \mathbf{v})$, we need only find the solution to the following Helmholtz wave equation [see Eq. (6.1-8)]:

$$\nabla^2 \varphi(\mathbf{r}) + k_0^2 n^2(\mathbf{r}) \varphi(\mathbf{r}) = 0, \tag{7.2-2}$$

where

$$k_0 = 2\pi f / c_0 = 2\pi / \lambda_0 \tag{7.2-3}$$

is the constant, reference wave number,

$$n(\mathbf{r}) = c_0 / c(\mathbf{r}) \tag{7.2-4}$$

is the *random* index of refraction,

$$c_0 = c(\mathbf{r}_0) = f \lambda_0 \tag{7.2-5}$$

is the constant, reference speed of sound at the source position $\mathbf{r}_0 = (x_0, y_0, z_0)$, and

$$\varphi(t, \mathbf{r}) = \varphi(\mathbf{r}) \exp(+j2\pi f t) \tag{7.2-6}$$

is the time-harmonic solution of Eq. (7.2-1) when $x_M(t, \mathbf{r})$ is set equal to zero, and $\varphi(\mathbf{r})$ is the solution of Eq. (7.2-2). Note that the wave number

$$k(\mathbf{r}) = 2\pi f / c(\mathbf{r}) \tag{7.2-7}$$

can be expressed as

$$k(\mathbf{r}) = k_0 n(\mathbf{r}). \tag{7.2-8}$$

Therefore, $k(\mathbf{r}_0) = k_0$ since $n(\mathbf{r}_0) = 1$. The index of refraction is commonly written as

$$n(\mathbf{r}) = n_D(\mathbf{r}) + n_R(\mathbf{r}) \tag{7.2-9}$$

or

$$n(\mathbf{r}) = n_D(\mathbf{r}) + \sigma(\mathbf{r}) n_{NR}(\mathbf{r}), \tag{7.2-10}$$

where $n_D(\mathbf{r})$ is the *deterministic component*, usually close to unity in value, $n_R(\mathbf{r})$ is the *random, zero mean component*, $\sigma(\mathbf{r})$ is the *standard deviation* of $n_R(\mathbf{r})$, and

$$n_{NR}(\mathbf{r}) = n_R(\mathbf{r}) / \sigma(\mathbf{r}) \tag{7.2-11}$$

is the *normalized random component* with zero mean and variance equal to unity. We shall work with Eq. (7.2-10) in this chapter. Note that the average value of $n(\mathbf{r})$ is equal to $n_D(\mathbf{r})$.

Let us assume that the speed of sound is only a function of the depth y, that is, $c(\mathbf{r}) = c(y)$, so that Eq. (7.2-2) reduces to

$$\nabla^2 \varphi(x, y, z) + k_0^2 n^2(y) \varphi(x, y, z) = 0, \tag{7.2-12}$$

where, from Eq. (7.2-10),

$$n(y) = n_D(y) + \sigma(y)n_{NR}(y). \tag{7.2-13}$$

By using the method of separation of variables and the WKB approximation, an approximate solution of Eq. (7.2-12) is given by (see Section 6.1)

$$\varphi(x, y, z) \approx \exp[-jk_X(x - x_0)]a_Y(y)\exp[+j\theta_Y(y)]\exp[-jk_Z(z - z_0)], \tag{7.2-14}$$

where

$$a_Y(y) = [k_Y(y)]^{-1/2}, \tag{7.2-15}$$

$$\theta_Y(y) = -\int_{y_0}^{y} k_Y(\zeta)\,d\zeta, \tag{7.2-16}$$

and

$$k_Y(y) = [k_0^2 n^2(y) - k_X^2 - k_Z^2]^{1/2}, \tag{7.2-17}$$

where k_X, $k_Y(y)$, and k_Z are the components of the propagation vector

$$\mathbf{k}(y) = k_X\hat{x} + k_Y(y)\hat{y} + k_Z\hat{z}. \tag{7.2-18}$$

Note that k_X and k_Z are constants, while $k_Y(y)$ is a function of the depth y. In addition, Eqs. (7.2-14) through (7.2-16) allow for a general source location with position vector $\mathbf{r}_0 = (x_0, y_0, z_0)$.

The WKB approximation given by Eqs. (7.2-15) and (7.2-16) is a valid solution for the depth dependence provided that (1) the transmitted frequencies f are high, (2) the sound-speed profile $c(y)$ is slowly changing, and (3) the depth interval from y_0 to y does *not* include any turning points [see Eq. (6.1-93)]. A turning point exists at $y = y_T$ if $k_Y(y_T) = 0$.

Since the reference propagation vector $\mathbf{k}_0$ can be expressed as

$$\mathbf{k}_0 = k_X\hat{x} + k_Y\hat{y} + k_Z\hat{z}, \tag{7.2-19}$$

$$k_0^2 = |\mathbf{k}_0|^2 = k_X^2 + k_Y^2 + k_Z^2, \tag{7.2-20}$$

where

$$k_X = k_0 u_0, \tag{7.2-21}$$

$$k_Y = k_0 v_0, \tag{7.2-22}$$

and

$$k_Z = k_0 w_0, \tag{7.2-23}$$

where

$$u_0 = \sin\theta_0 \cos\psi_0, \tag{7.2-24}$$

$$v_0 = \sin\theta_0 \sin\psi_0 = \cos\beta_0, \tag{7.2-25}$$

and

$$w_0 = \cos\theta_0 \tag{7.2-26}$$

are the direction cosines with respect to the positive X, Y, and Z axes, respectively. The angles (θ_0, ψ_0) (see Fig. 7.2-1) are the vertical and azimuthal spherical angles measured with respect to the positive Z and X axes, respectively, representing the *initial* directions of wave propagation. The angle β_0 (see Fig. 7.2-1) is measured with respect to the positive Y axis. From Eq. (7.2-20) we can write that

$$k_X^2 + k_Z^2 = k_0^2 - k_Y^2, \tag{7.2-27}$$

and substituting Eqs. (7.2-27) and (7.2-22) into Eq. (7.2-17) yields

$$k_Y(y) = k_Y\left\{1 + \frac{n^2(y) - 1}{v_0^2}\right\}^{1/2}. \tag{7.2-28}$$

Note that $k_Y(y_0) = k_Y$ since $n(y_0) = 1$. Therefore, the square root expression in Eq. (7.2-28) is responsible for changing the initial Y direction of propagation k_Y. If

$$|[n^2(y) - 1]/v_0^2| \ll 1, \tag{7.2-29}$$

then the square root expression in Eq. (7.2-28) can be approximated by the first two terms in a *binomial expansion*, yielding

$$k_Y(y) \approx k_Y + k_0^2[n^2(y) - 1]/(2k_Y). \tag{7.2-30}$$

It will be shown later that the binomial expansion criterion given by Eq. (7.2-29) can be related to the *critical angle of incidence* and, hence, *total*

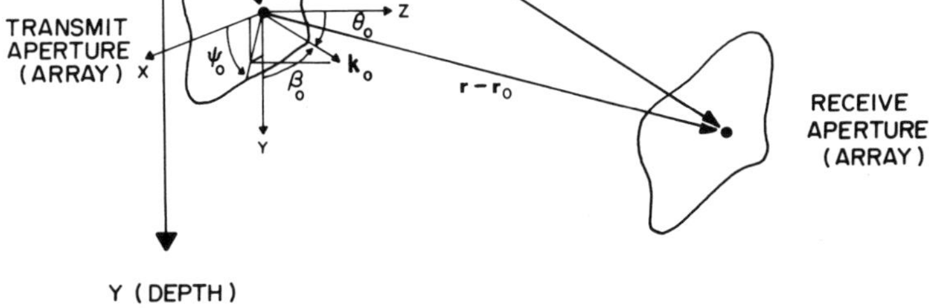

FIG. 7.2-1 The reference propagation vector $\mathbf{k}_0$ and associated angles.

reflection. The assumption concerning the absence of turning points will also be discussed later, but first let us return to the derivation of the transfer function.

Recall that we designated the vector $\mathbf{v}$ to represent the transmitted spatial frequencies as in $X_M(f, \mathbf{v})$ (see Fig. 1.1-3). If we let

$$\mathbf{v} = (f_X, f_Y, f_Z), \tag{7.2-31}$$

then

$$f_X = k_X/(2\pi) = u_0/\lambda_0, \tag{7.2-32}$$

$$f_Y = k_Y/(2\pi) = v_0/\lambda_0, \tag{7.2-33}$$

and

$$f_Z = k_Z/(2\pi) = w_0/\lambda_0 \tag{7.2-34}$$

are the spatial frequencies in the X, Y, and Z directions, respectively, where $c_0 = c(y_0) = f\lambda_0$. Therefore, if Eqs. (7.2-30) through (7.2-34) are substituted into Eqs. (7.2-14) through (7.2-16), then Eq. (7.2-6) becomes

$$\varphi(t, x, y, z) \approx \frac{\exp\{-j[k_0^2/(4\pi f_Y)] \int_{y_0}^{y} [n^2(\zeta) - 1]\, d\zeta\}}{\{2\pi f_Y + k_0^2[n^2(y) - 1]/(4\pi f_Y)\}^{1/2}} \times \exp(+j2\pi f t) \exp[-j2\pi \mathbf{v} \cdot (\mathbf{r} - \mathbf{r}_0)]. \tag{7.2-35}$$

Since the input to the communication channel is the time-harmonic plane wave

$$\exp(+j2\pi f t) \exp[-j2\pi \mathbf{v} \cdot (\mathbf{r} - \mathbf{r}_0)]$$

and since Eq. (7.2-35) represents the output from the channel at any time t and position $\mathbf{r}$ [see Eq. (2.1-21)],

$$H_M(f, \mathbf{v}; t, \mathbf{r}) = H_M(f, \mathbf{v}; \mathbf{r}) = H_M(f, f_Y; y), \tag{7.2-36}$$

where the random, time-invariant, space-variant transfer function of the ocean medium is given by

$$H_M(f, f_Y; y) = A_M(f, f_Y; y) \exp[+j\theta_M(f, f_Y; y)] \tag{7.2-37}$$

and where

$$A_M(f, f_Y; y) = \{2\pi f_Y + k_0^2[n^2(y) - 1]/(4\pi f_Y)\}^{-1/2}, \tag{7.2-38}$$

$$\theta_M(f, f_Y; y) = -[k_0^2/(4\pi f_Y)] \int_{y_0}^{y} [n^2(\zeta) - 1]\, d\zeta, \tag{7.2-39}$$

$$k_0 = 2\pi f/c_0, \tag{7.2-3}$$

and, from Eq. (7.2-13),

$$n^2(y) = n_D^2(y) + 2n_D(y)\sigma(y)n_{NR}(y) + \sigma^2(y)n_{NR}^2(y). \tag{7.2-40}$$

If the medium is "weakly irregular" or "weakly scattering," that is, if $\sigma(y)$ is very small compared to unity, so that

$$k_0\sigma(y)s_Y \ll 1, \tag{7.2-41}$$

where s_Y is the scale size of the random medium in the Y direction (i.e., the average distance over which the refractive index fluctuations remain correlated), then terms involving $\sigma^2(y)$ can be neglected and Eq. (7.2-40) reduces to

$$n^2(y) \approx n_D^2(y) + 2n_D(y)\sigma(y)n_{NR}(y). \tag{7.2-42}$$

Therefore, by using Eq. (7.2-29) to simplify Eq. (7.2-38) and upon substituting Eq. (7.2-42) into Eq. (7.2-39), one obtains the following:

$$A_M(f, f_Y; y) = (2\pi f_Y)^{-1/2} \tag{7.2-43}$$

and

$$\theta_M(f, f_Y; y) = \theta_{MD}(f, f_Y; y) + \theta_{MR}(f, f_Y; y), \tag{7.2-44}$$

where

$$\theta_{MD}(f, f_Y; y) = -[k_0^2/(4\pi f_Y)] \int_{y_0}^{y} [n_D^2(\zeta) - 1]\, d\zeta \tag{7.2-45}$$

is the deterministic or average component of the phase function and

$$\theta_{MR}(f, f_Y; y) = -[k_0^2/(2\pi f_Y)] \int_{y_0}^{y} n_D(\zeta)\sigma(\zeta)n_{NR}(\zeta)\, d\zeta \tag{7.2-46}$$

is the random component.

The medium transfer function H_M given by Eq. (7.2-37) with amplitude and phase functions A_M and θ_M specified by Eq. (7.2-43) and Eqs. (7.2-44) through (7.2-46), respectively, indicate that for a weakly irregular medium, the major effect of the medium is to *angle modulate* the transmitted field. Furthermore, if $n_D(y) \approx 1$, then $\theta_{MD}(f, f_Y; y) \approx 0$ [see Eq. (7.2-45)] and, as a result, the angle modulation is due strictly to the random fluctuations $\sigma(y)n_{NR}(y)$ of the index of refraction [see Eq. (7.2-46)]. Note that the angle-modulation process is often referred to as "scattering." Also, note that the transfer function derived in this section can be written as the product of two functions, one deterministic and the other random, that is,

$$H_M = [A_M \exp(+j\theta_{MD})] \exp(+j\theta_{MR}),$$

which agrees with the general practice of representing a field in a random medium as the product of a deterministic and a random function.

Let us now discuss the two important assumptions responsible for the derivation of the transfer function, namely, the aforementioned binomial expansion criterion given by Eq. (7.2-29) and the assumption regarding the absence of turning points.

Since

$$n(y) = c_0/c(y), \tag{7.2-47}$$

substituting Eqs. (7.2-25) and (7.2-47) into Eq. (7.2-29) yields

$$\beta_0 \ll \cos^{-1}\left\{\frac{|c^2(y) - c_0^2|^{1/2}}{c(y)}\right\}, \tag{7.2-48}$$

where β_0 is the angle of incidence of the reference propagation vector $\mathbf{k}_0$ (see Fig. 7.2-1). Now recall that the *critical angle of incidence* β_c associated with a time-harmonic plane wave incident upon a plane boundary between two fluid media is given by (see Fig. 7.2-2)

$$\sin\beta_c = c_1/c_2, \qquad c_2 > c_1, \tag{7.2-49}$$

or, equivalently,

$$\beta_c = \cos^{-1}\left\{\frac{(c_2^2 - c_1^2)^{1/2}}{c_2}\right\}, \qquad c_2 > c_1, \tag{7.2-50}$$

where c_2 *must* be greater than c_1 for β_c to exist. When the angle of incidence $\beta \geq \beta_c$, there is *total reflection*, and thus *no* transmission of energy into medium II. The speed of sound $c(y)$ generally increases with depth, for example, in the deep ocean. Also, $c(y)$ increases above and below the SOFAR channel axis. Therefore, when $c(y) > c_0$, the absolute value sign in Eq. (7.2-48)

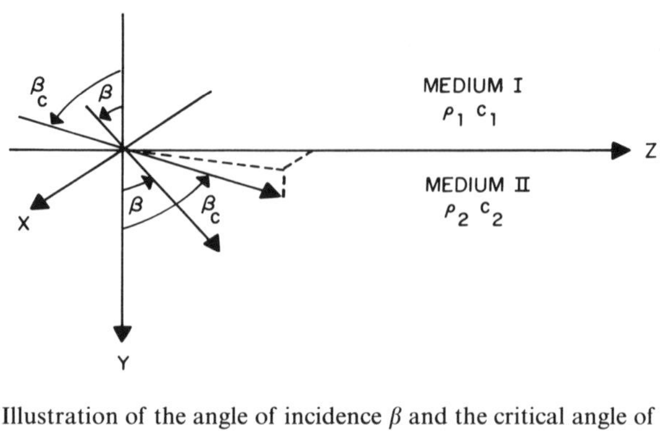

FIG. 7.2-2 Illustration of the angle of incidence β and the critical angle of incidence β_c.

can be removed, and by comparing Eq. (7.2-50) with the right-hand side of Eq. (7.2-48), the binomial expansion criterion indicates that the initial angles of transmission β_0 must be less than the "critical angle" in order to *avoid* total reflection, and thus passing through a turning point. This is consistent with the fact that the WKB approximation is invalid at a turning point.

7.2.2 Output Electrical Signal

Now that we have derived an ocean medium transfer function, let us demonstrate the use of the coupling equations presented in Section 7.1 by calculating the output electrical signal $y(t,\mathbf{r})$ from the receive aperture (array) as given by Eq. (7.1-12). As can be seen by inspecting the integrand of Eq. (7.1-12), we need expressions for both of the kernels $Y_{\mathrm{M}}(\eta,\mathbf{r})$ and $A_{\mathrm{R}}(\eta,\mathbf{r})$. Let us compute $Y_{\mathrm{M}}(\eta,\mathbf{r})$ first [see Eq. (7.1-18)].

Assume that the transmit aperture depicted in Fig. 1.1-2 is a planar array of $M \times N$ (odd) complex-weighted point sources, centered at (x_0, y_0, z_0) and parallel to the XY plane. In addition, assume that the complex weights are separable. Since in most practical situations an identical input electrical signal is applied to all elements in the transmit array *before* the complex weights, that is, since (see Example 3.2-2)

$$x(t,\mathbf{r}) \equiv x(t),$$

Eq. (7.1-7) reduces to

$$X(f,\boldsymbol{\alpha}) = X(f)\,\delta(\boldsymbol{\alpha}) \tag{7.2-51}$$

since $F_{\mathbf{r}}\{1\} = \delta(\boldsymbol{\alpha})$. Substituting Eq. (7.2-51) into Eq. (7.1-1) yields

$$X_{\mathrm{M}}(f,\mathbf{v}) = X(f) D_{\mathrm{T}}(f,\mathbf{v}), \tag{7.2-52}$$

where (see Section 4.5.1)

$$\begin{aligned} D_{\mathrm{T}}(f,\mathbf{v}) = \sum_{m=-(M-1)/2}^{(M-1)/2} \sum_{n=-(N-1)/2}^{(N-1)/2} & c_m d_n \exp(+j2\pi f_X m d_X) \\ & \times \exp(+j2\pi f_Y n d_Y) \exp(+j2\pi f_X x_0) \\ & \times \exp(+j2\pi f_Y y_0) \exp(+j2\pi f_Z z_0) \end{aligned} \tag{7.2-53}$$

is the far-field beam pattern of the transmit array, c_m and d_n are complex weights, d_X and d_Y are the interelement spacings in the X and Y directions, respectively, and the last three complex exponentials are phase factors that account for the array being centered at (x_0, y_0, z_0) instead of at the origin $(0,0,0)$.

Since the transfer function derived in Section 7.2.1 is time-invariant, Eq. (7.1-19) becomes

$$B_{\mathrm{M}}(f, \mathbf{v}; \phi, \mathbf{r}) = H_{\mathrm{M}}(f, \mathbf{v}; \mathbf{r})\, \delta(\phi), \tag{7.2-54}$$

and, as a result,

$$B_{\mathrm{M}}(f, \mathbf{v}; \eta - f, \mathbf{r}) = H_{\mathrm{M}}(f, \mathbf{v}; \mathbf{r})\, \delta(\eta - f). \tag{7.2-55}$$

Therefore, substituting Eqs. (7.2-52) and (7.2-55) into Eq. (7.1-18) yields

$$Y_{\mathrm{M}}(\eta, \mathbf{r}) = X(\eta) \int_{-\infty}^{\infty} D_{\mathrm{T}}(\eta, \mathbf{v}) H_{\mathrm{M}}(\eta, \mathbf{v}; \mathbf{r}) \exp(-j2\pi \mathbf{v} \cdot \mathbf{r})\, d\mathbf{v}, \tag{7.2-56}$$

and upon substituting Eq. (7.2-56) into Eq. (7.1-12), we obtain

$$y(t, \mathbf{r}) = \int_{-\infty}^{\infty} X(f) \int_{-\infty}^{\infty} D_{\mathrm{T}}(f, \mathbf{v}) H_{\mathrm{M}}(f, \mathbf{v}; \mathbf{r}) \exp(-j2\pi \mathbf{v} \cdot \mathbf{r})\, d\mathbf{v} \times A_{\mathrm{R}}(f, \mathbf{r}) \exp(+j2\pi f t)\, df, \tag{7.2-57}$$

where η was replaced by f since there is no frequency spreading in this example due to the time-invariant property of the transfer function. We need to specify the complex aperture function $A_{\mathrm{R}}(f, \mathbf{r})$ next.

Assume that the receive aperture depicted in Fig. 1.1-2 is a planar array of $M' \times N'$ (odd) complex-weighted point sources, centered at $(x_{\mathrm{R}}, y_{\mathrm{R}}, z_{\mathrm{R}})$ and parallel to the XY plane. In addition, assume that the complex weights are separable. Therefore, the receive aperture function is given by (see Section 4.5.1)

$$A_{\mathrm{R}}(f, \mathbf{r}) = \sum_{i=-(M'-1)/2}^{(M'-1)/2} \sum_{q=-(N'-1)/2}^{(N'-1)/2} c_i' d_q'\, \delta[x - (x_{\mathrm{R}} + i d_X')]\, \delta[y - (y_{\mathrm{R}} + q d_Y')] \times \delta(z - z_{\mathrm{R}}), \tag{7.2-58}$$

where c_i' and d_q' are complex weights and d_X' and d_Y' are the interelement spacings in the X and Y directions, respectively.

Upon substituting Eqs. (7.2-31), (7.2-36), (7.2-53), and (7.2-58) into Eq. (7.2-57) and recalling that $\mathbf{r} = (x, y, z)$, one obtains

$$\begin{aligned} y(t, x, y, z) = {} & \sum_{i=-(M'-1)/2}^{(M'-1)/2} \sum_{q=-(N'-1)/2}^{(N'-1)/2} c_i' d_q' \int_{-\infty}^{\infty} X(f) \\ & \times \sum_{m=-(M-1)/2}^{(M-1)/2} \sum_{n=-(N-1)/2}^{(N-1)/2} c_m d_n \int_{-\infty}^{\infty} \int_{-\infty}^{\infty} \int_{-\infty}^{\infty} H_{\mathrm{M}}(f, f_Y; y_{\mathrm{R}} + q d_Y') \\ & \times \exp[-j2\pi(f_X \Delta X_{im} + f_Y \Delta Y_{qn} + f_Z \Delta Z)]\, df_X\, df_Y\, df_Z \\ & \times \exp(+j2\pi f t)\, df\, \delta[x - (x_{\mathrm{R}} + i d_X')]\, \delta[y - (y_{\mathrm{R}} + q d_Y')]\, \delta(z - z_{\mathrm{R}}), \end{aligned} \tag{7.2-59}$$

where

$$\Delta X_{im} = x_R - x_0 + id'_X - md_X, \tag{7.2-60}$$

$$\Delta Y_{qn} = y_R - y_0 + qd'_Y - nd_Y, \tag{7.2-61}$$

and

$$\Delta Z = z_R - z_0. \tag{7.2-62}$$

To proceed further we make the following observations. From Eq. (7.2-20) we can write that

$$k_Z^2 = k_0^2 - (k_X^2 + k_Y^2), \tag{7.2-63}$$

and upon substituting Eq. (7.2-3) and Eqs. (7.2-32) through (7.2-34) into Eq. (7.2-63), we obtain

$$f_Z = \pm[(f/c_0)^2 - (f_X^2 + f_Y^2)]^{1/2}. \tag{7.2-64}$$

Thus, the spatial frequency f_Z is seen to be dependent on f, f_X, and f_Y. As a result, we can eliminate the integration with respect to f_Z in Eq. (7.2-59). Equation (7.2-64) can also be expressed in terms of the direction cosines u_0 and v_0 by using Eqs. (7.2-32) and (7.2-33), that is,

$$f_Z = \pm(f/c_0)[1 - (u_0^2 + v_0^2)]^{1/2} \tag{7.2-65}$$

or

$$f_Z = fw_0/c_0, \tag{7.2-66}$$

where

$$w_0 = \pm[1 - (u_0^2 + v_0^2)]^{1/2} \tag{7.2-67}$$

is the direction cosine with respect to the positive Z axis. If $u_0^2 + v_0^2 \leq 1$, then

$$w_0 = +[1 - (u_0^2 + v_0^2)]^{1/2}, \qquad u_0^2 + v_0^2 \leq 1, \tag{7.2-68}$$

which corresponds to an acoustic field propagating in the positive Z direction. However, if $u_0^2 + v_0^2 > 1$, then

$$w_0 = -[1 - (u_0^2 + v_0^2)]^{1/2}, \qquad u_0^2 + v_0^2 > 1 \tag{7.2-69}$$

or

$$w_0 = -j(u_0^2 + v_0^2 - 1)^{1/2}, \qquad u_0^2 + v_0^2 > 1, \tag{7.2-70}$$

which corresponds to an *evanescent wave*, that is, a decaying exponential.

If we now (1) eliminate the integration with respect to f_Z in Eq. (7.2-59), (2) change variables from the spatial frequencies f_X and f_Y to the direction cosines u_0 and v_0 by substituting Eqs. (7.2-32) and (7.2-33) into Eq. (7.2-59) and treat the frequency variable f as a constant, so that

$$df_X = d(fu_0/c_0) = (f/c_0)\,du_0$$

and

$$df_Y = d(fv_0/c_0) = (f/c_0)\,dv_0,$$

and (3) substitute Eqs. (7.2-66) and (7.2-68) into $\exp(-j2\pi f_Z \Delta Z)$, which appears in the integrand of Eq. (7.2-59), then Eq. (7.2-59) reduces to

$$\begin{aligned} y(t,x,y,z) = {} & \left(\frac{1}{c_0}\right)^2 \sum_{i=-(M'-1)/2}^{(M'-1)/2} \sum_{q=-(N'-1)/2}^{(N'-1)/2} c'_i d'_q \int_{-\infty}^{\infty} f^2 X(f) \\ & \times \int_{u_0=-1}^{+1} \int_{v_0=a_q}^{+1} \sum_{n=-(N-1)/2}^{(N-1)/2} d_n H_M(f,(fv_0/c_0); y_R + qd'_Y) \\ & \times \exp\{-j(2\pi f/c_0)[1-(u_0^2+v_0^2)]^{1/2}\,\Delta Z\} \\ & \times \exp[-j(2\pi f/c_0)v_0\,\Delta Y_{qn}]\,dv_0 \\ & \times \sum_{m=-(M-1)/2}^{(M-1)/2} c_m \exp[-j(2\pi f/c_0)u_0\,\Delta X_{im}]\,du_0 \exp(+j2\pi ft)\,df \\ & \times \delta[x-(x_R+id'_X)]\,\delta[y-(y_R+qd'_Y)]\,\delta(z-z_R), \quad u_0^2+v_0^2 \le 1, \end{aligned} \tag{7.2-71}$$

where

$$a_q = |n_D^2(y_R + qd'_Y) - 1|^{1/2} \tag{7.2-72}$$

was obtained from Eq. (7.2-29) by replacing $n^2(\cdot)$ with $n_D^2(\cdot)$. The transfer function H_M is given by Eqs. (7.2-37) and (7.2-43) through (7.2-46) with the exception that *the lower limit of integration* y_0 *in Eqs.* (7.2-45) *and* (7.2-46) *must be replaced by* $y_0 + nd_Y$, *which is the Y coordinate of a point source in the transmit planar array.*

Equation (7.2-71) is the desired result, that is, it represents the random output electrical signal at *each* element in the receive array in terms of the frequency spectrum of the transmitted electrical signal, the transmit and receive arrays, and the random transfer function of the ocean medium. If $y(t,x,y,z)$ given by Eq. (7.2-71) is complex, then we simply take the real part to obtain the real output electrical signal. Note that Eq. (7.2-71) pertains to the case of an acoustic field propagating in the positive Z direction since $u_0^2 + v_0^2 \le 1$. When $u_0^2 + v_0^2 > 1$,

$$y(t,x,y,z) = 0, \qquad u_0^2 + v_0^2 > 1 \tag{7.2-73}$$

since only rapidly decaying evanescent waves emanate from the transmit array.

Finally, consider the following term that appears in the integrand of Eq. (7.2-71):

$$H_M(f,(fv_0/c_0);\, y_R + qd'_Y)\exp\{-j(2\pi f/c_0)[1-(u_0^2+v_0^2)]^{1/2}\,\Delta Z\}. \quad (7.2\text{-}74)$$

If we now assume that the medium is *homogeneous*, that is, space-invariant, and nonrandom, then the deterministic and random components of the index of refraction become, respectively,

$$n_D(y) = 1 \quad (7.2\text{-}75)$$

and

$$n_R(y) = \sigma(y)n_{NR}(y) = 0. \quad (7.2\text{-}76)$$

Therefore, substituting Eqs. (7.2-75) and (7.2-76) into Eqs. (7.2-45) and (7.2-46), respectively, yields $\theta_{MD} = 0$ and $\theta_{MR} = 0$; and, as a result, the transfer function reduces to [see Eq. (7.2-37) and Eqs. (7.2-43) through (7.2-46)]

$$H_M(f,(fv_0/c_0);\, y_R + qd'_Y) = (k_0 v_0)^{-1/2}, \quad (7.2\text{-}77)$$

where k_0 is given by Eq. (7.2-3). Substituting Eq. (7.2-77) into Eq. (7.2-74) yields

$$(k_0 v_0)^{-1/2}\exp\{-j(2\pi f/c_0)[1-(u_0^2+v_0^2)]^{1/2}\,\Delta Z\}. \quad (7.2\text{-}78)$$

With the exception of the amplitude term, which is due to the WKB approximation, Eq. (7.2-78) is equivalent to the expression one obtains for a space-invariant transfer function in Fourier optics.

7.2.3 Coherence Function

Upon inspecting Eqs. (7.2-57) and (7.2-36), it can be seen that the autocorrelation function of the output electrical signal depends on the autocorrelation function of the transfer function

$$R_{H_M}(f,f',f_Y,f'_Y;y,y') = E\{H_M(f,f_Y;y)H_M^*(f',f'_Y;y')\}, \quad (7.2\text{-}79)$$

which is also known as the *coherence function.* Substituting Eqs. (7.2-37), (7.2-39), and (7.2-43) into Eq. (7.2-79) yields

$$R_{H_M}(f,f',f_Y,f'_Y;y,y') = [2\pi(f_Y,f'_Y)^{1/2}]^{-1}\,E\{\exp[+j(KI(y)+K'I(y'))]\}, \quad (7.2\text{-}80)$$

where

$$K = -k_0^2/(4\pi f_Y), \quad (7.2\text{-}81)$$

$$K' = +(k'_0)^2/(4\pi f'_Y), \quad (7.2\text{-}82)$$

$$I(y) = \int_{y_0}^{y} [n^2(\zeta) - 1]\, d\zeta, \tag{7.2-83}$$

$$I(y') = \int_{y_0}^{y'} [n^2(\zeta) - 1]\, d\zeta, \tag{7.2-84}$$

$$k_0 = 2\pi f/c_0, \tag{7.2-85}$$

$$k_0' = 2\pi f'/c_0, \tag{7.2-86}$$

and $n^2(\cdot)$ is given by Eq. (7.2-42). The expectation appearing in Eq. (7.2-80) is the *characteristic function* of the random quantity $KI(y) + K'I(y')$. If it is assumed that $I(y)$ is a *real, Gaussian random process*, which implies that the index of refraction is a real, Gaussian random process, then Eq. (7.2-80) can be written as

$$\begin{aligned} R_{H_M}(f, f', f_Y, f_Y'; y, y') = {} & [2\pi(f_Y f_Y')^{1/2}]^{-1} \exp\{+j[\theta_{MD}(f, f_Y; y) \\ & - \theta_{MD}(f', f_Y'; y')]\} \\ & \times \exp(-E\{\theta_{MR}^2(f, f_Y; y)\}/2) \\ & \times \exp(+E\{\theta_{MR}(f, f_Y; y)\theta_{MR}(f', f_Y'; y')\}) \\ & \times \exp(-E\{\theta_{MR}^2(f', f_Y'; y')\}/2, \end{aligned} \tag{7.2-87}$$

where

$$E\{\theta_{MR}^2(f, f_Y; y)\} = [k_0^2/(2\pi f_Y)]^2 \int_{y_0}^{y} \int_{y_0}^{y} n_D(\zeta) n_D(\zeta') \sigma(\zeta) \sigma(\zeta') R_{n_{NR}}(\zeta, \zeta')\, d\zeta\, d\zeta', \tag{7.2-88}$$

$$\begin{aligned} E\{\theta_{MR}^2(f', f_Y'; y')\} = {} & [(k_0')^2/(2\pi f_Y')]^2 \\ & \times \int_{y_0}^{y'} \int_{y_0}^{y'} n_D(\zeta) n_D(\zeta') \sigma(\zeta) \sigma(\zeta') R_{n_{NR}}(\zeta, \zeta')\, d\zeta\, d\zeta', \end{aligned} \tag{7.2-89}$$

$$\begin{aligned} E\{\theta_{MR}(f, f_Y; y)\theta_{MR}(f', f_Y'; y')\} = {} & +[k_0^2/(2\pi f_Y)][(k_0')^2/(2\pi f_Y')] \\ & \times \int_{y_0}^{y} \int_{y_0}^{y'} n_D(\zeta) n_D(\zeta') \sigma(\zeta) \sigma(\zeta') \\ & \times R_{n_{NR}}(\zeta, \zeta')\, d\zeta\, d\zeta', \end{aligned} \tag{7.2-90}$$

and

$$R_{n_{NR}}(y, y') = E\{n_{NR}(y) n_{NR}(y')\}. \tag{7.2-91}$$

If it is further assumed that the deterministic component $n_D(y)$ of the index of refraction is equal to unity and that the random component $n_R(y)$ and, hence, the normalized random component $n_{\text{NR}}(y)$, is wide-sense stationary, that is,

$$R_{n_{\text{NR}}}(y, y') = R_{n_{\text{NR}}}(\Delta y), \tag{7.2-92}$$

where $\Delta y = y - y'$, then [see Eq. (7.2-45)]

$$\theta_{\text{MD}}(f, f_Y; y) = 0, \tag{7.2-93}$$

$$\theta_{\text{MD}}(f', f_Y'; y') = 0, \tag{7.2-94}$$

and Eqs. (7.2-88) through (7.2-90) become, respectively,

$$\text{E}\{\theta_{\text{MR}}^2(f, f_Y; y)\} = (y - y_0)\left[\frac{k_0^2\sigma}{2\pi f_Y}\right]^2 \times \int_{-(y-y_0)}^{y-y_0}\left[1 - \frac{|\zeta|}{y - y_0}\right] R_{n_{\text{NR}}}(\zeta)\, d\zeta, \tag{7.2-95}$$

$$\text{E}\{\theta_{\text{MR}}^2(f', f_Y'; y')\} = (y' - y_0)\left[\frac{(k_0')^2\sigma}{2\pi f_Y'}\right]^2 \times \int_{-(y'-y_0)}^{y'-y_0}\left[1 - \frac{|\zeta|}{y' - y_0}\right] R_{n_{\text{NR}}}(\zeta)\, d\zeta, \tag{7.2-96}$$

and

$$\text{E}\{\theta_{\text{MR}}(f, f_Y; y)\theta_{\text{MR}}(f', f_Y'; y')\} = \frac{(k_0 k_0' \sigma)^2}{(2\pi f_Y)(2\pi f_Y')} \times \left\{(y - y')\int_{y-y'}^{y-y_0} R_{n_{\text{NR}}}(\zeta)\, d\zeta + (y' - y_0)\int_{-(y'-y_0)}^{y-y_0} R_{n_{\text{NR}}}(\zeta)\, d\zeta + \int_{-(y'-y_0)}^{0} \zeta R_{n_{\text{NR}}}(\zeta)\, d\zeta - \int_{y-y'}^{y-y_0} \zeta R_{n_{\text{NR}}}(\zeta)\, d\zeta\right\}, \tag{7.2-97}$$

where σ is the constant standard deviation of the real, Gaussian, zero-mean, wide-sense stationary, random component $n_{\text{R}}(y)$ of the index of refraction. The significance of the coherence function is discussed at the end of Section 7.1.

7.3 THE PARABOLIC EQUATION APPROXIMATION REVISITED

7.3.1 Transfer Function Derivation

A transfer function that models the bistatic communication channel geometry shown in Fig. 1.1-2 will be derived in this section using the parabolic equation approximation of the Helmholtz wave equation. This transfer function models the same physical situation as is described in the first paragraph of Section 7.2.1.

As was discussed in Section 7.2.1, since the coupling equations presented in Section 7.1 already allow for an arbitrary source distribution, we need not solve the wave equation given by Eq. (7.2-1) but, rather, we need only find the solution to the Helmholtz wave equation given by Eq. (7.2-2).

The parabolic equation approximation of the Helmholtz wave equation is given by [see Eq. (6.3-7)]

$$\frac{\partial}{\partial z} g(\mathbf{r}) = [-j/(2k_0)]\nabla_T^2 g(\mathbf{r}) - j(k_0/2)[n^2(\mathbf{r}) - 1]g(\mathbf{r}), \tag{7.3-1}$$

where ∇_T^2 is the transverse Laplacian given by Eq. (6.3-6). We shall use Eq. (7.3-1) as the starting point for the transfer function derivation.

From Eq. (7.2-10),

$$n^2(\mathbf{r}) = n_D^2(\mathbf{r}) + 2n_D(\mathbf{r})\sigma(\mathbf{r})n_{NR}(\mathbf{r}) + \sigma^2(\mathbf{r})n_{NR}^2(\mathbf{r}). \tag{7.3-2}$$

If the medium is "weakly irregular" or "weakly scattering," that is, if $\sigma(\mathbf{r})$ is very small compared to unity, so that

$$k_0\sigma(\mathbf{r})s_Z \ll 1, \tag{7.3-3}$$

where s_Z is the scale size of the random medium in the Z direction, then terms involving $\sigma^2(\mathbf{r})$ can be neglected. If it is further assumed that the deterministic component $n_D(\mathbf{r})$ of the index of refraction is equal to unity, then Eq. (7.3-2) reduces to

$$n^2(\mathbf{r}) \approx 1 + 2\sigma(\mathbf{r})n_{NR}(\mathbf{r}). \tag{7.3-4}$$

Substituting Eq. (7.3-4) into Eqs. (7.2-2) and (7.3-1) yields

$$\nabla^2\varphi(\mathbf{r}) + k_0^2[1 + 2\sigma(\mathbf{r})n_{NR}(\mathbf{r})]\varphi(\mathbf{r}) = 0 \tag{7.3-5}$$

and

$$\frac{\partial}{\partial z} g(\mathbf{r}) = [-j/(2k_0)]\nabla_T^2 g(\mathbf{r}) - jk_0\sigma(\mathbf{r})n_{NR}(\mathbf{r})g(\mathbf{r}), \tag{7.3-6}$$

respectively. The solution $\varphi(\mathbf{r})$ of the Helmholtz wave equation given by Eq. (7.3-5) is related to the solution $g(\mathbf{r})$ of the parabolic partial differential equation given by Eq. (7.3-6) by [see Eq. (6.3-1)]

$$\varphi(\mathbf{r}) = g(\mathbf{r})\exp(-jk_0 z). \tag{7.3-7}$$

The first term on the right-hand side of Eq. (7.3-6) accounts for free-space propagation in a uniform medium, while the second term accounts for scattering by the random medium.

An approximate solution of Eq. (7.3-6) is given by

$$g(\mathbf{r}) \approx g_{\mathrm{D}}(\mathbf{r}) g_{\mathrm{R}}(\mathbf{r}), \tag{7.3-8}$$

where $g_{\mathrm{D}}(\mathbf{r})$ and $g_{\mathrm{R}}(\mathbf{r})$ are the deterministic and random components, respectively, of the solution. In order to determine $g_{\mathrm{D}}(\mathbf{r})$, we set

$$\sigma(\mathbf{r}) n_{\mathrm{NR}}(\mathbf{r}) = n_{\mathrm{R}}(\mathbf{r}) = 0$$

in Eq. (7.3-6) and, as a result, we obtain

$$\frac{\partial}{\partial z} g(\mathbf{r}) = [-j/(2k_0)] \nabla_{\mathrm{T}}^2 g(\mathbf{r}). \tag{7.3-9}$$

An exact solution of Eq. (7.3-9) is given by $g(\mathbf{r}) = g_{\mathrm{D}}(\mathbf{r})$, where [see Eq. (6.3-9)]

$$\begin{aligned} g_{\mathrm{D}}(\mathbf{r}) = {} & \exp[-j2\pi f_X(x - x_0)] \exp[-j2\pi f_Y(y - y_0)] \\ & \times \exp\{+j[(2\pi)^2(f_X^2 + f_Y^2)/(2k_0)](z - z_0)\}, \end{aligned} \tag{7.3-10}$$

where

$$f_X = u_0/\lambda_0, \tag{7.3-11}$$

$$f_Y = v_0/\lambda_0, \tag{7.3-12}$$

and

$$f_Z = w_0/\lambda_0 \tag{7.3-13}$$

are the *transmitted spatial frequencies* (in cycles per meter) in the X, Y, and Z directions, respectively; k_0 and λ_0 are given by Eqs. (7.2-3) and (7.2-5), respectively; and

$$u_0 = \sin\theta_0 \cos\psi_0, \tag{7.3-14}$$

$$v_0 = \sin\theta_0 \sin\psi_0 = \cos\beta_0, \tag{7.3-15}$$

and

$$w_0 = \cos\theta_0 \tag{7.3-16}$$

are the direction cosines with respect to the positive X, Y, and Z axes, respectively. The angles (θ_0, ψ_0) (see Fig. 7.2-1) are the vertical and azimuthal spherical angles measured with respect to the positive Z and X axes, respectively, representing the *initial* directions of wave propagation. The angle β_0 (see Fig. 7.2-1) is measured with respect to the positive Y axis. The reference or transmitted propagation vector $\mathbf{k}_0$ is given by (see Fig. 7.2-1)

$$\mathbf{k}_0 = k_X\hat{x} + k_Y\hat{y} + k_Z\hat{z}, \tag{7.3-17}$$

where

$$k_X = k_0 u_0 = 2\pi f_X, \tag{7.3-18}$$

$$k_Y = k_0 v_0 = 2\pi f_Y, \tag{7.3-19}$$

and

$$k_Z = k_0 w_0 = 2\pi f_Z. \tag{7.3-20}$$

In order to determine the random component of the solution $g_{\mathbf{R}}(\mathbf{r})$, we ignore the first term in Eq. (7.3-6), or equivalently, we make the following additional assumptions that are analogous to Eq. (6.3-4):

$$\left|\frac{\partial^2}{\partial x^2} g(\mathbf{r})\right| \ll 2k_0 \left|\frac{\partial}{\partial z} g(\mathbf{r})\right| \tag{7.3-21}$$

and

$$\left|\frac{\partial^2}{\partial y^2} g(\mathbf{r})\right| \ll 2k_0 \left|\frac{\partial}{\partial z} g(\mathbf{r})\right|. \tag{7.3-22}$$

Using Eqs. (7.3-21) and (7.3-22), Eq. (7.3-6) reduces to

$$\frac{\partial}{\partial z} g(\mathbf{r}) = -jk_0 \sigma(\mathbf{r}) n_{\mathrm{NR}}(\mathbf{r}) g(\mathbf{r}), \tag{7.3-23}$$

which has an exact solution $g(\mathbf{r}) = g_{\mathbf{R}}(\mathbf{r})$, where

$$g_{\mathbf{R}}(x, y, z) = g(x, y, z_0) \exp\left\{-jk_0 \int_{z_0}^{z} \sigma(x, y, \zeta) n_{\mathrm{NR}}(x, y, \zeta)\, d\zeta\right\}. \tag{7.3-24}$$

However, as mentioned previously, since the coupling equations presented in Section 7.1 already account for an arbitrary source distribution as a function of time and space, the initial field distribution term $g(x, y, z_0)$ can be dropped from Eq. (7.3-24). Thus, for our purposes, we shall use the following expression for the random component $g_{\mathbf{R}}(\mathbf{r})$:

$$g_{\mathbf{R}}(x, y, z) = \exp\left\{-jk_0 \int_{z_0}^{z} \sigma(x, y, \zeta) n_{\mathrm{NR}}(x, y, \zeta)\, d\zeta\right\}. \tag{7.3-25}$$

Therefore, in summary, Eq. (7.3-8) is an approximate solution of the parabolic partial differential equation given by Eq. (7.3-6), where $g_{\mathbf{D}}(\mathbf{r})$ and $g_{\mathbf{R}}(\mathbf{r})$ are given by Eqs. (7.3-10) and (7.3-25), respectively, provided that the assumptions represented by Eqs. (6.3-4), (7.3-4), (7.3-21), and (7.3-22) are valid; and, in addition, that the cross product terms

$$\frac{1}{k_0} \frac{\partial}{\partial x} g_{\mathbf{D}}(\mathbf{r}) \frac{\partial}{\partial x} g_{\mathbf{R}}(\mathbf{r}) \ll 1 \tag{7.3-26}$$

and

$$\frac{1}{k_0}\frac{\partial}{\partial y}g_{\mathrm{D}}(\mathbf{r})\frac{\partial}{\partial y}g_{\mathrm{R}}(\mathbf{r}) \ll 1 \tag{7.3-27}$$

so that they too can be neglected. The additional requirements imposed by Eqs. (7.3-26) and (7.3-27) are a consequence of substituting Eqs. (7.3-8), (7.3-10), and (7.3-25) into Eq. (7.3-6) and determining which additional terms must be small so that the left-hand side of Eq. (7.3-6) is approximately equal to the right-hand side. We are now in a position to obtain an expression for the ocean medium transfer function.

The solution of the Helmholtz wave equation given by Eq. (7.3-5) is approximately equal to

$$\varphi(\mathbf{r}) \approx g_{\mathrm{D}}(\mathbf{r})g_{\mathrm{R}}(\mathbf{r})\exp(-jk_0 z), \tag{7.3-28}$$

where Eq. (7.3-8) was substituted into Eq. (7.3-7). The corresponding time-harmonic solution is given by

$$\varphi(t,\mathbf{r}) \approx g_{\mathrm{D}}(\mathbf{r})g_{\mathrm{R}}(\mathbf{r})\exp(-jk_0 z)\exp(+j2\pi ft), \tag{7.3-29}$$

where Eq. (7.3-28) was substituted into Eq. (7.2-6). Substituting Eqs. (7.3-10) and (7.3-25) into Eq. (7.3-29) and multiplying the result by

$$\exp[-j2\pi f_Z(z-z_0)]\exp[+j2\pi f_Z(z-z_0)] = 1 \tag{7.3-30}$$

yields

$$\varphi(t,\mathbf{r}) \approx \exp[-j2\pi(\hat{f}_Z - f_Z)(z-z_0)]\exp\left\{-jk_0\int_{z_0}^{z}\sigma(x,y,\zeta)n_{\mathrm{NR}}(x,y,\zeta)\,d\zeta\right\}$$
$$\times \exp(+j2\pi ft)\exp[-j2\pi\mathbf{v}\cdot(\mathbf{r}-\mathbf{r}_0)], \tag{7.3-31}$$

where

$$\mathbf{v} = (f_X, f_Y, f_Z), \tag{7.3-32}$$

$$\hat{f}_Z = \hat{w}_0/\lambda_0 = f\hat{w}_0/c_0, \tag{7.3-33}$$

and

$$\hat{w}_0 = 1 - [(u_0^2 + v_0^2)/2] \tag{7.3-34}$$

or

$$\hat{w}_0 = 1 - [(\sin^2\theta_0)/2], \tag{7.3-35}$$

which is an approximation or estimate of the direction cosine w_0. Equations (7.3-34) and (7.3-35) are, in fact, *binomial expansions* of w_0, valid so long as

$$u_0^2 + v_0^2 = \sin^2\theta_0 \ll 1 \tag{7.3-36}$$

or

$$\sin^2 \theta_0 \leq 0.1, \tag{7.3-37}$$

which implies

$$\theta_0 \leq 18.0^\circ. \tag{7.3-38}$$

Thus, $\varphi(t, \mathbf{r})$ as given by Eq. (7.3-31) is based on the assumption that the initial angles of transmission θ_0 are small, that is, that θ_0 satisfies Eq. (7.3-38) (see Fig. 6.3-1). In addition, it is also assumed that $s_Z/\lambda_0 \gg 1$, so that *forward scatter* predominates. This assumption is in conjunction with the form of the solution given by Eq. (7.3-7), which represents a wave propagating in the positive Z direction, and with Eq. (6.3-4). The "parabolic approximation" represented by Eq. (6.3-4) is responsible for neglecting $\partial^2 g(\mathbf{r})/\partial z^2$ in favor of $\partial g(\mathbf{r})/\partial z$. The first-order partial derivative with respect to z implies that only one independent solution in the Z direction is allowed, namely, forward-scattered waves propagating in the positive Z direction are accounted for, while backscattered waves propagating in the negative Z direction are neglected.

Since the input to the communication channel is the time-harmonic plane wave

$$\exp(+j2\pi ft)\exp[-j2\pi \mathbf{v} \cdot (\mathbf{r} - \mathbf{r}_0)],$$

and since Eq. (7.3-31) represents the output from the channel at any time t and position $\mathbf{r}$ [see Eq. (2.1-21)],

$$H_{\mathrm{M}}(f, \mathbf{v}; t, \mathbf{r}) = H_{\mathrm{M}}(f, \mathbf{v}; \mathbf{r}) = H_{\mathrm{M}}(f, f_Z; \mathbf{r}), \tag{7.3-39}$$

where the random, time-invariant, space-variant transfer function of the ocean medium is given by

$$H_{\mathrm{M}}(f, f_Z; \mathbf{r}) = \exp[+j\theta_{\mathrm{MD}}(f, f_Z; \mathbf{r})] \exp[+j\theta_{\mathrm{MR}}(f, f_Z; \mathbf{r})], \tag{7.3-40}$$

where

$$\theta_{\mathrm{MD}}(f, f_Z; \mathbf{r}) = -2\pi(\hat{f}_Z - f_Z)(z - z_0) \tag{7.3-41}$$

is the deterministic phase component and

$$\theta_{\mathrm{MR}}(f, f_Z; \mathbf{r}) = -k_0 \int_{z_0}^{z} \sigma(x, y, \zeta) n_{\mathrm{NR}}(x, y, \zeta)\, d\zeta \tag{7.3-42}$$

is the random phase component.

The medium transfer function given by Eq. (7.3-40) indicates that the major effect of the medium is to *angle modulate* the transmitted field. Furthermore, if $\hat{f}_Z - f_Z \approx 0$, then $\theta_{\mathrm{MD}}(f, f_Z; \mathbf{r}) \approx 0$ [see Eq. (7.3-14)] and, as a result, the angle modulation is due strictly to the random fluctuations $\sigma(\mathbf{r}) n_{\mathrm{NR}}(\mathbf{r})$ of the index of refraction [see Eq. (7.3-42)]. Note that the angle-modulation process is often referred to as "scattering." Also, note that the transfer function derived in this section is written as the product of two functions, one deterministic and the other random, that is,

$$H_M = \exp(+j\theta_{MD})\exp(+j\theta_{MR}),$$

which agrees with the general practice of representing a field in a random medium as the product of a deterministic and a random function. Note that the form of Eq. (7.3-40) is a direct consequence of the assumed form of Eq. (7.3-8).

Finally, it should be emphasized that the transfer function specified by Eqs. (7.3-40) through (7.3-42) was based, in part, on the assumption given by Eq. (7.3-3), which allowed neglect of terms involving $\sigma^2(\mathbf{r})$. Equation (7.3-3) restricts the upper limit of integration z appearing in Eq. (7.3-42) to be on the order of the scale size s_Z of the random medium in the Z direction. Although the random component of the index of refraction angle modulates or "scatters" the propagating wave by only a small amount (i.e., the wave is "weakly scattered" or "singly scattered") within a distance s_Z, this does not mean that the theory is restricted to weak scatter only. For longer distances of wave propagation, one need only divide the overall distance z into *elementary layers* or *phase screens* of thickness s_Z and then integrate from z_0 to s_Z, s_Z to $2s_Z$, etc. Some authors suggest that the thickness of the elementary layers should be $10s_Z$.

7.3.2 Output Electrical Signal

Now that we have derived an ocean medium transfer function, let us demonstrate the use of the coupling equations presented in Section 7.1 by calculating the output electrical signal $y(t, \mathbf{r})$ from the receive aperture (array). In the case of a time-invariant, space-variant transfer function, $y(t, \mathbf{r})$ is given by [see Eq. (7.2-57)]

$$y(t, \mathbf{r}) = \int_{-\infty}^{\infty} X(f) \int_{-\infty}^{\infty} D_T(f, \mathbf{v}) H_M(f, \mathbf{v}; \mathbf{r}) \exp(-j2\pi\mathbf{v} \cdot \mathbf{r})\, d\mathbf{v} \times A_R(f, \mathbf{r}) \exp(+j2\pi f t)\, df, \tag{7.3-43}$$

where $X(f)$ is the frequency spectrum of the transmitted electrical signal, $D_T(f, \mathbf{v})$ is the far-field transmit directivity function (beam pattern), $H_M(f, \mathbf{v}; \mathbf{r})$ is the random, time-invariant, space-variant transfer function of the ocean medium, and $A_R(f, \mathbf{r})$ is the complex receive aperture function.

Assume that the transmit aperture depicted in Fig. 1.1-2 is a planar array of $M \times N$ (odd) complex-weighted point sources centered at (x_0, y_0, z_0) and parallel to the XY plane. In addition, assume that the complex weights are separable. Therefore (see Section 4.5.1)

$$D_{\mathrm{T}}(f,\mathbf{v}) = \sum_{m=-(M-1)/2}^{(M-1)/2} \sum_{n=-(N-1)/2}^{(N-1)/2} c_m d_n \exp(+j2\pi f_X m d_X) \exp(+j2\pi f_Y n d_Y) \times \exp(+j2\pi f_X x_0) \exp(+j2\pi f_Y y_0) \times \exp(+j2\pi f_Z z_0) \tag{7.3-44}$$

is the far-field beam pattern of the transmit array, c_m and d_n are complex weights, d_X and d_Y are the interelement spacings in the X and Y directions, respectively, and the last three complex exponentials are phase factors that account for the array being centered at (x_0, y_0, z_0) instead of at the origin $(0, 0, 0)$.

Next, assume that the receive aperture depicted in Fig. 1.1-2 is a planar array of $M' \times N'$ (odd) complex-weighted point sources centered at $(x_{\mathrm{R}}, y_{\mathrm{R}}, z_{\mathrm{R}})$ and parallel to the XY plane. In addition, assume that the complex weights are separable. Therefore, the receive aperture function is given by (see Section 4.5.1)

$$A_{\mathrm{R}}(f,\mathbf{r}) = \sum_{i=-(M'-1)/2}^{(M'-1)/2} \sum_{q=-(N'-1)/2}^{(N'-1)/2} c'_i d'_q \,\delta[x - (x_{\mathrm{R}} + i d'_X)]\,\delta[y - (y_{\mathrm{R}} + q d'_Y)] \times \delta(z - z_{\mathrm{R}}), \tag{7.3-45}$$

where c'_i and d'_q are complex weights and d'_X and d'_Y are the interelement spacings in the X and Y directions, respectively.

Upon substituting Eqs. (7.3-32), (7.3-39), (7.3-44), and (7.3-45) into Eq. (7.3-43) and recalling that $\mathbf{r} = (x, y, z)$, one obtains

$$\begin{aligned} y(t,x,y,z) = {} & \sum_{i=-(M'-1)/2}^{(M'-1)/2} \sum_{q=-(N'-1)/2}^{(N'-1)/2} c'_i d'_q \int_{-\infty}^{\infty} X(f) \sum_{m=-(M-1)/2}^{(M-1)/2} \sum_{n=-(N-1)/2}^{(N-1)/2} c_m d_n \\ & \times \int_{-\infty}^{\infty}\int_{-\infty}^{\infty}\int_{-\infty}^{\infty} H_{\mathrm{M}}(f, f_Z; x_{\mathrm{R}} + i d'_X, y_{\mathrm{R}} + q d'_Y, z_{\mathrm{R}}) \\ & \times \exp[-j2\pi(f_X \Delta X_{im} + f_Y \Delta Y_{qn} + f_Z \Delta Z)]\, df_X\, df_Y\, df_Z \\ & \times \exp(+j2\pi f t)\, df\, \delta[x - (x_{\mathrm{R}} + i d'_X)]\,\delta[y - (y_{\mathrm{R}} + q d'_Y)]\,\delta(z - z_{\mathrm{R}}), \end{aligned} \tag{7.3-46}$$

where

$$\Delta X_{im} = x_{\mathrm{R}} - x_0 + i d'_X - m d_X, \tag{7.3-47}$$

$$\Delta Y_{qn} = y_{\mathrm{R}} - y_0 + q d'_Y - n d_Y, \tag{7.3-48}$$

and

$$\Delta Z = z_{\mathbf{R}} - z_0. \tag{7.3-49}$$

In order to proceed further we make the following observations. From Eq. (7.3-17) we can write that

$$k_Z^2 = k_0^2 - (k_X^2 + k_Y^2), \tag{7.3-50}$$

where $k_0^2 = |\mathbf{k}_0|^2$, and upon substituting Eq. (7.2-3) and Eqs. (7.3-18) through (7.3-20) into Eq. (7.3-50), we obtain

$$f_Z = \pm[(f/c_0)^2 - (f_X^2 + f_Y^2)]^{1/2}. \tag{7.3-51}$$

Thus, the spatial frequency f_Z is seen to be dependent on f, f_X, and f_Y. As a result, we can eliminate the integration with respect to f_Z in Eq. (7.3-46). Equation (7.3-51) can also be expressed in terms of the direction cosines u_0 and v_0 by using Eqs. (7.3-11) and (7.3-12), that is,

$$f_Z = \pm(f/c_0)[1 - (u_0^2 + v_0^2)]^{1/2} \tag{7.3-52}$$

or

$$f_Z = f w_0/c_0, \tag{7.3-53}$$

where

$$w_0 = \pm[1 - (u_0^2 + v_0^2)]^{1/2} \tag{7.3-54}$$

is the direction cosine with respect to the positive Z axis. If $u_0^2 + v_0^2 \leq 1$, then

$$w_0 = +[1 - (u_0^2 + v_0^2)]^{1/2}, \qquad u_0^2 + v_0^2 \leq 1, \tag{7.3-55}$$

which corresponds to an acoustic field propagating in the positive Z direction. However, if $u_0^2 + v_0^2 > 1$, then

$$w_0 = -[1 - (u_0^2 + v_0^2)]^{1/2}, \qquad u_0^2 + v_0^2 > 1, \tag{7.3-56}$$

or

$$w_0 = -j(u_0^2 + v_0^2 - 1)^{1/2}, \qquad u_0^2 + v_0^2 > 1, \tag{7.3-57}$$

which corresponds to an *evanescent wave*, that is, a decaying exponential.

If we now (1) eliminate the integration with respect to f_Z in Eq. (7.3-46), (2) change variables from the spatial frequencies f_X and f_Y to the direction cosines u_0 and v_0 by substituting Eqs. (7.3-11) and (7.3-12) into Eq. (7.3-46) and treat the frequency variable f as a constant, so that

$$df_X = d(fu_0/c_0) = (f/c_0)\,du_0$$

and

$$df_Y = d(fv_0/c_0) = (f/c_0)\,dv_0,$$

and (3) substitute Eqs. (7.3-53) and (7.3-55) into $\exp(-j2\pi f_Z \Delta Z)$, which appears in the integrand of Eq. (7.3-46), then Eq. (7.3-46) becomes

$$\begin{aligned} y(t,x,y,z) = {} & \left(\frac{1}{c_0}\right)^2 \sum_{i=-(M'-1)/2}^{(M'-1)/2} \sum_{q=-(N'-1)/2}^{(N'-1)/2} c'_i d'_q \int_{-\infty}^{\infty} f^2 X(f) \\ & \times \int_{u_0=-1}^{1} \int_{v_0=-v_{0_{\max}}}^{v_{0_{\max}}} \sum_{m=-(M-1)/2}^{(M-1)/2} \sum_{n=-(N-1)/2}^{(N-1)/2} c_m d_n \\ & \times H_M(f,(f/c_0)[1-(u_0^2+v_0^2)]^{1/2}; x_R + id'_X, y_R + qd'_Y, z_R) \\ & \times \exp\{-j(2\pi f/c_0)[1-(u_0^2+v_0^2)]^{1/2}\,\Delta Z\} \\ & \times \exp[-j(2\pi f/c_0)(u_0 \Delta X_{im} + v_0 \Delta Y_{qn})]\, dv_0\, du_0 \\ & \times \exp(+j2\pi f t)\, df\, \delta[x-(x_R + id'_X)]\, \delta[y-(y_R + qd'_Y)] \\ & \times \delta(z - z_R), \qquad u_0^2 + v_0^2 \le 1, \end{aligned} \tag{7.3-58}$$

where

$$v_{0_{\max}} = \sin\theta_{0_{\max}} \tag{7.3-59}$$

and [see Eq. (7.3-38)]

$$\theta_{0_{\max}} \le 18.0^\circ. \tag{7.3-60}$$

The limits of integration with respect to v_0 were obtained by noting that

$$\frac{\pi}{2} - \theta_{0_{\max}} \le \beta_0 \le \frac{\pi}{2} + \theta_{0_{\max}},$$

and since $v_0 = \cos\beta_0$ [see Eq. (7.3-15) and Fig. 7.2-1],

$$\cos\left(\frac{\pi}{2} + \theta_{0_{\max}}\right) = -\sin\theta_{0_{\max}}$$

and

$$\cos\left(\frac{\pi}{2} - \theta_{0_{\max}}\right) = \sin\theta_{0_{\max}},$$

and therefore

$$-\sin\theta_{0_{\max}} \le v_0 \le \sin\theta_{0_{\max}}.$$

The transfer function H_M is given by Eqs. (7.3-40) through (7.3-42), where $\hat{f}_Z$ is given by Eq. (7.3-33) and $\hat{w}_0$ by Eq. (7.3-34).

Equation (7.3-58) is the desired result, that is, it represents the random output electrical signal at *each* element in the receive array in terms of the frequency spectrum of the transmitted electrical signal, the transmit and receive arrays, and the random transfer function of the ocean medium. If $y(t, x, y, z)$ is complex, then simply take the real part to obtain the real output electrical signals. Note that Eq. (7.3-58) pertains to the case of an acoustic field propagating in the positive Z direction since $u_0^2 + v_0^2 \leq 1$. When $u_0^2 + v_0^2 > 1$,

$$y(t, x, y, z) = 0, \qquad u_0^2 + v_0^2 > 1, \tag{7.3-61}$$

since only rapidly decaying evanescent waves emanate from the transmit array.

Finally, consider the following term that appears in the integrand of Eq. (7.3-58):

$$\begin{aligned} &H_M(f, (f/c_0)[1 - (u_0^2 + v_0^2)]^{1/2}; x_R + id'_X, y_R + qd'_Y, z_R) \\ &\quad \times \exp\{-j(2\pi f/c_0)[1 - (u_0^2 + v_0^2)]^{1/2}\,\Delta Z\}. \end{aligned} \tag{7.3-62}$$

If Eqs. (7.3-40), (7.3-41), (7.3-33), (7.3-34), (7.3-53), and (7.3-55) are substituted into Eq. (7.3-62), we obtain

$$\begin{aligned} &\exp\{-j(2\pi f/c_0)(1 - [(u_0^2 + v_0^2)/2])\,\Delta Z\} \\ &\quad \times \exp[+j\theta_{MR}(f, (f/c_0)[1 - (u_0^2 + v_0^2)]^{1/2}; x_R + id'_X, y_R + qd'_Y, z_R)], \end{aligned} \tag{7.3-63}$$

where θ_{MR} is given by Eq. (7.3-42). If we now assume that the medium is *homogeneous*, that is, space-invariant, and nonrandom, then the random component of the index of refraction is zero and, as a result [see Eqs. (7.2-11) and (7.3-42)], $\theta_{MR} = 0$. With $\theta_{MR} = 0$, Eq. (7.3-63) becomes

$$\exp\{-j(2\pi f/c_0)(1 - [(u_0^2 + v_0^2)/2])\,\Delta Z\}. \tag{7.3-64}$$

It is interesting to note that Eq. (7.3-64) is equivalent to the expression that one obtains in Fourier optics for a space-invariant transfer function that describes the effects of propagation in the region of Fresnel diffraction close to the Z axis, that is, where θ_0 does not exceed 18°. Also, note that Eq. (7.3-64) can be obtained directly by taking the spatial Fourier transform with respect to x and y of the impulse response or Green's function specified by Eqs. (6.3-28) and (6.3-21).

7.3.3 Coherence Function

Upon inspecting Eqs. (7.3-43) and (7.3-39), it can be seen that the autocorrelation function of the output electrical signal depends on the autocorrelation function of the transfer function

$$R_{H_M}(f, f', f_Z, f'_Z; \mathbf{r}, \mathbf{r}') = E\{H_M(f, f_Z; \mathbf{r}) H_M^*(f', f'_Z; \mathbf{r}')\}, \tag{7.3-65}$$

which is also known as the *coherence function*. Substituting Eqs. (7.3-40) and

(7.3-42) into Eq. (7.3-65) yields

$$R_{H_M}(f, f', f_Z, f_Z'; \mathbf{r}, \mathbf{r}') = \exp\{+j[\theta_{MD}(f, f_Z; \mathbf{r}) - \theta_{MD}(f', f_Z'; \mathbf{r}')]\} \times \mathrm{E}\{\exp[+j(KI(\mathbf{r}) + K'I(\mathbf{r}'))]\}, \qquad (7.3\text{-}66)$$

where

$$K = -k_0 = -2\pi f/c_0, \qquad (7.3\text{-}67)$$

$$K' = +k_0' = +2\pi f'/c_0, \qquad (7.3\text{-}68)$$

$$I(\mathbf{r}) = \int_{z_0}^{z} \sigma(x, y, \zeta) n_{NR}(x, y, \zeta)\, d\zeta, \qquad (7.3\text{-}69)$$

and

$$I(\mathbf{r}') = \int_{z_0}^{z'} \sigma(x', y', \zeta') n_{NR}(x', y', \zeta')\, d\zeta', \qquad (7.3\text{-}70)$$

where θ_{MD} is given by Eq. (7.3-41) and the random component of the index of refraction $n_R(\mathbf{r}) = \sigma(\mathbf{r}) n_{NR}(\mathbf{r})$ was assumed to be *real*. The expectation appearing in Eq. (7.3-66) is the *characteristic function* of the random quantity $KI(\mathbf{r}) + K'I(\mathbf{r}')$. If it is assumed that $I(\mathbf{r})$ is a *real, Gaussian random process,* which implies that $n_R(\mathbf{r})$ and, hence, $n_{NR}(\mathbf{r})$, is a real, Gaussian random process, then Eq. (7.3-66) can be written as

$$\begin{aligned} R_{H_M}(f, f', f_Z, f_Z'; \mathbf{r}, \mathbf{r}') &= \exp\{+j[\theta_{MD}(f, f_Z; \mathbf{r}) - \theta_{MD}(f', f_Z'; \mathbf{r}')]\} \\ &\times \exp(-\mathrm{E}\{\theta_{MR}^2(f, f_Z; \mathbf{r})\}/2 \\ &\times \exp(+\mathrm{E}\{\theta_{MR}(f, f_Z; \mathbf{r})\theta_{MR}(f', f_Z'; \mathbf{r}')\}) \\ &\times \exp(-\mathrm{E}\{\theta_{MR}^2(f', f_Z'; \mathbf{r}')\}/2), \end{aligned} \qquad (7.3\text{-}71)$$

where

$$\mathrm{E}\{\theta_{MR}^2(f, f_Z; \mathbf{r})\} = k_0^2 \int_{z_0}^{z}\int_{z_0}^{z} \sigma(x, y, \zeta)\sigma(x, y, \zeta') \times R_{n_{NR}}(x, x, y, y, \zeta, \zeta')\, d\zeta\, d\zeta', \qquad (7.3\text{-}72)$$

$$\mathrm{E}\{\theta_{MR}^2(f', f_Z'; \mathbf{r}')\} = (k_0')^2 \int_{z_0}^{z'}\int_{z_0}^{z'} \sigma(x', y', \zeta')\sigma(x', y', \zeta'') \times R_{n_{NR}}(x', x', y', y', \zeta', \zeta'')\, d\zeta'\, d\zeta'', \qquad (7.3\text{-}73)$$

$$\mathrm{E}\{\theta_{MR}(f, f_Z; \mathbf{r})\theta_{MR}(f', f_Z'; \mathbf{r}')\} = k_0 k_0' \int_{z_0}^{z}\int_{z_0}^{z'} \sigma(x, y, \zeta)\sigma(x', y', \zeta') \times R_{n_{NR}}(x, x', y, y', \zeta, \zeta')\, d\zeta\, d\zeta', \qquad (7.3\text{-}74)$$

and

$$R_{n_{\mathrm{NR}}}(x, x', y, y', z, z') = \mathrm{E}\{n_{\mathrm{NR}}(x, y, z) n_{\mathrm{NR}}(x', y', z')\}. \tag{7.3-75}$$

If it is further assumed that the random component $n_{\mathrm{R}}(\mathbf{r})$ and, hence, the normalized random component $n_{\mathrm{NR}}(\mathbf{r})$ is wide-sense stationary, that is,

$$R_{n_{\mathrm{NR}}}(x, x', y, y', z, z') = R_{n_{\mathrm{NR}}}(\Delta x, \Delta y, \Delta z), \tag{7.3-76}$$

where $\Delta x = x - x'$, $\Delta y = y - y'$, and $\Delta z = z - z'$, then Eqs. (7.3-72) through (7.3-74) become, respectively,

$$\mathrm{E}\{\theta_{\mathrm{MR}}^2(f, f_Z; \mathbf{r})\} = (k_0 \sigma)^2 (z - z_0) \int_{-(z-z_0)}^{z-z_0} \left[1 - \frac{|\zeta|}{z - z_0}\right] R_{n_{\mathrm{NR}}}(0, 0, \zeta)\, d\zeta, \tag{7.3-77}$$

$$\mathrm{E}\{\theta_{\mathrm{MR}}^2(f', f_Z'; \mathbf{r}')\} = (k_0' \sigma)^2 (z' - z_0) \int_{-(z'-z_0)}^{z'-z_0} \left[1 - \frac{|\zeta|}{z' - z_0}\right] R_{n_{\mathrm{NR}}}(0, 0, \zeta)\, d\zeta, \tag{7.3-78}$$

and

$$\begin{aligned} \mathrm{E}\{\theta_{\mathrm{MR}}(f, f_Z; \mathbf{r}) \theta_{\mathrm{MR}}(f', f_Z'; \mathbf{r}')\} &= k_0 k_0' \sigma^2 \\ &\times \left\{ (z - z') \int_{z-z'}^{z-z_0} R_{n_{\mathrm{NR}}}(\Delta x, \Delta y, \zeta)\, d\zeta + (z' - z_0) \int_{-(z'-z_0)}^{z-z_0} R_{n_{\mathrm{NR}}}(\Delta x, \Delta y, \zeta)\, d\zeta \right. \\ &\left. + \int_{-(z'-z_0)}^{0} \zeta R_{n_{\mathrm{NR}}}(\Delta x, \Delta y, \zeta)\, d\zeta - \int_{z-z'}^{z-z_0} \zeta R_{n_{\mathrm{NR}}}(\Delta x, \Delta y, \zeta)\, d\zeta \right\}, \end{aligned} \tag{7.3-79}$$

where σ is the constant standard deviation of the real, Gaussian, zero-mean, wide-sense stationary, random component $n_{\mathrm{R}}(\mathbf{r})$ of the index of refraction. The significance of the coherence function is discussed at the end of Section 7.1.

BIBLIOGRAPHY

R. H. Clarke, Sound propagation in a variable ocean. *J. Sound and Vibration* **34**, 457–477 (1974).

R. H. Clarke, Acoustic and electromagnetic waves propagating in a tenuous random medium. In *Aspects of Signal Processing*, Vol. I (edited by G. Tacconi), pp. 61–67. D. Reidel, Dordrecht, Holland, 1977.

S. M. Flatté, ed., *Sound Transmission through a Fluctuating Ocean.* Cambridge Univ. Press, London and New York, 1979.

J. D. Gaskill, *Linear Systems, Fourier Transforms, and Optics.* Wiley, New York, 1978.

J. W. Goodman, *Introduction to Fourier Optics.* McGraw-Hill, New York, 1968.

A. Ishimaru, *Wave Propagation and Scattering in Random Media*, Vol. I. Academic Press, New York, 1978.

D. Mintzer, Wave propagation in a randomly inhomogeneous medium. I. *J. Acoust. Soc. Am.* **25**, 922–927 (1953).

D. Mintzer, Wave propagation in a randomly inhomogeneous medium. II. *J. Acoust. Soc. Am.* **25**, 1107–1111 (1953).

D. Mintzer, Wave propagation in a randomly inhomogeneous medium. III. *J. Acoust. Soc. Am.* **26**, 186–190 (1954).

C. B. Officer, *Introduction to the Theory of Sound Transmission.* McGraw-Hill, New York, 1958.

J. A. Stratton, *Electromagnetic Theory.* McGraw-Hill, New York, 1941.

B. J. Uscinski, *The Elements of Wave Propagation in Random Media.* McGraw-Hill, New York, 1977.

L. J. Ziomek, *Linear Time-Variant Space-Variant Filters and the WKB Approximation.* Naval Postgraduate School Technical Report 62-83-058, October 1983, 93 pages. Also available from DTIC, AD Number A 134886.

L. J. Ziomek, Linear Time-Invariant Space-Variant Filters and the Parabolic Equation Approximation. *Signal Processing* (1985), in press.

Answers to Problems

CHAPTER 3

3-6 (a) $$D_N(f, \theta, \psi) = \frac{\sin[(\pi L/\lambda)\sin\theta\sin\psi]}{(\pi L/\lambda)\sin\theta\sin\psi}$$

(b) $$D_N(f, \theta, \psi) = \frac{\sin[(\pi L/\lambda)\cos\theta]}{(\pi L/\lambda)\cos\theta}$$

3-11 $$D(f, f_X) = -j\frac{L^2}{2}\pi f_X\left[\frac{\sin(\pi L f_X/2)}{\pi L f_X/2}\right]^2$$

3-12 (a) $$D(f, f_X) = \frac{4h}{(2\pi f_X)^2(L/2 - d)}\sin\left\{\frac{\pi f_X(L + 2d)}{2}\right\}\sin\left\{\frac{\pi f_X(L - 2d)}{2}\right\},$$

where $f_X = u/\lambda$ and $u = \sin\theta\cos\psi$;

(b) $$D(f, f_X) = \frac{L}{2}\left[\frac{\sin(\pi L f_X/2)}{\pi L f_X/2}\right]^2,$$

which is the far-field unnormalized beam pattern of the triangular amplitude window;

(c) $D(f, f_X) = \sin(\pi L f_X)/(\pi f_X)$,

which is the far-field unnormalized beam pattern of the rectangular amplitude window.

3-15 Hanning:

$$D_N(f, f_X) = \frac{\sin(\pi L f_X)}{\pi L f_X[1 - (L f_X)^2]}$$

Hamming:

$$D_N(f, f_X) = \frac{[0.08(L f_X)^2 - 0.54]}{0.54[(L f_X)^2 - 1]}\,\frac{\sin(\pi L f_X)}{\pi L f_X}$$

Blackman:

$$D_N(f, f_X) = \frac{[1.68 - 0.18(Lf_X)^2]}{0.42[(Lf_X)^2 - 1][(Lf_X)^2 - 4]} \frac{\sin(\pi L f_X)}{\pi L f_X}$$

3-16 (a) 83.3°, 52.6°, 30.8°
(b) 15.3°, 10.2°, 6.1°
(c) 7.6°, 5.1°, 3.0°

3-17 triangular: $\Delta u = 1.276\lambda/L$
cosine: $\Delta u = 1.189\lambda/L$
Hanning: $\Delta u = 1.441\lambda/L$
Hamming: $\Delta u = 1.303\lambda/L$
Blackman: $\Delta u = 1.644\lambda/L$

3-18 triangular:
(a) 146.3°, 79.3°, 45.0°
(b) 22.1°, 14.7°, 8.8°
(c) 11.0°, 7.3°, 4.4°
cosine:
(a) 126.2°, 73.0°, 41.8°
(b) 20.5°, 13.7°, 8.2°
(c) 10.2°, 6.8°, 4.1°
Hanning:
(a) invalid, 92.2°, 51.2°
(b) 25.0°, 16.6°, 9.9°
(c) 12.4°, 8.7°, 5.0°
Hamming:
(a) 155.5°, 81.3°, 46.0°
(b) 22.5°, 15.0°, 9.0°
(c) 11.2°, 7.5°, 4.5°
Blackman:
(a) invalid, 110.6°, 59.1°
(b) 28.6°, 18.9°, 11.3°
(c) 14.2°, 9.4°, 5.7°

3-19

ψ' (deg)	$\Delta\psi$ (deg) REC	TRI	COS	HAN	HAM	BLK
90	3.0	4.4	4.1	5.0	4.5	5.7
60	3.5	5.1	4.7	5.7	5.2	6.5
45	4.3	6.2	5.8	7.0	6.3	8.0
30	6.1	8.8	8.2	9.9	9.0	11.3
0	26.5	31.8	30.7	33.8	32.1	36.1

3-21 (a) near-field since $\pi(L/2)^2/\lambda = 52.4$ m
(b) $-17.809x + 1.05x^2$
(c) far-field
$-38.897x$
(d) near-field
$-34.953y + 1.05y^2$
far-field
$-41.712y$

3-22 (a) $D_N(f, f_X, f_Y) = \dfrac{\sin(\pi L_X f_X)}{\pi L_X f_X} \dfrac{\sin(\pi L_Y f_Y)}{\pi L_Y f_Y}$
(b) $D_N(f, f_X, f_Y) = \left[\dfrac{\sin(\pi L_X f_X/2)}{\pi L_X f_X/2}\right]^2 \left[\dfrac{\sin(\pi L_Y f_Y/2)}{\pi L_Y f_Y/2}\right]^2$
(c) $D_N(f, f_X, f_Y) = \dfrac{\sin[\pi L_X(f_X - f_X')]}{\pi L_X(f_X - f_X')} \left[\dfrac{\sin[\pi L_Y(f_Y - f_Y')/2]}{\pi L_Y(f_Y - f_Y')/2}\right]^2$
(d) $D_N(f, f_X, f_Y) = \left[\dfrac{\sin(\pi L_X f_X/2)}{\pi L_X f_X/2}\right]^2 \dfrac{\cos(\pi L_Y f_Y)}{[1 - (2L_Y f_Y)^2]}$
(e) $D_N(f, f_X, f_Y) = \dfrac{\cos[\pi L_X(f_X - f_X')]}{\{1 - [2L_X(f_X - f_X')]^2\}} \dfrac{\cos[\pi L_Y(f_Y - f_Y')]}{\{1 - [2L_Y(f_Y - f_Y')]^2\}}$

3-24 (a) near-field since $\pi[(L_X/2)^2 + (L_Y/2)^2]/\lambda = 102.1$ m
(b) $+6.196x - 26.838y + 0.436(x^2 + y^2)$
(c) far-field
$-23.28x - 30.894y$

CHAPTER 4

4-2 (a) $S_N(f_X) = \dfrac{1}{N} \dfrac{\sin(\pi f_X N d)}{\sin(\pi f_X d)}$
(b) $S_N(f_X) = \dfrac{4}{(N-1)^2} \dfrac{\sin^2(\pi f_X[(N-1)/2]d)}{\sin^2(\pi f_X d)}$
(c) $\Delta u = 0.255$
(d) $\Delta u = 0.449$
(e) $\Delta\theta = 14.7°$
(f) $\Delta\theta = 25.9°$

4-3 (c) $\Delta u = 0.162$
(d) $\Delta u = 0.26$
(e) $\Delta\theta = 9.3°$
(f) $\Delta\theta = 14.9°$

4-4 (a) $a_0 = 1.0$
$a_{-1} = a_1 = 0.916$
$a_{-2} = a_2 = 0.694$
$a_{-3} = a_3 = 0.544$

(b) $a_0 = 1.0$
$a_{-1} = a_1 = 0.840$
$a_{-2} = a_2 = 0.479$
$a_{-3} = a_3 = 0.159$
(c) $\Delta\theta = 16.4^\circ$
$\Delta\theta = 20.6^\circ$

4-5 (a) $a_{-1} = a_1 = 1.0$
$a_{-2} = a_2 = 0.684$
$a_{-3} = a_3 = 0.296$
(b) $\Delta\theta = 22.1^\circ$

4-7 (b) $\theta_1 = -(\pi d/\lambda)\sin\theta'\cos\psi'$
$\theta_{-1} = -\theta_1$
(c) $\tau_1 = -[d/(2c)]\sin\theta'\cos\psi'$
$\tau_{-1} = -\tau_1$

4-8 (b) same as 4-7(b)
(c) same as 4-7(c)

4-9 (b) $\theta_1 = -(\pi d/\lambda)\sin\theta'\cos\psi'$
$\theta_0 = 0$
$\theta_{-1} = -\theta_1$
(c) $\tau_1 = -[d/(2c)]\sin\theta'\cos\psi'$
$\tau_0 = 0$
$\tau_{-1} = -\tau_1$

4-15 (a) near-field since $\pi(L/2)^2/\lambda = 70.686$ m
(b) $\theta_n = -0.644n + 1.25 \times 10^{-3}n^2$
(c) far-field
$\theta_n = +1.437n$
(d) 61
(e) near-field
$\theta_n = -0.795n + 1.25 \times 10^{-3}n^2$
far-field
$\theta_n = -2.703n$

4-17 (a) $S_N(f,\theta,\psi) = \dfrac{\sin[(5\pi d_Y/\lambda)\sin\theta\sin\psi]}{5\sin[(\pi d_Y/\lambda)\sin\theta\sin\psi]}$
(b) $S_N(f,\theta,\psi) = \dfrac{\sin^2[(2\pi d_Y/\lambda)\sin\theta\sin\psi]}{4\sin^2[(\pi d_Y/\lambda)\sin\theta\sin\psi]}$
(c) $\theta_{-2} = (4\pi d_Y/\lambda)\sin\theta'\sin\psi'$
$\theta_{-1} = (2\pi d_Y/\lambda)\sin\theta'\sin\psi'$
$\theta_0 = 0$
$\theta_1 = -\theta_{-1}$
$\theta_2 = -\theta_{-2}$

(d) $\tau_{-2} = (2d_Y/c)\sin\theta'\sin\psi'$
$\tau_{-1} = (d_Y/c)\sin\theta'\sin\psi'$
$\tau_0 = 0$
$\tau_1 = -\tau_{-1}$
$\tau_2 = -\tau_{-2}$
(e) $d_Y < 0.03$ m

4-18 $S(f_X, f_Y) = 2\,\mathrm{Re}\sum_{m=0}^{(M-1)/2}\sum_{n=0}^{(N-1)/2} c_{mn}\exp[+j2\pi(f_X m d_X + f_Y n d_Y)]$,
where $c_{00} \triangleq a_{00}/2$.

4-19 $D_N(f, f_X, f_Y) = \dfrac{\sin(3\pi f_X d_X)}{3\sin(\pi f_X d_X)}\dfrac{\sin^2(2\pi f_Y d_Y)}{4\sin^2(\pi f_Y d_Y)}\dfrac{2J_1[2\pi a(f_X^2 + f_Y^2)^{1/2}]}{2\pi a(f_X^2 + f_Y^2)^{1/2}}$,

where

$$f_X = u/\lambda = \sin\theta\cos\psi/\lambda$$

and

$$f_Y = v/\lambda = \sin\theta\sin\psi/\lambda$$

4-20 $D_N(f, f_X, f_Y) = \dfrac{\sin^2(2\pi f_X d_X)}{4\sin^2(\pi f_X d_X)}\dfrac{\sin(\pi L_X f_X)}{\pi L_X f_X}\dfrac{\sin(\pi L_Y f_Y)}{\pi L_Y f_Y}$

4-21 (a) $S(f_X, f_Y) = 1.0 + 2\cos[2\pi(f_X d_X + f_Y d_Y)]$
(b) $S(f_X, f_Y) = 2\{\cos[2\pi(f_X d_X + f_Y d_Y)] - 1.0\}$

4-22 $D(f, \theta, \psi) = -4Q\sin[(\pi d/\lambda)\sin\theta\cos\psi]\sin[(\pi d/\lambda)\sin\theta\sin\psi]$

4-23 (a) near-field since $\pi[(L_X/2)^2 + (L_Y/2)^2]/\lambda = 67.034$ m
(b) $\theta_{mn} = -0.739m - 0.850n + 2.36 \times 10^{-3}(m^2 + n^2)$
(c) far-field
$\theta_{mn} = +1.352m - 2.772n$
(d) $M = 27$, $N = 41$

CHAPTER 5

5-7 (a) $T = 1.0$ sec
$b = -2356.194$ rad/sec^2
$A = \sqrt{2}$
(b) $|b|T/\pi = 750.0$ Hz

5-9 (a) $s = 0.997$
$\tau = 1.335$ sec
$\phi = -28.2$ Hz
$|\mathbf{V}_S|/c = 0.00163$
$\phi = -28.3$ Hz

(b) $s = 1.00283$
$\tau = 1.331$ sec
$\phi = +28.3$ Hz
$|\mathbf{V}_S|/c = 0.0024$
$\phi = +28.3$ Hz

5-10 (a) $s = 1.004725$
$\tau = 2.660$ sec
$\phi = +70.9$ Hz
(b) $s = 0.995297$
$\tau = 2.673$ sec
$\phi = -70.5$ Hz

Index

P

Q

R

S